D1759334

3 0116 00494 5240

This book is due for return not later than the
last date stamped below, unless recalled sooner.

THE HILBERT TRANSFORM OF SCHWARTZ DISTRIBUTIONS AND APPLICATIONS

PURE AND APPLIED MATHEMATICS

A Wiley-Interscience Series of Texts, Monographs, and Tracts

Founded by RICHARD COURANT
Editor Emeritus: PETER HILTON
Editors: MYRON B. ALLEN III, DAVID A. COX,
      HARRY HOCHSTADT, PETER LAX, JOHN TOLAND

A complete list of the titles in this series appears at the end of this volume.

THE HILBERT TRANSFORM OF SCHWARTZ DISTRIBUTIONS AND APPLICATIONS

J. N. PANDEY
Carleton University

A Wiley-Interscience Publication
JOHN WILEY & SONS, INC.
New York • Chichester • Brisbane • Toronto • Singapore

This text is printed on acid-free paper.

Copyright © 1996 by John Wiley & Sons, Inc.

All rights reserved. Published simultaneously in Canada.

Reproduction or translation of any part of this work beyond
that permitted by Section 107 or 108 of the 1976 United
States Copyright Act without the permission of the copyright
owner is unlawful. Requests for permission or further
information should be addressed to the Permissions Department,
John Wiley & Sons, Inc., 605 Third Avenue, New York, NY
10158-0012.

Library of Congress Cataloging in Publication Data:
Pandey, J. N.
    The Hilbert transform of Schwartz distributions and applications /
by J. N. Pandey.
        p.   cm. — (Pure and applied mathematics)
    Includes bibliographical references.
    ISBN 0-471-03373-1 (cloth : alk. paper)
    1. Hilbert transform.   2. Schwartz distributions.   I. Title.
II. Series: Pure and applied mathematics (John Wiley & Sons :
Unnumbered)
QA432.P335   1996
515′.782—dc20   95-18944

10 9 8 7 6 5 4 3 2 1

To my parents
(Pandit Chandrika Pandey and Shrimati Chameli Devi),
as well as to all of the 329 passengers and crew members
of Air India flight number 182 which crashed on June 23, 1985
near the Irish coast.

CONTENTS

PREFACE

For the past two decades I have been researching the Hilbert transform of Schwartz distributions. I and my colleagues have arrived at many new results. These results form the basis of this book which will be of interest not only to mathematicians but also to engineers and applied scientists. My objective is to demonstrate the wide applicability of Hilbert transform techniques. This book may be used either as a graduate-level textbook on the Hilbert transform of Schwartz distributions and periodic distributions or as a research monograph.

The Hilbert transform $(Hf)(x) = \frac{1}{\pi}(P) \int_{-\infty}^{\infty} \frac{f(t)}{x-t} dt$ arises in many fields such as

-
   i. Signal processing (the Hilbert transform of periodic functions)
  ii. Metallurgy (Griffith crack problem and the theory of elasticity)
 iii. Dirichlet boundary value problems (potential theory)
  iv. Dispersion relation in high energy physics, spectroscopy, and wave equations
   v. Wing theory
  vi. The Hilbert problem
 vii. Harmonic analysis

The Hilbert problem during the last four decades has received considerable attention in metallurgical problems, namely in the Griffith crack problem in the theory of elasticity. Sneddon and Lowengrub who have been pioneers of applying the finite Hilbert transform in the theory of elasticity state: "The major development of the present century in the field of two-dimensional elasticity has been Muskhelishvili's work on the complex form of the two-dimensional equations due to G. B. Kolsov." Consequently a fair amount of treatment of classical as well as distributional Hilbert problems has been incorporated in the book. In particular, Chapter 2 is devoted to the classical Hilbert problem, whereas Chapter 3 and Chapter 6 cover distributional Hilbert problems.

The singular nature of the kernel $\frac{1}{\pi(x-t)}$ of the Hilbert transform has made the work on the Hilbert transform very difficult to accomplish and in turn the work on the Hilbert transform of distributions has suffered. Nevertheless, the problem of the Hilbert transform of distributions has received the attention of many mathematicians who had started working on the Hilbert transform of various subspaces of Schwartz distributions. Among them are Laurent Schwartz [87], Gel'fand and Shilov [44], Horvath [5], Bremermann [9], Jones [53], Lauwerier [58], Tillmann [97, 63], Beltrami and Wohlers [6], Orton [72], Mitrovic [61, 62, 63], and Carmichael [15, 16]. The approach to the Hilbert transform of distributions that I have developed with my colleagues is the simplest and the most effective. It is easily accessible to applied scientists despite the fact that I have used a fairly advanced treatment in this book.

Among many new results that I wish to point out are the inversion formula for the n-dimensional Hilbert transform $H^2 f = (-1)^n f$, $n > 1$ and a new definition for the Hilbert transform of periodic functions with period 2τ:

$$(Hf)(x) = \frac{1}{\pi} \lim_{N \to \infty} (P) \int_{-N}^{N} \frac{f(t)}{x - t} dt \tag{i}$$

$$= \frac{1}{2\tau}(P) \int_{-\tau}^{\tau} f(x - t) \cot\left(\frac{t\pi}{2\tau}\right) dt \tag{ii}$$

This identity is true at least for the class of functions $f \in L_{2\tau}^p$. My definition of the Hilbert transform of periodic functions is a generalization of the Hilbert transform of periodic functions with period 2π, defined as

$$(Hf)(x) = \frac{1}{2\pi}(P) \int_{-\pi}^{\pi} f(x - t) \cot \frac{t}{2} dt \tag{iii}$$

Definition (iii) was widely used by Butzer, Nessel, Oppenheim, Schaefer, and many others, and to the best of my knowledge, there has been no formula or definition for the Hilbert transform of periodic functions with period other than 2π. I also believe that the definition of the Hilbert transform of periodic functions in the form (i) will be especially useful to people working in signal processing for computational purposes. From definition (i), which is the definition for the Hilbert transform of functions, a unified theory of the Hilbert transform of periodic as well as nonperiodic functions can be developed.

In Chapter 7 I develop the theory of the Hilbert transform of periodic distributions and also the approximate Hilbert transform of periodic distributions. I use this theory to find a harmonic function $U(x, y)$ which is periodic in x with period 2τ vanishes as $y \to \infty$, uniformly $\forall x \in \mathbb{R}$ and tends to a periodic distribution f (with period 2τ) as $y \to 0^+$, in the weak distributional sense. The uniqueness of the solution is also proved.

My discussion proceeds from a Paley-Wiener type of theorem (Theorem 6.18) which gives the characterization of functions or generalized functions whose Fourier transform vanishes over certain orthants or the union of orthants of $\mathbb{R}^n$.

In Chapter 5 I also give a generalization of the Hilbert problem

$$F_+(x) - F_-(x) = f(x)$$

in higher dimensions and solve it. In Section 6.7 I calculate the p-norm $\|H\|_p$ of the Hilbert transform operator $H: L^p(\mathbb{R}^n) \to L^p(\mathbb{R}^n)$, $p > 1$. In Theorem 6.3 I give a characterization of bounded linear operators on $L^p(\mathbb{R}^n)$, $p > 1$, which commute with translation and dilatation.

Another highlight of the book is the very elegant treatment of the one-dimensional Hilbert transform of distributions in D'_{L^p}, $p > 1$, in Chapter 3. Chapter 3 will be especially useful to applied scientists.

The book assumes that the reader has a background in the elements of functional analysis. Chapter 1 essentially deals with the prerequisite materials for the theory of distributions and Fourier transform.

Chapter 2 presents the Riemann-Hilbert problem and gives the background material to the study of the Hilbert transform. It includes sections on the appearance of the Hilbert transform in wing theory, in the theory of elasticity, in spectroscopy, and in high-energy physics.

Chapter 3 discusses the Hilbert transform of Schwartz distributions in D'_{L^p} and related boundary value problems.

Chapter 4 considers the Hilbert transform of Schwartz distributions in D'. It also discusses a Gel'fand and Shilov technique for the Hilbert transform of generalized functions and an improvement to their techniques.

Chapter 5 deals with n-dimensional Hilbert transform and the approximation technique in evaluating the Hilbert transform and the inversion formulas. The Hilbert transform of distribution in $D'_{L^p}(\mathbb{R}^n)$ is also covered, and many applications are given.

Chapter 6 considers the applications of the Hilbert transform to Riemann-Hilbert problems (classical as well as distributional). Many other related results are presented. One among many is the derivation of a Paley-Wiener theorem.

Chapter 7 deals with the periodic distributions and their Hilbert transforms.

With the firm belief that perfection never comes without practice I have included numerous examples in every chapter.

I wish to acknowledge the assistance of Professor E. L. Koh of the University of Regina, and of Professor S. A. Naimpally and Dr. James Bondar of my department who were very kind and patient in going through various chapters of the manuscript and gave me very useful suggestions. I want to express my sincere gratitude to Professor Angelo Mingarelli of my department who very patiently entered the graphic designs on my manuscript and helped me consult CDRAM (Math Reviews) for the preparation of the manuscript.

I further wish to thank Mr. Andrew E. Dabrowski a former student at Carleton, Mr. Sanjay Varma of the Mehta Research Institute Allahabad India, Ms. Nalini Sreeshylan of the Institute of Sciences Bangalore India, Mr. K. P. Sivaraman of the Tata Institute of Fundamental Research, Bombay, Mrs. Shelly Bereznin, Mr. Ibrahim Farah, and Mrs. Diane Berezowski of Carleton University for their help in the typing of the manuscript in its various forms.

The major part of the typing was done by Mrs. Diane Berezowski who modified all the chapters typed by others and unified them into a single TEX scheme along with her own typing. She never lost her temper despite the many changes I had asked her to incorporate in the manuscript. I am grateful to her for her patience and for her active and prompt cooperation.

In addition I would like to thank Drs. Q. M. Tariq and Ehab Bassily of my department who very patiently prepared the Index and the Notation Index of this book.

I also wish to acknowledge a grant from the Natural Sciences and Engineering Research Council of Canada in support of my research. I wish to thank the Tata Institute of Fundamental Research, Bombay, the Indian Institute of Science, Bangalore, and the Mehta Research Institute, Allahabad, for financially supporting my visits to their institutions, where a considerable part of the research work on this monograph was completed. In particular, I express my gratitude to Professor H. S. Mani, Director, Mehta Research Institute of Mathematics and Mathematical Physics for encouraging me to use the facilities of his institute.

My debt to my wife, Krishna for her constant support and cheerfulness under difficult circumstances when the manuscript was under preparation, is so profound as to defy description.

J. N. PANDEY

Ottawa, Canada

1

SOME BACKGROUND

1.1. FOURIER TRANSFORMS AND THE THEORY OF DISTRIBUTIONS

This chapter discusses some very important properties of the Fourier transform of functions that will be useful in developing the theory of the Fourier transform of distributions. It also develops some basic results concerning topological vector spaces, in particular, locally convex spaces, and extends these results to develop a theory of distributions and tempered distributions.

Definition. Let f be a function of a real variable t defined on the real line. Then its Fourier transform $F(w)$ is defined by the relation

$$(\mathcal{F}f)(w) = F(w) = \int_{-\infty}^{\infty} f(t)e^{iwt}\,dt \tag{1.1}$$

provided that the integral exists.

There are many variations on definition (1.1). Some authors add the factor $\frac{1}{2\pi}$ or $\frac{1}{\sqrt{2\pi}}$ outside the integral sign, and some take the kernel of the Fourier transform as e^{-iwt} in place of the kernel e^{iwt}. Some authors including L. Schwartz have written the kernel of the Fourier transform $e^{2\pi iwt}$. But these variations matter little.

The inverse Fourier transform of f in our case will be defined as

$$\mathcal{F}^{-1}f(t) = \frac{1}{2\pi} \int_{-\infty}^{\infty} f(w)e^{-iwt}\,dw \tag{1.2}$$

provided that the integral exists.

Example 1. Let

$$f(t) = \begin{cases} 1, & -1 \le t \le 1 \\ 0 & \text{elsewhere} \end{cases}$$

Then

$$(\mathcal{F}f)(w) = \int_{-1}^{1} 1 e^{itw} dt = \begin{cases} \dfrac{2\sin w}{w}, & w \neq 0 \\ 2 & \text{when } w = 0 \end{cases}$$

Note that the function $f(t) \in L^1$ but $(\mathcal{F}f)(w) \notin L^1$.

Theorem 1. Let $f \in L^1$. Then

  i. $F(w) = \displaystyle\int_{-\infty}^{\infty} f(t)e^{iwt} dt$ is well defined $\forall\, w \in \mathbb{R}$.

  ii. $F(w)$ is uniformly continuous and bounded on $\mathbb{R}$.

  iii. $F(w) \to 0$ as $|w| \to \infty$.

Proof. (i) $\int_{-\infty}^{\infty} |f(t)e^{iwt}|\, dt \leq \int_{-\infty}^{\infty} |f(t)|\, dt$. Clearly $f(t)e^{iwt}$ is a measurable function of t. Therefore $f(t)e^{iwt}$ is absolutely integrable, and it is integrable for each $w \in \mathbb{R}$. Hence $F(w)$ is well defined for each $w \in \mathbb{R}$.

(ii) $|F(w)| \leq \int_{-\infty}^{\infty} |f(t)e^{iwt}|\, dt \leq \int_{-\infty}^{\infty} |f(t)|\, dt$. $F(w)$ is uniformly bounded. Now we prove that $F(w)$ is uniformly continuous on R. Choosing N large enough so that for an arbitrary $\epsilon > 0$, we have

$$\int_{N}^{\infty} |f(t)|\, dt + \int_{-\infty}^{-N} |f(t)|\, dt < \frac{\epsilon}{4} \tag{1.3}$$

A simple calculation shows that

$$\Delta F = F(w + \Delta w) - F(w)$$

$$= \int_{-N}^{N} f(t)[e^{i(w+\Delta w)t} - e^{iwt}]\, dt$$

$$+ \left(\int_{-\infty}^{-N} + \int_{N}^{\infty} \right) f(t)e^{iwt}[e^{i\Delta wt} - 1]\, dt \tag{1.4}$$

Now denote the first integral in the right hand side of (1.4) by I and the second pair of integrals by J:

$$|I| \leq \int_{-N}^{N} |f(t)|\, |e^{i\Delta wt} - 1|\, dt$$

By virtue of the uniform continuity of $(e^{i\Delta wt} - 1)$, we can choose δ small enough so that

$$|I| < \frac{\epsilon}{2} \quad \text{whenever} \quad |\Delta w| < \delta \tag{1.5}$$

δ being independent of t (and w as well).

$$|J| \leq \left(\int_{-\infty}^{\infty} + \int_{-\infty}^{-N} \right) 2|f(t)| \, dt \leq \frac{\epsilon}{2} \tag{1.6}$$

Combining (1.5) and (1.6), we have

$$|\Delta F| \leq \epsilon \quad \text{whenever} \quad |\Delta w| < \delta$$

This proves the uniform continuity of $F(w)$ over the real line.

(iii) The space $\mathcal{D}(\mathbb{R}^n)$ of infinitely differentiable functions with compact support on $\mathbb{R}^n$ is dense in $L^p(\mathbb{R}^n)$ $p \geq 1$, and the identity map from $\mathcal{D}(\mathbb{R}^n)$ to $L^p(\mathbb{R}^n)$ is continuous [67, 101]. Let now $\varphi \in \mathcal{D}(\mathbb{R})$ be such that

$$\int_{-\infty}^{\infty} |f(t) - \varphi(t)| \, dt < \frac{\epsilon}{2}.$$

Then

$$F(w) = \int_{-\infty}^{\infty} [f(t) - \varphi(t)] e^{iwt} \, dt + \int_{-\infty}^{\infty} \varphi(t) e^{iwt} \, dt \tag{1.7}$$

Denoting the two integrals in the right hand side of (1.7) by J_1 and J_2, respectively, we see that

$$|J_1| \leq \int_{-\infty}^{\infty} |f(t) - \varphi(t)| \, dt < \frac{\epsilon}{2} \tag{1.8}$$

A simple integration by parts shows that

$$J_2 \to 0 \text{ as } |w| \to \infty$$

Therefore there exists a $k > 0$ such that

$$|J_2| < \frac{\epsilon}{2} \, \forall \, |w| > k \tag{1.9}$$

Combining (1.8) and (1.9), we get for $\epsilon > 0$, that there exists a constant $k > 0$ such that $|F(w)| < \epsilon \, \forall \, |w| > k$. Since ϵ is arbitrary our result is proved. $\square$

Theorem 2. Let $f \in L^1$ and $F(w)$ be the Fourier transform of f. Assume that $F(w) \in L^1$. Then

$$f(t) = \frac{1}{2\pi} \int_{-\infty}^{\infty} F(w) e^{-iwt} \, dw \qquad \text{a.e.} \tag{1.10}$$

The equality (1.10) holds at all points of continuity of f. Proof of this inversion formula for the Fourier transform can be found in many books on integral transforms.

The Fourier transform of a function $f(t)$ defined from $\mathbb{R}^n \to \mathbb{R}$ is defined as $F(w) = \int_{\mathbb{R}^n} f(t) e^{it \cdot w} \, dt$, provided that the integral exists. Here $t = (t_1, t_2, \ldots, t_n)$ and

$w = (w_1, w_2, \ldots, w_n)$, $t \cdot w = t_1 w_1 + t_2 w_2 + \cdots + t_n w_n$. Theorems analogous to Theorems 1 and 2 are valid. The inversion formula analogous to (1.10) is

$$f(t) = \left(\frac{1}{2\pi}\right)^n \int_{\mathbb{R}^n} F(w)e^{-iw \cdot t} dw \qquad \text{a.e.}$$

which is also valid if f and F both $\in L^1(\mathbb{R}^n)$.

1.2. FOURIER TRANSFORMS OF L^2 FUNCTIONS

1.2.1. Fourier Transforms of Some Well-known Functions

Consider

$$\mathcal{F}(e^{-|t|}) = \int_{-\infty}^{\infty} e^{-|t|} e^{iwt} dt$$

$$= 2 \int_{0}^{\infty} \cos wt e^{-t}$$

$$= \frac{2}{w^2 + 1}$$

$$\mathcal{F}[h(t-1) - h(t-2)] = \int_{1}^{2} e^{iwt} dt$$

where $h(t)$ is Heaviside's unit function,

$$= \frac{e^{iwt}}{iw}\bigg|_{1}^{2} = \left[\frac{e^{2iw} - e^{iw}}{iw}\right]$$

$$\mathcal{F}\left[\frac{1}{1+t^2}\right] = \int_{-\infty}^{\infty} \frac{e^{itw}}{1+t^2} dt$$

$$= \pi[e^{-w}h(w) + e^{+w}h(-w)].$$

In a different category from the above is

$$\mathcal{F}(1) = \int_{-\infty}^{\infty} e^{iwt} dt$$

which does not exist in the classical sense but does exist in the distributional sense, as will be proved later.

We can verify that our inversion formula as stated before is valid in the case of these functions. For example,

$$\mathcal{F}^{-1}[e^{-w}h(w) + e^{w}h(-w)]\pi = \frac{1}{2\pi} \int_{0}^{\infty} \pi e^{-w} e^{-iwt} dw + \frac{1}{2\pi} \pi \int_{-\infty}^{0} e^{w} e^{-iwt} dw$$

$$= \frac{1}{2} \left[\frac{1}{1+it} + \frac{1}{1-it} \right]$$

$$= \frac{1}{1+t^2}$$

Theorem 3. Let $f(t)$ be continuous, and let $f'(t)$ be piecewise continuous on the real line such that $\lim_{|t| \to \infty} f(t) = 0$, and $f(t)$ is Fourier transformable $\forall\, w \in \mathbb{R}$. Then $f'(w)$ is also Fourier transformable $\forall\, w \in \mathbb{R}$, and

$$\mathcal{F}(f')(w) = (-iw)(\mathcal{F}f)(w)$$

Proof. Consider the operations

$$\mathcal{F}(f')(w) = \int_{-\infty}^{\infty} f'(t) e^{iwt}\, dt$$

$$= e^{iwt} f(t) \Big|_{-\infty}^{\infty} - \int_{-\infty}^{\infty} iw e^{iwt} f(t)\, dt$$

$$= (-iw) \int_{-\infty}^{\infty} e^{iwt} f(t)\, dt$$

$$= (-iw)\mathcal{F}(f)(w)$$

These operations can be justified by integrating between $-M$ and N and letting $M, N \to \infty$. $\square$

Corollary 1. If f' is continuous on $\mathbb{R}$ and is Fourier transformable, and if $\lim_{|t| \to \infty} f(t) = 0$, then $f(t)$ is also Fourier transformable and $(\mathcal{F}f')(w) = (-iw)(\mathcal{F}f)(w)$.

Corollary 2. If $f^{(n)}(t)$ is Fourier transformable and is continuous such that $\lim_{t \to \pm\infty} f^{(k)}(t) = 0$ for $k = 0, 1, 2, \ldots, n-1$, then $f^{(n-1)}, f^{(n-2)}, \ldots, f', f$ are all Fourier transformable and

$$\mathcal{F}(f^{(k)})(w) = (-iw)^k (\mathcal{F}f)(w), \qquad k = 1, 2, \ldots, n$$

Corollary 3. If f is continuously differentiable up to order n such that $\lim_{|t| \to \infty} f^{(k)}(t) = 0$ for each $k = 0, 1, 2, \ldots, n-1$ and $f(t)$ is Fourier transformable, then each of the derivatives $f', f'', \ldots, f^{(n)}$ is Fourier transformable and

$$\mathcal{F}(f^{(k)})(w) = (-iw)^k (\mathcal{F}f)(w), \qquad k = 1, 2, 3, \ldots, n.$$

Corollary 4. If $f, f', f'', \ldots, f^{(n-1)}$ are all continuous and $f^{(n)}$ is piecewise continuous in any aribtrary, finite closed interval of $\mathbb{R}$ and if $\lim_{|t| \to \infty} f^{(k)}(t) = 0$ for each $k = 0, 1, 2, \ldots, n-1$ and $f(t)$ is Fourier transformable, then $f^{(k)}(t)$ is Fourier transformable and $\mathcal{F}(f^{(k)})(w) = (-iw)^k (\mathcal{F}f)(w)$ for each $k = 1, 2, 3, \ldots, n$.

For functions defined over a finite measure space, every $f \in L^2(X)$ belongs to $L^1(X)$, but this result is not true here in general. Let us consider

$$f(t) = \begin{cases} \dfrac{\sin t}{t}, & t \neq 0 \\ 1, & t = 0 \end{cases}$$

This function $f \notin L^1(\mathbb{R})$, but it does belong to $L^2(\mathbb{R})$. There are functions that belong to $L^1(\mathbb{R})$ and $L^2(\mathbb{R})$ as well. For example $e^{-|t|} \in L^1(\mathbb{R}) \cap L^2(\mathbb{R})$.

The Fourier transform of functions belonging to $L^2(\mathbb{R})$ does not necessarily exist in general in the pointwise sense. Also, if $f \in L^2(\mathbb{R})$, then the truncated function $f(t)\chi_{[-a,a]} \to f(t)$ in $L^2(\mathbb{R})$ as $a \to \infty$. Since $f(t)\chi_{[-a,a]} \in L^1(\mathbb{R}) \cap L^2(\mathbb{R})$, the space of functions belonging to $L^1(\mathbb{R}) \cap L^2(\mathbb{R})$ forms a dense subset of $L^2(\mathbb{R})$. The question now arises as to how the Fourier transform of $f \in L^2(\mathbb{R})$ is to be defined.

Using the above-mentioned density property, Plancherel proved the following well-known theorem [3, p. 91], which is called the *Plancheral theorem*.

Theorem 4. Let $f \in L^2(\mathbb{R})$. Then there exists a function $\hat{f}(w) \in L^2(\mathbb{R})$ such that

$$\left\| \hat{f}(w) - \int_{-a}^{a} f(t)e^{iwt}\,dt \right\|_2 \to 0 \qquad \text{as } a \to \infty \tag{1.11}$$

that is,

$$\hat{f}(w) = \text{l.i.m.}_{a\to\infty} \int_{-a}^{a} f(t)e^{iwt}\,dt$$

Moreover

$$\left\| f(x) - \frac{1}{2\pi} \int_{-a}^{a} \hat{f}(w)e^{-iwx}\,dw \right\|_2 \to 0 \qquad \text{as } a \to \infty \tag{1.12}$$

that is,

$$f(x) = \text{l.i.m.}_{a\to\infty} \frac{1}{2\pi} \int_{-a}^{a} \hat{f}(w)e^{-iwx}\,dw$$

For a measure theoretic and modern proof see Rudin [84] on the real and complex analysis.

It is further proved that

$$\hat{f}(w) = \frac{d}{dw} \int_{-\infty}^{\infty} \frac{e^{iwt} - 1}{it} f(t)\,dt \qquad \text{a.e.} \tag{1.13}$$

and

$$f(x) = \frac{d}{dx} \int_{-\infty}^{\infty} \frac{e^{-iwx} - 1}{-iw} \hat{f}(w)\,dw \qquad \text{a.e.} \tag{1.14}$$

A very elementary proof of the fact that (1.13) and (1.14) are eqivalent to (1.11) and (1.12), respectively, is given by Akhiezer and Glazman in their work on the theory of linear operators in Hilbert space [3, pp. 75–76]. These concepts of the Fourier transform were further developed by Titchmarch [99] for L^p functions $1 < p \leq 2$.

Theorem 5. Titchmarch's Theorem. Let $f \in L^p(\mathbb{R})$, $1 < p \leq 2$. Then there exists a function $\hat{f}(\xi) \in L^q(\mathbb{R})$ where $\frac{1}{p} + \frac{1}{q} = 1$ such that

$$\left\| \hat{f}(\xi) - \int_{-N}^{N} f(t)e^{it\xi}\,dt \right\|_q \to 0 \qquad \text{as } N \to \infty$$

Furthermore

$$\left\| f(x) - \frac{1}{2\pi} \int_{-N}^{N} \hat{f}(\xi)e^{-i\xi x}\,d\xi \right\|_p \to 0 \qquad \text{as } N \to \infty.$$

The Fourier receprocity relation also holds in the sense that

$$\hat{f}(\xi) = \frac{d}{d\xi} \int_{-\infty}^{\infty} f(t)\frac{e^{i\xi t} - 1}{it}\,dt \qquad \text{a.e.}$$

$$f(x) = \frac{d}{dx} \int_{-\infty}^{\infty} \hat{f}(\xi)\frac{e^{-i\xi x} - 1}{-i\xi}\,d\xi \qquad \text{a.e.}$$

Also

$$\|\hat{f}\|_q \leq K(p) \left\{ \int_{-\infty}^{\infty} |f(x)|^p dx \right\}^{1/(p-1)},$$

where $K(p)$ is a constant depending upon p. Thus the Fourier integral operator $\mathcal{F}$ is a bounded linear operator from L^p to L^q. The work of generalizing the Fourier transform of functions was continued by Laurent Schwartz [87] who put forward the theory of the Fourier transform of tempered distributions and L. Ehrenpreis [36] who brought forward his theory of the Fourier transform of Schwartz distributions.

1.3. CONVOLUTION OF FUNCTIONS

Let f and g be complex-valued functions defined on the real line, which we denote by $\mathbb{R}$. Then their convolution $(f * g)(x)$ is defined by

$$(f * g)(x) = \int_{-\infty}^{\infty} f(x - y)g(y)\,dy$$

provided that the above integral exists. At the set of points where the convolution exists, we are able to define a new function $(f * g)(x)$. Since many of the properties of the convolution defined above are similar to the product, we call the convolution

$(f * g)(x)$ a *convolution product of two functions* f and g. It is a simple exercise to show that $(f * g)(x) = (g * f)(x)$. I will now give some results in the form of theorems that demonstrate the existence of the convolution. The proof of the following theorem is found in Rudin [84, pp. 146–147].

Theorem 6. Let $f, g \in L^1(-\infty, \infty)$. Then

    i. $(f * g)(x) = \displaystyle\int_{-\infty}^{\infty} f(x - y)g(y)\, dy$ exists and is finite a.e.

    ii. The function $(f * g)(x) \in L^1(-\infty, \infty)$.

    iii. $\|f * g\|_1 \leq \|f\|_1 \|g\|_1$,

where

$$\|f\|_1 = \int_{-\infty}^{\infty} |f(x)|\, dx$$

Proof. There is no loss of generality in assuming that f and g are Borel measurable. Clearly, if f and g are Lebesgue measurable, then there exist Borel measurable functions f_0 and g_0, respectively, defined on the real line such that $f = f_0$ a.e. and $g = g_0$ a.e. Borel measurable functions are necessarily Lebesgue measurable, so we may assume that f and g are Borel measurable functions. Also the value of an integral remains unchanged by changing the values of the integrand at a set of points of measure zero. Now define

$$F(x, y) = f(x - y)g(y)$$

We want to first show that the function $F(x, y)$ is a Borel function in $\mathbb{R}^2$. For a set $E \in \mathbb{R}$, let there be a set $\tilde{E} \in \mathbb{R}^2$ defined by

$$\tilde{E} = \{(x, y) : x - y \in E\}$$

Since $x - y$ is a continuous function of (x, y), $\tilde{E}$ must be open whenever E is. It is very easy to verify that the collection of all $E \in \mathbb{R}$ for which $\tilde{E}$ (as defined above) is a Borel set forms a σ-algebra on $\mathbb{R}$. Again, if V is an open set in $\mathbb{R}$ and f is a Borel function on $\mathbb{R}$, the set $E = \{x : f(x) \in V\}$ is a Borel set in R. Therefore

$$\{(x, y) : f(x - y) \in V\} = \{(x, y) : x - y \in E\} = \tilde{E}$$

is a Borel set in $\mathbb{R}^2$. Hence the function $(x - y) \rightarrow f(x - y)$ is a Borel function. The function $(x, y) \rightarrow g(y)$ is also a Borel function in $\mathbb{R}^2$. Therefore the product $f(x - y)g(y)$ is a Borel function on $\mathbb{R}^2$. Now

$$\int_{-\infty}^{\infty} dy \int_{-\infty}^{\infty} |F(x, y)|\, dx = \int_{-\infty}^{\infty} |g(y)|\, dy \int_{-\infty}^{\infty} |f(x - y)|\, dx$$

$$= \|f\|_1 \|g\|_1$$

We have used the translation invariance property of the Lebesgue measure to show that

$$\int_{-\infty}^{\infty} |f(x - y)| dx = \|f\|_1$$

So $F(x, y) \in L^1(\mathbb{R}^2)$. Therefore in view of the Fubini's theorem $(f * g)(x) = \int_{-\infty}^{\infty} f(x - y)g(y) dy$ exists for almost all $x \in \mathbb{R}$, and $(f * g)(x) \in L^1(\mathbb{R})$. This proves (i) and (ii) together. Now

$$\|f * g\|_1 = \int_{-\infty}^{\infty} |(f * g)(x)| dx \leq \int_{-\infty}^{\infty} \left[\int_{-\infty}^{\infty} |F(x, y)| dy\right] dx$$

$$= \int_{-\infty}^{\infty} \left(\int_{-\infty}^{\infty} |F(x, y)| dx\right) dy = \|f\|_1 \|g\|_1$$

This proves (iii). $\square$

Corollary 5. Let $f, g \in L^1(\mathbb{R})$. Then

$$\mathcal{F}(f * g)(w) = (\mathcal{F}f)(w)(\mathcal{F}g)(w)$$

where F is the Fourier transformation operator.

Proof.

$$\mathcal{F}(f * g)(w) = \int_{-\infty}^{\infty} \left(\int_{-\infty}^{\infty} f(x - y)g(y) dy\right) e^{iwx} dx$$

Then by Fubini's theorem we get

$$= \int_{-\infty}^{\infty} \left(\int_{-\infty}^{\infty} f(x - y)e^{iwx} dx\right) g(y) dy$$

Now

$$\int_{-\infty}^{\infty} f(x - y)e^{iwx} dx = e^{iwy} \int_{-\infty}^{\infty} f(x - y)e^{iw(x-y)} dx$$

$$= e^{iwy} \int_{-\infty}^{\infty} f(t)e^{iwt} dt$$

By the translation invariance property of the Lebesgue measure,

$$\mathcal{F}(f * g)(w) = \int_{-\infty}^{\infty} e^{iwy} \int_{-\infty}^{\infty} f(t)e^{iwt} dt \, g(y) dy$$

$$= \int_{-\infty}^{\infty} g(y)e^{iwy} dy \int_{-\infty}^{\infty} f(t)e^{iwt} dt$$

$$= (\mathcal{F}f)(w)(\mathcal{F}g)(w)$$

An excellent proof of the following theorem is given by Hewitt and Stromberg [48, p. 397]. □

Theorem 7. For $1 < p < \infty$, let $f \in L^1(\mathbb{R})$ and $g \in L^p(\mathbb{R})$. Then for almost all $x \in \mathbb{R}$, $f(x - y)g(y)$ and $f(y)g(x - y)$ as functions of y, $\in L^1(\mathbb{R})$. For all such x define

$$(f * g)(x) = \int_{\mathbb{R}} f(x - y)g(y)\, dy$$

and

$$(g * f)(x) = \int_{\mathbb{R}} g(x - y)f(y)\, dy$$

Then

$$(f * g)(x) = (g * f)(x) \qquad \text{a.e.}$$

and

$$(f * g)(x) \in L^p(\mathbb{R})$$

further $\|f * g\|_p \leq \|f\|_1 \|g\|_p$.

Proof. Let $q = \frac{p}{p-1}$, and let $h \in L^q(\mathbb{R})$. Then each of the functions $f(x - y)$, $g(y)$, $h(x)$, are Borel measurable in $\mathbb{R}^2$, and so also are their products taken two at a time and the function $f(x - y)\, g(y)\, h(x)$. Now using Fubini's theorem, translation invariance of the Lebesgue measure and Holder's inequality we have

$$\int_{-\infty}^{\infty} \int_{-\infty}^{\infty} |f(x - y)g(y)h(x)|\, dy\, dx$$

$$= \int_{-\infty}^{\infty} |h(x)| \int_{-\infty}^{\infty} |f(x - y)g(y)|\, dy\, dx$$

$$= \int_{-\infty}^{\infty} |h(x)| \int_{-\infty}^{\infty} |f(t)g(x - t)|\, dt\, dx$$

$$= \int_{-\infty}^{\infty} |f(t)| \int_{-\infty}^{\infty} |g(x - t)h(x)|\, dx\, dt$$

$$\leq \int_{-\infty}^{\infty} |f(t)| \|g(x - t)\|_p \|h(x)\|_q\, dt$$

$$= \|g\|_p \|h\|_q \|f\|_1 < \infty$$

It follows that

$$\int_{-\infty}^{\infty} |f(x - y)g(y)| \, dy$$

and

$$\int_{-\infty}^{\infty} |f(t)g(x - t)| \, dt$$

are both finite for almost all $x \in \mathbb{R}$, and so $(f * g)(x)$ and $(g * f)(x)$ are finite a.e.

$$\left| \int_{-\infty}^{\infty} (f * g)(x)h(x) \, dy \right| = \left| \int_{-\infty}^{\infty} (g * f)(x) \, h(x) \, dx \right|$$

$$\leq \|g\|_p \, \|f\|_1 \, \|h\|_q$$

Since the dual of $L^q(\mathbb{R})$ is $L^p(\mathbb{R})$, $(f * g)(x) \in$ the dual of $L^q(\mathbb{R})$, $(f * g)(x) \in L^p$. Now in the duality notation we have

$$|\langle (f * g)(x), h(x) \rangle| \leq \|g\|_p \, \|f\|_1 \, \|h\|_q$$

$$\|(f * g)(x)\|_p = \sup_{\|h\|_q = 1} \frac{|\langle (f * g)(x), h(x) \rangle|}{\|h(x)\|_q} \leq \|g\|_p \, \|f\|_1$$

$(f * g)(x) \in L^p(-\infty, \infty)$ and

$$\|f * g\|_p \leq \|f\|_1 \, \|g\|_p$$

The fact that $(f * g)(x) = (g * f)(x)$ a.e. is also proved by using the above duality results. This is because, by Fubini's theorem, we now have

$$\langle (f * g)(x), h(x) \rangle = \langle (g * f)(x), h(x) \rangle \qquad \forall \, h \in L^q(-\infty, \infty)$$

Therefore

$$(f * g)(x) = (g * f)(x) \qquad \text{in } L^p(-\infty, \infty)$$

and so a.e. $\square$

There are many interesting results on the convolution of classical functions. The interested reader may consult Hewitt and Stromberg [48] and Treves [101]. Treves [101] gives the proofs of Theorems 7 and 8 using Holder's inequality and the density property of the space $\mathcal{D}(\mathbb{R}^n)$ into the space $L^p(\mathbb{R}^n)$, $p \geq 1$. This is a very interesting approach.

1.3.1. Differentiation of the Fourier Transform

Theorem 8. Let f be a function defined on the real line such that $f(t)$ and $t\, f(t)$ are both absolutely integrable on the real line. Then

$$\frac{d}{dw}(\mathcal{F}f)(w) = \mathcal{F}[it\, f(t)]$$

The proof is simple. The theorem can be proved by using the classical result on the switch of the order of integration and differentation, as shown by Titchmarsh [100]. Theorems 6, 7, and 8 are also valid in $\mathbb{R}^n$ and the proofs of these theorems can be obtained in the same way.

I will develop some elementary properties of distributions and then will return to the theory of the Fourier transform of distributions. The discussion will throw some light on the work done by Schwartz and L. Ehrenpreis in this direction.

1.4. THEORY OF DISTRIBUTIONS

Before I turn to the Schwartz theory of distributions, I will consider some prerequisites to it. I will assume that the reader is familiar with elements of functional analysis.

Topological Space. Let X be a nonempty set. A collection τ of subsets of X is said to be a topology on X if the following properties (axioms) are satisfied by τ:

1. $\phi \in \tau$ and $X \in \tau$.
2. If $u_1, u_2, \ldots, u_n$ belong to τ, then $\bigcap_{i=1}^{n} u_i \in \tau$.
3. Let $\{u_\alpha\}_{\alpha \in A}$ be an arbitrary collection of the members of τ, then $\bigcup_{\alpha \in A} u_\alpha \in \tau$.

It is understood that the collection $\{u_\alpha\}_{\alpha \in A}$ in property 3 can be finite or infinite (countable or uncountable).

If τ is a topology on X then members of τ are called *open sets* in X, and the pair (X, τ) together forms a topological space. Often the topological space (X, τ) is simply denoted by X.

Example 2. Let $X = \{0, 1, 2, 4\}$, and let τ be the collection of subsets of X given by

$$\tau = \{\phi\},\ \{0, 1\},\ \{1, 2\},\ \{1, 0, 2\},\ \{0, 1, 2, 4\}$$

Prove that the collection τ does not form a topology on X:

$$\{0, 1\} \cap \{1, 2\} = \{1\} \notin \tau$$

So τ does not form a topology on X. But the collection

$$\tau = \{\phi\},\ \{0, 1, 2, 4\},\ \{1, 2\}$$

does form a topology on X. Verification of axioms 1, 2, and 3 is easy.

Now let X and Y be two topological spaces, and let $f : X \to Y$. The mapping f is said to be continuous if $f^{-1}(V)$ is an open set in X for every open set V in Y. The mapping f is said to be continuous at a point $x \in X$ if the inverse image of every open set containing $f(x)$ in Y is an open set in X.

A nonempty set X is said to be a metric space if there is a distance function ρ defined on it such that

1. $0 \leq \rho(x, y) < \infty \; \forall \, x, y \in X$.
2. $\rho(x, y) = 0$ iff $x = y$.
3. $\rho(x, y) = \rho(y, x) \; \forall \, x, y \in X$.
4. $\rho(x, y) \leq \rho(x, z) + \rho(z, y)$ (triangle inequality).

The set $\{y \in X : \rho(x, y) < r\}$ is defined to be an open ball with center at x and radius $r > 0$. Let τ be the collection of all sets $E \in X$ that are arbitrary unions of open balls. Then τ is a topology on X. It is very easy to verify that τ satisfies the axioms 1, 2, and 3 for a topology. Therefore a metric space is a topological space.

1.4.1. Topological Vector Spaces

A topological vector space (or TVS) X is a vector space equipped with a topology τ such that the operation of addition from $X \times X \to X$ and the scalar multiplication from $\mathbb{C} \times X \to X$ are both continuous. Thus a topological vector space is a vector space whose topology is compatible with its linear structure. A normed linear space X is a topological vector space. If $x_\alpha \to x_0$, $y_\alpha \to y_0$, then

$$\|x_\alpha + y_\alpha - (x_0 + y_0)\| \leq \|x_\alpha - x_0\| + \|y_\alpha - y_0\| \to 0$$

Therefore $x_\alpha \to x_0, y_\alpha \to y_0 \Rightarrow (x_\alpha + y_\alpha) \to x_0 + y_0$. If $c_\alpha \to c_0$ and $x_\alpha \to x_0$, then

$$\|c_\alpha x_\alpha - c_0 x_0\| = \|(c_\alpha - c_0)x_\alpha\| + \|c_0(x_\alpha - x_0)\|$$
$$\leq |c_\alpha - c_0| \, \|x_\alpha\| + |c_0| \, \|x_\alpha - x_0\|$$

Since $x_\alpha \to x_0$, it can be shown that there exists $M > 0$ such that $\|x_\alpha\| \leq M, \forall \, x_\alpha \in$ a neighborhood of x_0. Therefore

$$\|c_\alpha x_\alpha - c_0 x_0\| \leq |c_\alpha - c_0|M + |c_0| \, \|x_\alpha - x_0\| \to 0$$

as $c_\alpha \to c_0$ and $x_\alpha \to x_0$; then $c_\alpha \to c_0$ and $x_\alpha \to x_0 \Rightarrow c_\alpha x_\alpha \to c_0 x_0$.

A subset A of a topological vector space X is said to be bounded if for any neighborhood U of $0 \in X$ there exists a constant $k > 0$ such that $kA \subset U$. A neighborhood of a point is defined to be a set containing an open set that contains the point. Equivalently a subset V of a TVS X is said to be bounded if by suitable contraction it can be contained in any neighborhood of zero. More precisely for a given neighborhood U of zero there exists a $\lambda > 0$ such that

$$V \subset \lambda U$$

Definition. A subset A of a vector space V over $\mathbb{C}$ (field of complex numbers) is said to be convex if $x, y \in A$ and $\forall \alpha, \beta \in R_+$, such that $\alpha + \beta = 1$ implies that $\alpha x + \beta y \in A$.

Example 3. $\{x : |x - x_0| < r\}$ on $\mathbb{R}^n$ is convex where

$$|x - x_0| = \left[\sum_{i=1}^{n}(x_i - x_{0_i})^2\right]^{1/2}$$

Example 4. $\{x : \|x\| \le b\}$ in a Banach space is convex. Also the set

$$\left\{x : \left[\sum_{i=1}^{n}(x_i - x_{0i})^2\right]^{1/2} \le b\right\}$$

in $\mathbb{R}^n$ is convex.

Example 5. Intersection of any family of convex set is a convex set. A subset A of a vector space V is said to be balanced if $\lambda A \subset A$ for all $\lambda \in \mathbb{C}$ such that $|\lambda| \le 1$. Taking $\lambda = 0$, we see that 0 belongs to any balanced set A. For example,

$$\left\{x : |x| = \sqrt{x_1^2 + x_2^2 + \cdots + x_n^2} \le \rho\right\}$$

is a balanced set of $\mathbb{R}^n$, but

$$\left\{x : \sqrt{(x_1 - 1)^2 + (x_2 - 1)^2 + \cdots + (x_n - 1)^2} \le \rho\right\}$$

is not a balanced set of $\mathbb{R}^n$ if $\rho < \sqrt{n}$.

On $V = \mathbb{R}$ the closed interval $[1, 2]$ is convex but not balanced. One can verify that the intersection of a family of convex and balanced subsets in a vector space V is a convex and balanced subset of V.

A subset A of a vector space V is said to be absorbing if for every $x \in V$, there exists a $\lambda > 0$ such that $x \in \lambda A$. Note that an absorbing set must contain the origin. The set $[-2, 0, 5]$ is an absorbing set in $\mathbb{R}$, but the set $[1, 2, 5]$ is not an absorbing subset of $\mathbb{R}$.

Any neighborhood of the origin in a topological vector space V is an absorbing subset of V. A family of open sets in a topological space (X, τ) is said to form a basis of the topology if every open set in X is the union of sets of the family. A family of neighborhoods of a point x in a topological space is said to form a *neighborhood basis* if every neighborhood of x contains some set of the family. Two neighborhood bases at the point x are said to be *equivalent* if each neighborhood in one base contains a neighborhood in the other base. Let T_1 and T_2 be two topological bases in X. Then T_1 and T_2 are said to be equivalent if each open set in T_1 contains an open set of T_2 and each open set of T_2 contains an open set of T_1. For every $x \in X$, let B_x be a neighborhood basis at x, and let $\mathcal{B} = \bigcup_{x \in X} B_x$. Then obviously $\mathcal{B}$ forms a basis for the topology on X. The topological space X satisfies first countability axiom if there is a countable neighborhood basis at every point of $x \in X$. If there is a countable basis for X, then X is said to satisfy the second countability axiom.

Consider a topological vector space X, and define a translation mapping $x \to x + x_0$. Clearly it is a mapping from X onto itself and is continuous, and its inverse mapping $x \to x - x_0$ is also a continuous mapping from X onto itself. Therefore the translation mapping is a homeomorphism from X onto itself. It maps an open set into an open set. Once we know the neighborhood basis at the origin, we know the neighborhood basis at any point $x_0 \in X$ by simply making use of the translation mapping. Thus the topology of a topological vector space can be completely determined by knowing a neighborhood basis at the origin. If T is a linear functional on a topological vector space X, then the continuity of T at any arbitrary point $x \Leftrightarrow$ continuity of T at 0. The proof is very simple.

A subset Y of X is defined to be a topological subset of X if Y is equipped with the topology induced by X on Y, which really means that a set $M \subset Y$ is said to be open set in Y iff there exists an open set $B \subset X$ such that

$$M = Y \cap B$$

Let there be two topologies τ_1 and τ_2 on a set Y. We say that the topology τ_1 is stronger than the topology τ_2 if every element (open set) of τ_2 is contained in τ_1. We say that two topologies τ_1 and τ_2 on X are equivalent if τ_1 is stronger than τ_2 and also τ_2 is stronger than τ_1.

Definition. A topological vector space X is said to be locally convex and called a locally convex space if there exists a neighborhood basis at 0 in X whose every element is a convex set.

A functional γ on a vector space X is said to be a seminorm if it satisfies the following conditions:

 i. $\gamma(\varphi) \geq 0 \; \forall \; \varphi \in X$.
 ii. $\gamma(\alpha\varphi) = |\alpha| \gamma(\varphi) \; \forall \; \varphi \in X$ and $\alpha \in \mathbb{C}$.
 iii. $\gamma(\varphi + \psi) \leq \gamma(\varphi) + \gamma(\psi)$.

In fact the condition i is redundant. If we put $\alpha = 0$ in condition ii, we get $\gamma(0) = 0$. Likewise, by putting $\varphi = \varphi - \psi$ in condition iii and interchanging the roles of φ and ψ, we get $\gamma(\varphi - \psi) \geq |\gamma(\varphi) - \gamma(\psi)|$. Thus γ is a nonnegative functional on X.

There may be a nonzero $x \in V$ such that $\gamma(x) = 0$. Then the seminorm with the additional condition that $\gamma(x) = 0 \Rightarrow x = 0$ becomes a norm.

Theorem 9. Let γ be a seminorm on a linear space X. Then the γ-ball $B_\gamma(r) = \{x \in X : \gamma(x) < r\}$ is (i) convex, (ii) balanced, and (iii) absorbing.

Proof. (i) Convex: If $x, y \in B_\gamma(r)$, then for $\alpha \in [0, 1]$,

$$\gamma(\alpha x + (1 - \alpha)y) \leq \gamma(\alpha x) + \gamma((1 - \alpha)y)$$
$$\leq |\alpha|\gamma(x) + |1 - \alpha|\gamma(y)$$
$$\leq \alpha\gamma(x) + (1 - \alpha)\gamma(y)$$
$$< \alpha r + (1 - \alpha)r$$
$$< r$$

(ii) balanced: Let $\alpha \in \mathbb{C}$ such that $|\alpha| \leq 1$ and $x \in B_\gamma(r)$. Then

$$\gamma(\alpha x) = |\alpha|\gamma(x) < r$$
$$\bigcup_\alpha \alpha B_\gamma(r) \subset B_\gamma(r) \qquad \text{if } |\alpha| \leq 1$$

(iii) absorbing: $\lambda 0 \in B_\gamma(r)$. If $\gamma(x) = 0$ for $x \neq 0$, then $x \in B_\gamma(r)$. Therefore for a nonzero $x \in V$ such that $\gamma(x) \neq 0$, choose $\lambda_x = \frac{r}{\gamma(x)}$. Now $\gamma(\lambda_x x) < r$, that is, $\lambda_x x \in B_\gamma(r)$. $\square$

Let E be an absorbing set in a vector space V, and let x be any point in V. Define a functional μ_E by $\mu_E(x) = \inf\{\lambda > 0 : \lambda^{-1}x \in E\} = \inf\{\lambda > 0 : x \in \lambda E\}$. Clearly $\mu_E(0) = 0$ in that, if λ_i is a sequence of positive number tending to zero, then $\inf_i \lambda_i = 0$. Therefore

$$0 \leq \mu_E(0) \leq \inf\{\lambda_i > 0 : 0 \in \lambda_i E\}$$
$$0 \leq \mu_E(0) \leq 0.$$

Hence $\mu_E(0) = 0$. The functional $\mu_E : X \to [0, \infty)$ as defined above is called the *Minkowski functional* of E [1].

Theorem 10. Let E be a convex balanced and absorbing subset of a vector space V. Then the Minkowski functional $\mu_E(x)$ of the set E is a seminorm over the vector space V.

Proof. For any $x, y \in V$, let

$$\mu(x) = \inf\{\lambda > 0 : x \in \lambda E\}$$
$$\mu(y) = \inf\{\mu > 0 : y \in \mu E\}$$

We can choose nonnegative numbers λ and μ satisfying

$$\mu_E(x) \leq \lambda \leq \mu_E(x) + \epsilon \quad \text{and} \quad x \in \lambda E$$
$$\mu_E(y) \leq \mu \leq \mu_E(y) + \epsilon \quad \text{and} \quad y \in \mu E$$

By virtue of convexity of E,

$$\frac{1}{\lambda + \mu}(x + y) = \frac{\lambda}{\lambda + \mu}\lambda^{-1}x + \frac{\mu}{\lambda + \mu}\mu^{-1}y \in E$$

Therefore

$$\mu_E(x + y) \leq \lambda + \mu \leq \mu_E(x) + \epsilon + \mu_E(y) + \epsilon$$

Since ϵ is an arbitrary positive number, we have

$$\mu_E(x + y) \leq \mu_E(x) + \mu_E(y).$$

Again,

$$\mu_E(x) \leq \lambda \leq \mu_E(x) + \epsilon, \qquad x \in \lambda E$$

If c is a complex number (excluding the trivial cases $c = 0$) and η is an arbitrary positive number $< \mu_E(x) \neq 0$, then by definition

$$x \notin (\mu_E(x) - \eta)E$$

Therefore

$$cx \notin (|c|\mu_E(x) - |c|\eta)E$$

If $cx \in \lambda E$ for $\lambda > 0$, then $\lambda > (|c|\mu_E(x) - |c|\eta)$. Therefore $\mu_E(cx) = \inf \lambda \geq |c|\mu_E(x) - |c|\eta$. Since η is arbitrary, we get

$$\mu_E(cx) \geq \mu_E(x)|c| \qquad (1.15)$$

Also

$$cx \in \lambda|c|E \; \forall \; \lambda \qquad \text{satisfying } \lambda > \mu_E(x)$$

$$\mu_E(cx) \leq \inf \lambda|c| = |c| \inf \lambda = |c|\mu_E(x) \qquad (1.16)$$

Combining the two inequalities (1.15) and (1.16), we have

$$\mu_E(cx) = |c|\mu_E(x)$$

The result follows from (1.16) if $\mu_E(x) = 0$. $\square$

Definition. The family $P = \{p_\alpha\}_{\alpha \in A}$ of seminorms defined over a topological vector space X is said to be separating if for each nonzero $x \in X$ there is at least one seminorm $p_\beta \in P$ such that $p_\beta(x) \neq 0$.

A TVS X is said to be Hausdorff if every two different points in X have disjoint neighborhoods.

Theorem 11. The topology of a locally convex topological vector space X is equivalent to the topology generated by a metric if and only if X is Hausdorff and has a countable local base.

Proof. (i) Assume that X is metrizable, namely that there exists a metric $\rho(x, y)$ that generates the topology of X. Then the set of balls $\{B_n\}_{n=1}^{\infty}$ where $B_n = \{x \in X : \rho(0, x) < \frac{1}{n}\}$ are convex, balanced, and absorbing and form a countable base at 0 for a Hausdorff topology. (ii) If X is Hausdorff and has a countable local base $\{B_n\}_{n=1}^{\infty}$ at 0 such that each of B_n is convex, balanced, and absorbing, define the seminorm p_i on the space X to be the Minkowski functional of B_i. Set

$$\rho(x, y) = \sum_{i=1}^{\infty} \frac{1}{2^i} \frac{p_i(x - y)}{1 + p_i(x - y)} \qquad \forall\, x, y \in X$$

Now

$$\frac{a}{1 + a} \le \frac{b}{1 + b} \qquad \text{if } 0 \le a \le b$$

$$\frac{p_i(x - y)}{1 + p_i(x - y)} = \frac{p_i[x - z + z - y]}{1 + p_i[x - z + z - y]}$$

$$\le \frac{p_i(x - z) + p_i(z - y)}{1 + p_i(x - z) + p_i(z - y)}$$

$$\le \frac{p_i(x - z)}{1 + p_i(x - z)} + \frac{p_i(z - y)}{1 + p_i(z - y)}$$

and

$$\rho(x, y) \le \rho(x, z) + \rho(z, y)$$

Therefore ρ is subadditive and $\rho(x, y) = 0 \Rightarrow x = y$ as $\{p_i\}_{i=1}^{\infty}$ is separating.

Next to show that $\rho(x, y) = \rho(y, x) \ge 0$ is trivial. The fact that the topology generated by $\rho(x, y)$ is compatible with the local structure of X follows from the properties of seminorms p_i. Now let

$$U_n = \left\{ x \in X : d(0, x) \le \frac{i}{2^n} \right\}$$

$\rho : X \times X \to R$ is continuous as the series representing $\rho(x, y)$ converges uniformly. Since each of the terms $\frac{p_n(x-y)}{1+p_n(x-y)}$ is continuous, the metric $\rho(x, y)$ to which this series converges must be continuous function of x, y. So

$$U_n = \left\{ x \in X : \rho(0, x) < \frac{1}{2^n} \right\}$$

is open in (X, B). Clearly $U_{n+1} \subset B_n$ because, if $x \notin B_n$, then

$$p_n(x) \geq 1 \Rightarrow \rho(0, x) \geq \frac{1}{2^n} \frac{p_n(x)}{1 + p_n(x)} \geq \frac{1}{2^n} \frac{1}{1 + 1}$$

$$\rho(0, x) \geq \frac{1}{2^{n+1}} \qquad \text{(a contradiction)}$$

Therefore $\{U_n\}$ forms a local base for the topology of (X, B). $\square$

Theorem 12. Let V be a topological vector space. A subset B of V is bounded if and only if every sequence contained in B is bounded in V.

Proof. If B is a bounded set in V, then every sequence in B is bounded (obvious).

Now assume that every sequence of B is bounded in V. We wish to show that B is bounded. Assume that B is not bounded. Then corresponding to any positive integer n and a neighborhood U of zero, we can find an element $x_n \in B$ such that $x_n \notin nU$. Therefore we are able to find a sequence $\{x_n\}_{n=1}^{\infty}$ in B that is not bounded, and this is a contradiction. $\square$

There are situations where the topology of a topological vector spaces can be generated by a norm. The topology generated by a norm will be equivalent to the topology of the topological vector space under consideration. Such topological vector spaces are said to be *normable*.

Any ball in a normed linear space is bounded. So in a normed linear space there exist balls which are neighborhoods of the zero element that are at the same time bounded. This property is the characteristic of normable spaces at least among Hausdorff locally convex spaces.

Some topological vector spaces are not metrizable. An example of this type of a TVS is the Schwartz testing function space D equipped with the usual topology, which I will discuss in the next section. An excellent proof of this fact is given by Shilov [88]. Since D is not metrizable, it is not normable either. Theorem 13 below gives a characterization of locally convex spaces that are normable. We will need the following lemmas to prove it:

Lemma 1. (i) Let U be a convex subset of a vector space V. Then the set U_0 defined by $U_0 = \bigcup_{|\lambda| \leq \epsilon} \lambda U$ is also a convex subset of the vector space V. (ii) U_0 is balanced.

Proof. If x and y are two elements of the set U, then for $0 < \alpha < 1$, $\alpha x + (1 - \alpha)y \in U$. Therefore $\lambda \alpha x + \lambda(1 - \alpha)y \in \lambda U$. Let $\lambda_1 x$ and $\lambda_2 y$ be two representative elements of $\lambda_1 U$ and $\lambda_2 U$, respectively, for $0 < |\lambda_1|, |\lambda_2|$ both $\leq \epsilon$. Our claim is that the element

$$\alpha \lambda_1 x + (1 - \alpha)\lambda_2 y$$

belongs to the set

$$(\alpha \lambda_1 + (1 - \alpha)\lambda_2)U$$

where $0 < \alpha < 1$. If we put

$$\Lambda[\beta x + (1 - \beta)y] = \alpha\lambda_1 x + (1 - \alpha)\lambda_2 y$$

then

$$\left.\begin{array}{l} \Lambda\beta = \alpha\lambda_1 \\ \Lambda(1 - \beta) = (1 - \alpha)\lambda_2 \end{array}\right\} \Rightarrow \Lambda = \alpha\lambda_1 + (1 - \alpha)\lambda_2$$

Further note that if $|\lambda_1|, |\lambda_2| \leq \epsilon$, then

$$|\alpha\lambda_1 + (1 - \alpha)\lambda_2| \leq \epsilon$$

Therefore U_0 is a convex set.

   (ii) Let $k \in \mathbb{C}$ such that $|k| \leq 1$. if $\lambda \in C$ such that $|\lambda| \leq \epsilon$. Then $k\lambda \in \mathbb{C}$, satisfying $|k\lambda| \leq \epsilon$.

$$kU_0 = \bigcup_{|\lambda| \leq \epsilon} k\lambda U$$

$$= \bigcup_{\substack{|\mu| \leq \epsilon \\ \mu = k\lambda}} \mu U \subset U_0$$

Therefore U_0 is balanced. $\square$

Corollary 6. Let X be a locally convex Hausdorff TVS. Then there exists a subset U_0 of X that is convex, balanced, and absorbing.

Proof. Since X is a locally convex TVS, every neighborhood of 0 contains a convex neighborhood of 0. Again the mapping

$$\alpha x : \mathbb{C} \times X \to X$$

is continuous. Let V be a neighborhood of zero. There exists $\epsilon > 0$ and a neighborhood U of 0 such that

$$\alpha x \in V \; \forall \; \alpha \qquad \text{satisfying } |\alpha| \leq \epsilon \text{ and } x \in U$$

that is, $\alpha U \in V$. The set U is chosen to be a convex set by choice as X is locally convex. Now let

$$U_0 = \bigcup_{|\alpha| \leq \epsilon} \alpha U$$

Each αU is a neighborhood of 0, so U_0 is a neighborhood of 0 as well. Therefore U_0 is absorbing, as a neighborhood of zero in a TVS is absorbing. By Lemma 1, U_0 is convex, balanced, and absorbing. $\square$

Theorem 13. Let X be a locally convex and Hausdorff TVS. The topology of X can be generated by a norm if and only if its zero vector has a bounded neighborhood.

Proof. Let X be normable with a norm. Then the unit ball $\{x : \|x\| < 1\}$ is a bounded neighborhood of 0. If U is a bounded neighborhood of 0 in a TVS X that is Hausdorff and locally convex, we can find an open neighborhood U_0 of zero contained in U such that U_0 is convex balanced, absorbing, and bounded. Let p be the Minkowski functional defined on U_0. By the mode of construction of the Minkowski functional, if $p(x) = 0$, then $x \in \lambda U_0$ for any $\lambda > 0$. Now, for any neighborhood of 0, we can choose $\lambda > 0$ appropriately so that λU_0 is contained in the neighborhood of zero. So x belong to every neighborhood of zero. Since X is Hausdorff two points 0, and x cannot be contained in every neighborhood of zero unless $x = 0$. Therefore $p(x) = 0 \Rightarrow x = 0$. So p is a norm. $\square$

For further details about the normable and metrizable TVS one can look into Treves [103].

Definition. The support of a complex-valued function f defined over a set Ω is the closure of the set $\{x \in \Omega : \varphi(x) \neq 0\}$ in the topological space Ω. That is, it is the smallest closed set containing the set $\{x \in \Omega : \varphi(x) \neq 0\}$. Define a function $f(x)$ over the interval $[0, 1]$ by

$$f(x) = \begin{cases} 1, & \text{where } x \text{ is rational} \\ 0, & \text{where } x \text{ is irrational} \end{cases} \tag{1.17}$$

Here the support of $f \equiv \operatorname{supp} f = $ the closed interval $[0, 1]$. If, however, we restrict to the case that $f(x)$ is defined over the interval $(-1, 1)$ by (1.17), then $\operatorname{supp} f = (-1, 1)$ and not $[-1, 1]$. Now consider the infinitely differentiable function $\varphi(x)$ such that

$$\varphi(x) = \begin{cases} \exp \frac{-1}{1-|x|^2} & \text{when } |x| < 1 \\ 0 & \text{when } |x| \geq 1 \end{cases} \tag{1.18}$$

Then the $\operatorname{supp} \varphi$ is the unit ball $\{x \in \mathbb{R}^n : |x| \leq 1\}$.

1.4.2. Locally Convex Spaces

Let Ω be an open subset of $\mathbb{R}^n$. We denote the set of functions defined on Ω and continuously differentiable up to order m by $C^{(m)}(\Omega)$. Now assume that K is a compact subset of Ω. Then the set of functions that are continuously differentiable up to order m and have support contained in the set K will be denoted by $C_K^{(m)}(\Omega)$. It is easy to see that $C^{(m)}(\Omega)$ and $C_K^{(m)}(\Omega)$ both are vector spaces and that $C_K^{(m)}(\Omega) \subset C^{(m)}(\Omega)$.

The space of infinitely differentiable complex-valued functions defined over Ω will be denoted by $C^\infty(\Omega) = \bigcap_{m \geq 0} C^{(m)}(\Omega)$. The set $C_K^\infty(\Omega)$ stands for the set of infinitely differentiable complex-valued functions defined over Ω and having support contained in K. The space $C_0^\infty(\Omega)$ will stand for the set of complex-valued and infinitely differentiable functions defined over Ω having compact supports in Ω. Recall that a topological space is a locally convex space if its topology has a local base whose members are convex sets. Each of the foregoing vector spaces can be equipped with a topology that will make them locally convex TVS. As an example consider the space $C_K^{(m)}(\Omega)$. Define the set of seminorms $\{\gamma_k\}_{k=0}^m$ by

$$\gamma_k(\varphi) = \sup_{\substack{x \in K \\ |\alpha| \le k}} |\partial^\alpha \varphi| \tag{1.19}$$

The collection of seminorms $\{\gamma_k\}_{k=0}^m$ is a separating collection of seminorms. The space $C_k^{(m)}(\Omega)$ becomes a locally convex Hausdorff TVS when it is equipped with the topology generated by the sequence of seminorms $\{\gamma_k\}_{k=0}^\infty$ defined by (1.19) as follows: We define the open neighborhood basis about the origin as the collection of balloons centered at the origin by

$$\{\varphi : \gamma_{\nu_k}(\varphi) < \epsilon_k, \ k = 1, 2, \ldots, r\}$$

where $\epsilon_1, \epsilon_2, \ldots, \epsilon_r$ are arbitrary set of positive numbers and $\nu_1, \nu_2, \ldots, \nu_r$ are integers lying between 0 to $|m|$, and likewise r is an integer lying between 0 and m. Denote this class of balloons by B_0. Since the space $C_k^{(m)}(\Omega)$ is a vector space, the open neighborhood basis about a point ψ will be given by $B_0 + \psi$, which we denote by B_ψ. Then $\bigcup_{x \in C_k^m(\Omega)} B_x$ forms a basis for the topology of the space $C_K^m(\Omega)$ with which we equip it. One can verify that the space $C_K^m(\Omega)$ is a locally convex, Hausdorff, and "complete (sequentially complete)" TVS. Convergence to zero of a sequence $\{\varphi_\nu\}_{\nu=1}^\infty$ in this topology is equivalent to the uniform convergence of each of the derivative sequence $\{\varphi_\nu^{(i)}\}_{\nu=1}^\infty$ to zero for each $|i| = 0, 1, 2, \ldots, m$ over K and so uniformly over Ω. This space is also metrizable because its topology can be generated by the metric

$$\rho(\varphi, \psi) = \sum_{|i|=0}^m \frac{1}{2^i} \frac{\gamma_i(\varphi - \psi)}{1 + \gamma_i(\varphi - \psi)} \tag{1.20}$$

For details of the basis of a topology, the reader is referred to Bremmerman [8], Friedman [42], and Kelley [55], and Zemanian [110]. The set of continuous linear functionals over the space $C_K^{(m)}(\Omega)$ is denoted by $(C_K^m(\Omega))'$. Obviously it forms a vector space. If $f, g \in (C_K^m(\Omega))'$ and α, β are complex constants, we define the functional $\alpha f + \beta g$ by

$$\langle \alpha f, \varphi \rangle = \alpha \langle f, \varphi \rangle$$

$$\langle f + g, \varphi \rangle = \langle f, \varphi \rangle + \langle g, \varphi \rangle \qquad \forall \ \varphi \in C_K^m(\Omega)$$

An example of a function belonging to the dual space $C_K^{(m)}(\Omega)$ is the functional $\delta(x - a)$ where

$$\langle \delta(x - a), \varphi \rangle = \varphi(a) \qquad \forall \ \varphi \in C_K^{(m)}(\Omega)$$

1.4.3. Schwartz Testing Function Space: Its Topology and Distributions

Let Ω be an open subset of $\mathbb{R}^n$ and $K_1, K_2, K_3, \ldots$ be compact subsets of Ω such that

$$K_1 \subset K_2 \subset \cdots \subset K_n \ldots$$

and

$$\bigcup_{i=1}^{\infty} K_i = \Omega$$

Such a construction is always possible [Yosida, 108].

We have already discussed the space $C_K^m(\Omega)$ and the space $C_0^\infty(\Omega)$. When the space $C_0^\infty(\Omega)$, which consists of infinitely differentiable complex-valued functions defined over a subset Ω of $\mathbb{R}^n$ having compact supports in Ω, is equipped with the topology as described by Laurent Schwartz [87], the space is named as $\mathcal{D}(\Omega)$. Clearly

$$\mathcal{D}_{K_1}(\Omega) \subset \mathcal{D}_{K_2}(\Omega) \subset \mathcal{D}_{K_3}(\Omega) \ldots$$

and

$$\mathcal{D}(\Omega) = \bigcup_{i=1}^{\infty} \mathcal{D}_{K_i}(\Omega)$$

Bremermann [8] describes briefly the topology of the space $\mathcal{D}(\mathbb{R}^n)$ described by Laurent Schwartz [87] as follows:

Let ϵ_j be a monotone sequence of positive numbers tending to zero, and let $\{m_j\}$ be a sequence of positive integers monotonically increasing to ∞. The neighborhood basis B of 0 is defined as follows: B consists of the sets

$$V(\{m_j\}, \{\epsilon_j\}) \tag{1.21}$$
$$= \{\varphi : \varphi \in \mathcal{D}(\mathbb{R}^n), |\partial^\alpha \varphi(t)| < \epsilon_j; \ \forall \ |\alpha| \leq m_j, \ t \notin \Omega_j\}$$

where $\Omega_0 = \emptyset$ and $\Omega_j = \{t : \|t\| < j\}$. B forms a neighborhood basis at the function $\varphi \equiv 0$ and thus generates a topology for $\mathcal{D}(\mathbb{R}^n)$. If $\{\varphi_\nu\}_{\nu=1}^{\infty}$ is a sequence in $\mathcal{D}(\mathbb{R}^n)$ tending to zero as $\nu \to \infty$ in the topology of $\mathcal{D}(\mathbb{R}^n)$ as described above, then there exists a compact subset K of $\mathbb{R}^n$ such that (1) the support of each of the elements $\varphi_\nu(t)$ is contained in K, and (2) each of the sequences $\{\varphi_\nu^{(k)}(t)\}_{\nu=1}^{\infty}$ tends to zero uniformly over K and so over $\mathbb{R}^n$ as $\nu \to \infty$, for each $k = 0, 1, 2, \ldots$. The uniformity here is assumed with respect to t and not with respect to k. The proof of this fact is easily obtained by the method of contradiction. The convergence in $\mathcal{D}(\mathbb{R}^n)$ is very strong. There exist sequences of functions in $\mathcal{D}(\mathbb{R}^n)$ converging to zero along with all their derivatives uniformly on $\mathbb{R}^n$, yet the sequence may not converge to zero in $\mathcal{D}(\mathbb{R}^n)$ as the following example shows.

Let φ be the function as defined by (1.18). Then $\varphi \in \mathcal{D}(\mathbb{R}^n)$. Define a sequence $\{\varphi_\nu(t)\}_{\nu=1}^\infty$ by

$$\varphi_\nu(t) = \frac{1}{\nu} \varphi(t).$$

Then $\varphi_\nu(t) \to 0$ in $\mathcal{D}(\mathbb{R}^n)$ as $\nu \to \infty$. But the sequence

$$\psi_\nu(t) = \frac{1}{\nu} \varphi \left(\frac{t}{\nu} \right)$$

$$= \frac{1}{\nu} \varphi \left(\frac{t_1}{\nu}, \frac{t_2}{\nu}, \ldots, \frac{t_n}{\nu} \right)$$

does not converge in $\mathcal{D}(\mathbb{R}^n)$ as $\nu \to \infty$. This is because there may not exist a compact set $K \in \mathbb{R}^n$ that contains the supports of each of $\psi_\nu(t)$. There is, however, an equivalent way of describing the topology of the space $\mathcal{D}$ (as described by Schwartz [87], Zemanian [110], and Yosida [108]). We will now explain it.

Let K_1, K_2, K_3 be the sequence of compact subsets of $\Omega \in \mathbb{R}^n$ such that

$$K_1 \subset K_2 \subset K_3 \ldots$$

and

$$\bigcup_{i=1}^\infty K_i = \Omega$$

Recall that

$$\mathcal{D}_{K_1}(\Omega) \subset \mathcal{D}_{K_2}(\Omega) \subset \mathcal{D}_{K_3}(\Omega) \ldots$$

and

$$\mathcal{D}(\Omega) = \bigcup_{i=1}^\infty \mathcal{D}_{K_i}(\Omega)$$

The topology on each of the spaces $\mathcal{D}_K(\Omega)$ is described by the sequence of seminorms $\{\gamma_i(\varphi)\}_{i=0}^\infty$ as defined by (1.19) and the space $\mathcal{D}_K(\Omega)$ is metrizable by the metric (1.20). It can be easily seen that the topology of $\mathcal{D}_{K_i}(\Omega)$ is the same as that induced on $\mathcal{D}_{K_i}(\Omega)$ by $\mathcal{D}_{K_{i+1}}(\Omega)$ for each $i = 1, 2, 3, 4, \ldots$.

A natural way to topologize the space $\mathcal{D}(\Omega)$ is that if V is an open set in $\mathcal{D}(\Omega)$, then $V \cap \mathcal{D}_K(\Omega)$ is an open set in $\mathcal{D}_K(\Omega)$ for any compact set $K \in \Omega$. Therefore, if V is an open set containing the origin in the space $\mathcal{D}(\Omega)$, then $V \cap \mathcal{D}_{K_i}(\Omega)$ is defined to be an open subset of $\mathcal{D}_{K_i}(\Omega)$ containing the origin for each $i = 1, 2, 3, \ldots$. When topologized in this way $\mathcal{D}(\Omega)$ is called as the inductive limit of $\mathcal{D}_{K_i}(\Omega)$'s (see Yosida [108, p. 28], Shaefer [85, pp. 56–57]). When $\mathcal{D}(\Omega)$ is topologized in this way, the embedding of $\mathcal{D}_K(\Omega)$ in $\mathcal{D}(\Omega)$ is continuous, and this is a very natural thing to expect.

Theorem 14. Let $\mathcal{D}(\Omega)$ be the inductive limit of $\mathcal{D}_{K_i}(\Omega)$. That is, K_i's are compact subsets of $\Omega \in \mathbb{R}^n$ and

$$\Omega = \bigcup_{i=1}^{\infty} K_i$$

$K_1 \subset K_2 \subset K_3 \subset \Omega$. Then the convergence of a sequence $\{\varphi_k\}$ to 0 in $\mathcal{D}(\Omega)$, namely $\lim_{k \to \infty} \varphi_k = 0$ in $\mathcal{D}(\Omega)$ means that

  i. There exists a compact subset K of Ω such that supp $\varphi_k \subseteq K \; \forall \, k = 1, 2, 3, \ldots$.
  ii. For any differential operator $\mathcal{D}^p$, the sequence $\{\mathcal{D}^p f_k(x)\}_{k=1}^{\infty}$ converges to 0 uniformly on K.

Proof (by Yosida [108]). We need prove only (i), since (ii) is merely a trivial consequence of (i). Assume that (i) is not true. Therefore there exists a sequence $\{x^{(k)}\}_{k=1}^{\infty}$ of points of Ω having no limit points in Ω and a subsequence $\{\varphi_{k_j}(x)\}$ of $\{\varphi_k(x)\}$ such that $\varphi_{k_j}(x_j) \neq 0$. Then the seminorm

$$p(\varphi) = \sum_{j=1}^{\infty} 2 \sup_{x \in K_j - K_{j-1}} \left| \frac{\varphi(x)}{\varphi_{k_j}(x_j)} \right|$$

$$x_j \in K_j - K_{j-1}$$

$$K_0 = \phi$$

defines a neighborhood $U = \{\varphi \in C_0^{\infty}(\Omega); p(\varphi) \leq 1\}$ of 0 of $\mathcal{D}(\Omega)$. This is because

$$U \cap \mathcal{D}_{K_1}(\Omega)$$

$$\subset \left\{ \varphi : \varphi \in \mathcal{D}_{K_1} : \|\varphi\| \leq \frac{|\varphi_{k_1}(x_1)|}{2} \right\}$$

and

$$U \cap \mathcal{D}_{K_2}(\Omega)$$

$$\subset \left\{ \varphi : \varphi \in \mathcal{D}_{K_2} : \|\varphi\| \leq \max \frac{(|\varphi_{k_1}(x_1)|, |\varphi_{k_2}(x_2)|)}{2} \right\}$$

and so on, where $\|\varphi\| = \sup_{x \in \mathbb{R}} |\varphi(x)|$.

Therefore U as described above is a neighborhood of 0 in $\mathcal{D}(\Omega)$. But none of the φ_{k_j} belong to U as $p(\varphi_k(x)) \geq 2$, so φ_{k_j} is not eventually in U. Therefore φ_{k_j} does not tend to zero, a contradiction. $\square$

I will now show that the space $\mathcal{D} = \mathcal{D}(\Omega)$ is not metrizable. The following counterexample is a special case of the counterexample given by Shilov [88]. Define a double sequence function $\{\varphi_{\nu,\mu}\}_{\nu,\mu=1}^{\infty}$ in $\mathcal{D}$ as

$$\varphi_{\nu,\mu}(t) = \begin{cases} \frac{1}{\mu}e^{-\frac{1}{\nu^2-t^2}}, & |t| < \nu \\ 0, & |t| \geq \nu \end{cases}$$

Clearly $\varphi_{\nu,\mu}(t) \in \mathcal{D}$, and the support of $\varphi_{\nu,\mu}(t)$ is contained in the interval $[-\nu, \nu]$. Assume that ρ is the metric describing the topology of the space $\mathcal{D}$. Now consider the following sequence of sequences

$$\varphi_{1,1}, \varphi_{1,2}, \varphi_{1,3} \ldots \to 0 \quad \text{in} \quad \mathcal{D}$$

$$\varphi_{2,1}, \varphi_{2,2}, \varphi_{2,3} \ldots \to 0 \quad \text{in} \quad \mathcal{D}$$

$$\varphi_{3,1}, \varphi_{3,2}, \varphi_{3,3} \ldots \to 0 \quad \text{in} \quad \mathcal{D}$$

Choose an element φ_{1,k_1} from the first horizontal sequence so that $\rho(0, \varphi_{1,k_1}) < \epsilon$. Make sure that φ_{1,k_1} lies to the right of $\varphi_{1,1}$ in the first horizontal sequence. Choose an element φ_{2,k_2} from the second horizontal sequence so that $\rho(0, \varphi_{2,k_2}) < \frac{\epsilon}{2}$, taking care that φ_{2,k_2} lies to the right of the element $\varphi_{2,2}$. Thus we are able to choose a sequence $\{\varphi_{\nu,k_\nu}\}_{\nu=1}^{\infty}$ such that

$$\rho(0, \varphi_{\nu,k_\nu}) < \frac{\epsilon}{2^\nu}, \qquad \nu = 1, 2, 3, \ldots$$

The sequence $\{\varphi_{\nu,k_\nu}\}_{\nu=1}^{\infty}$ tends to zero in $\mathcal{D}$ as $\nu \to \infty$. But this is a contradiction, for the support of φ_{ν,k_ν} is $[-\nu, \nu]$. So as $\nu \to \infty$, the support of the sequence φ_{ν,k_ν} gets unbounded. Therefore the sequence $\{\varphi_{\nu,k_\nu}\}_{\nu=1}^{\infty}$ does not converge in $\mathcal{D}$. This contradicts our assumption that the space $\mathcal{D}$ is metrizable.

A distribution is a continuous linear functional on $\mathcal{D}(\mathbb{R}^n)$. When $n = 1$, we denote the testing function space by $\mathcal{D}$. Sometimes a distribution is also called a *generalized function*. I now present some examples of distributions.

Definition. A regular distribution is a distribution generated by a locally integrable function f, and this distribution is also denoted by f,

$$\langle f, \varphi \rangle = \int_{-\infty}^{\infty} f(x)\varphi(x)\,dx$$

Put differently, the distribution f is the functional when operating upon the testing function φ generates the same number as the integral

$$\int_{-\infty}^{\infty} f(x)\varphi(x)\,dx \equiv \int_{a}^{b} f(x)\varphi(x)\,dx$$

where the support of $\varphi(x)$ is the interval $[a, b]$. The support of φ varies with φ. Linearity of this functional is trivial. To verify continuity, let $\varphi_\nu \to 0$ in $\mathcal{D}$ as $\nu \to \infty$, and let the supports of all φ_ν be contained in an interval $[A, B]$. Then

$$|\langle f, \varphi_\nu \rangle| = \left| \int_A^B f(x)\varphi_\nu(x)\, dx \right| \leq \epsilon \int_A^B |f(x)|\, dx$$

$$|\varphi_\nu(x)| \leq \epsilon \qquad \text{as } \nu \to \infty$$

Since ϵ is arbitrary, it follows that

$$\langle f, \varphi_\nu \rangle \to 0 \qquad \text{as } \nu \to \infty.$$

The example of a regular distribution over $\mathcal{D}(\mathbb{R}^n)$ can be given similarly. A distribution that cannot be generated by a locally integrable function is called a *singular distribution*. The Dirac δ function is a very simple example of a singular distribution,

$$\langle \delta(t - a), \varphi(t) \rangle = \varphi(a)$$

Linearity and continuity of this distribution are trivial. $pv\frac{1}{x}$ is another singular distribution, and it is defined as follows:

$$\left\langle pv\frac{1}{x}, \varphi(x) \right\rangle = (P) \int_{-\infty}^{\infty} \frac{\varphi(x)}{x}\, dx. \tag{1.22}$$

The fact that the right-hand side in (1.22) exists is very easy to prove.
Let the support of φ be contained in the interval $[-a, a]$. Then

$$\left\langle pv\frac{1}{x}, \varphi(x) \right\rangle = (P) \int_{-a}^{a} \frac{\varphi(x)}{x}\, dx$$

$$= (P) \int_{-a}^{a} \frac{\varphi(x) - \varphi(0)}{x}\, dx + (P) \int_{-a}^{a} \frac{\varphi(0)}{x}\, dx$$

$$= (P) \int_{-a}^{a} \frac{\varphi(x) - \varphi(0)}{x}\, dx$$

Now define a C^∞ function $\psi(x)$ by

$$\psi(x) = \begin{cases} \frac{\varphi(x) - \varphi(0)}{x}, & \text{if } x \neq 0 \\ \varphi'(0), & \text{if } x = 0 \end{cases} \tag{1.23}$$

Therefore

$$\left\langle pv\frac{1}{x}, \varphi(x) \right\rangle = \int_{-a}^{a} \psi(x)\, dx \tag{1.24}$$

Now

$$\psi(x) = \frac{1}{x} \int_0^x \varphi'(t)\, dt$$

Thus we obtain

$$|\psi(x)| \le \sup_t |\varphi'(t)| \tag{1.25}$$

The linearity of $pv\frac{1}{x}$ follows from (1.24), and the continuity from (1.25). Therefore $pv\frac{1}{x} \in \mathcal{D}'$.

Pseudofunction $1_+(t)t^{-3/2}$ where

$$1_+(t) = \begin{cases} 1, & t \ge 0 \\ 0, & t < 0 \end{cases}$$

The generalized function pseudofunction $1_+(t)t^{-3/2}$ is also denoted by $Pf1_+(t)t^{-3/2}$. The function $1_+(t)t^{-3/2}$ does not define a regular distribution because the integral $\int_0^\infty t^{-3/2}\varphi(t)\,dt$, $\varphi \in \mathcal{D}$ is divergent in general. We therefore generate a distribution from $1_+(t)t^{-3/2}$ by using Hadamard's technique of splitting a divergent integral into convergent and divergent parts. The convergent part of the integral $\int_0^\infty t^{-3/2}\varphi(t)\,dt$ denotes the expression

$$\langle Pf1_+(t)t^{-3/2}, \varphi(t)\rangle \qquad \forall\, \varphi \in \mathcal{D} \tag{1.26}$$

If b is large enough so that $\varphi(t) = 0$ for $t > b$, then the expression in (1.26) becomes

$$\lim_{\epsilon \to 0_+} Fp \int_\epsilon^b t^{-3/2}\varphi(t)\,dt$$

$$= \lim_{\epsilon \to 0_+} Fp \int_\epsilon^b t^{-3/2}[\varphi(0) + t\psi(t)]\,dt \qquad \text{from (1.23)}$$

$$= \lim_{\epsilon \to 0_+} Fp \left[2\frac{\varphi(0)}{\sqrt{\epsilon}} - \frac{2\varphi(0)}{\sqrt{b}} + \int_\epsilon^b \frac{\psi(t)}{\sqrt{t}}\,dt \right]$$

We throw away the divergent part and define

$$Fp \int_0^\infty t^{-3/2}\varphi(t)\,dt = \int_0^b \frac{\psi(t)}{\sqrt{t}}\,dt - \frac{2\varphi(0)}{\sqrt{b}}$$

Therefore

$$\langle Pf1_+(t)t^{-3/2}, \varphi(t)\rangle = \int_0^b \frac{\psi(t)}{\sqrt{t}}\,dt - \frac{2\varphi(0)}{\sqrt{b}}$$

Now the linearity of this functional is obvious. For $\{\varphi_\nu\} \to 0$ in $\mathcal{D}$, assume that the support of $\varphi_\nu(t)$ is contained in the interval $[-b, b]$. Therefore

$$\langle Pf1_+(t)t^{-3/2}, \varphi_\nu(t)\rangle = \int_0^b \frac{\psi_\nu(t)}{\sqrt{t}} - \frac{2\varphi_\nu(0)}{\sqrt{b}}$$

$$\left| \int_0^b \frac{\psi_\nu(t)}{\sqrt{t}}\,dt \right| \le \sup_x |\varphi_\nu'(x)|\, 2\sqrt{b} \to 0 \qquad \text{as } \nu \to \infty$$

This proves the continuity of $Pf1_+(t)t^{-3/2}$. Similarly we can define many other pseudofunctions by using the technique of Hadamard. Some of the well-known pseudofunctions are $Pf1_+(t)t^{-(2n+1)/2}$, $Pf1_+(t)\ln t$, $Pf\frac{1_+(t)}{t}$, and so on. The details can be seen in Zemanian [109]. A distribution f is said to be zero over an open set Ω if $\langle f, \varphi \rangle = 0$ for all $\varphi \in \mathcal{D}(\mathbb{R}^n)$ whose support is contained in Ω. Two distributions f and g are said to be equal over an open set Ω if

$$\langle f, \varphi \rangle = \langle g, \varphi \rangle \qquad \forall \, \varphi \in \mathcal{D}(\mathbb{R}^n)$$

with support in Ω. The support of a distribution f is the smallest closed set outside which the distribution vanishes. The union of all open sets over which a distribution vanishes is called the *null set* of the distribution. We can see that the support of a distribution is the complement of its null set. The support of $\delta(t)$ is the set $\{0\}$, and the support of the regular distribution $1_+(t-1)\,1_+(2-t)$ is the closed interval $[1, 2]$.

Theorem 15. If a distribution is equal to zero on every set of a collection of open sets, then it is equal to zero on the union of these sets.

Proof of this theorem is rather complex. Readers interested in the proof may look into Zemainian [109].

1.4.4. The Calculus of Distribution

The power of distributional analysis rests in large part on the facts that every distribution possesses derivatives of all orders and that differentiation is a continuous operator in this theory. Distributional differentiation commutes with many operations such as limiting operations, infinite summation, and integration.

Definition. We say that a sequence $\{f_n\}$ of distributions converges to a distribution f weakly or in the weak topology of $\mathcal{D}'(\Omega)$ if the sequence $\langle f_k, \varphi \rangle$ converges in the topology of C to $\langle f, \varphi \rangle$ as $k \to \infty$. Since $\langle \sin nx, \varphi(x) \rangle = \int_{-\infty}^{\infty} \sin nx\varphi(x)\, dx \to 0$ as $n \to \infty$. Therefore the regular distribution $\sin nx \to 0$ weakly as $n \to \infty$. The regular distribution generated by $\frac{\sin nx}{\pi x} \to \delta(x)$ as $n \to \infty$. This is an example of a case where a sequence of regular distributions converges to a distribution but the corresponding limit of $\frac{\sin nx}{\pi x}$ in the sense of function as $n \to \infty$ does not exist.

Consider the sequence

$$f_\nu(t) = \begin{cases} 0, & |t| \ge \frac{1}{\nu} \\ \nu^2, & |t| < \frac{1}{\nu} \end{cases}$$

$\lim\langle f_\nu, \varphi \rangle$ fails to exist whenever $\varphi(0) \ne 0$. This sequence is an example of an absolutely integrable function that tends to another absolutely integrable function but the distributional limit does not exist. Now consider another example of a directed sequence

$$f_\nu(t) = \begin{cases} \frac{\nu}{2}, & |t| \le \frac{1}{\nu} \\ 0, & |t| > \frac{1}{\nu} \end{cases}$$

$$\langle f_\nu, \varphi \rangle = \frac{\nu}{2} \int_{-1/\nu}^{1/\nu} \varphi(t)\, dt \to \varphi(0) = \langle \delta(t), \varphi \rangle$$

as $\nu \to \infty$. That is, $f_\nu \to \delta(t)$ in $\mathcal{D}'$ but the function sequence $f_\nu(t) \to 0$ a.e. as $\nu \to \infty$. In this example both limits exist, but they do not correspond.

Theorem 16. The space $\mathcal{D}'(\Omega)$ is weakly complete. Let $\{f_\nu\}_{\nu=1}^{\infty}$ be a sequence of distributions in $\mathcal{D}'(\Omega)$ such that $\langle f_\nu(t), \varphi(t) \rangle$, $\varphi \in \mathcal{D}(\Omega)$ is a Cauchy sequence in C. Then there exists a distribution f such that $\lim_{\nu \to \infty} \langle f_\nu(t), \varphi(t) \rangle = \langle f, \varphi \rangle \; \forall \; \varphi \in D(\Omega)$.

Proof. Since $\langle f_\nu, \varphi \rangle$ is a Cauchy sequence in C it must converge. Let us define a functional f by

$$\lim_{\nu \to \infty} \langle f_\nu, \varphi \rangle = \langle f, \varphi \rangle \tag{1.27}$$

This functional f defined by (1.27) is linear, and we now proceed to prove its continuity on $\mathcal{D}(\Omega)$.

Let K be a compact subset of Ω, and let I be an open set containing K. The space

$$\mathcal{D}_K(I) = \mathcal{D}_K(\Omega) \subset \mathcal{D}(\Omega)$$

The topology of the space $\mathcal{D}_K(\Omega)$ is generated by the separating collection of seminorms

$$\gamma_m(\varphi) = \sup_{\substack{|\alpha| \le m \\ t \in I}} |\partial^\alpha \varphi(t)|$$

The topology of $\mathcal{D}_K(\Omega)$ can likewise be generated by the norm

$$\|\varphi\| = \sum_{m=0}^{\infty} \frac{1}{2^m} \frac{\gamma_m(\varphi)}{1 + \gamma_m(\varphi)}$$

which we denote by $\|\varphi\|$.

The space $\mathcal{D}_K(I)$ is really a Banach space, and each of the sequences $\langle f_\nu, \varphi \rangle$ is bounded. Therefore, by the principle of uniform boundedness, $\|f_\nu\|$ is bounded. In other words, there exists a constant $c > 0$ satisfying $\|f_\nu\| \le c$. Now, using the property of a NLS, we have

$$|\langle f_\nu, \varphi \rangle| \le \|f_\nu\|\, \|\varphi\|$$

$$|\langle f_\nu, \varphi \rangle| \le c\|\varphi\|$$

The sequence f_ν is bounded over $\mathcal{D}_K(\Omega)$

$$\langle f, \varphi \rangle = \lim \langle f_\nu, \varphi \rangle$$

and therefore

$$|\langle f, \varphi \rangle| \leq c \|\varphi\|$$

So f is bounded on $\mathcal{D}_K(\Omega)$; f is continuous on $\mathcal{D}_K(\Omega)$. Since K is arbitrary compact subset of Ω, it follows from the definition of the inductive limit topology of $\mathcal{D}(\Omega)$ that f is continuous on $\mathcal{D}(\Omega)$. Thus $\mathcal{D}'(\Omega)$ is weakly complete. $\square$

Definition. We say that a sequence of distributions $\{f_\nu\}_{\nu=1}^\infty$ tends to f in the strong topology of $\mathcal{D}'(\Omega)$ if for every bounded set B of $\mathcal{D}(\Omega)$, $\langle f_\nu, \varphi \rangle \to \langle f, \varphi \rangle$ uniformly $\forall \varphi \in B$ where B is an arbitrary bounded set of $\mathcal{D}(\Omega)$. One can immediately observe that convergence of a sequence of distributions in the strong topology of $\mathcal{D}'(\Omega)$ implies its convergence in the weak topology of $\mathcal{D}'(\Omega)$.

1.4.5. Distributional Differentiation

The distributional derivative $\partial^\alpha f$ of a distribution f is defined as a distribution that assigns the same number to a $\varphi \in \mathcal{D}$ as f assigns to $(-1)^{|\alpha|}\partial^\alpha \varphi$. It is given by the formula

$$\langle \partial^\alpha f, \varphi \rangle = \langle f, (-1)^{|\alpha|}\partial^\alpha \varphi \rangle \qquad \forall \varphi \in \mathcal{D}(\Omega). \tag{1.28}$$

Recall that the expression $\partial^\alpha \varphi$ stands for partial derivative of φ with respect to $x_1, x_2, \ldots, x_n$ and of order $|\alpha| = \alpha_1 + \alpha_2 + \cdots + \alpha_n$. The definition (1.28) is coined in analogy to the integration by parts in the classical analysis. Let

$$h(t) = \begin{cases} 1, & t \geq 0 \\ 0, & t < 0 \end{cases}$$

Then $h(t)$ is locally integrable and the distributional derivative of $h(t)$ is $\delta(t)$. For

$$\langle Dh(t), \varphi \rangle = \langle h(t), -\varphi^{(1)}(t) \rangle$$

$$= -\int_0^\infty \varphi'(t)\, dt = \varphi(0) = \langle \delta, \varphi \rangle$$

we have

$$Dh(t) = \delta(t)$$

1.5. PRIMITIVE OF DISTRIBUTIONS

The primitive of a distribution f is another distribution $f^{(-1)}$ such that $Df^{(-1)} = f$. The primitive of $\delta(t) = h(t) + c$. There are infinitely many primitives of a distribution, and any two of them differ by a constant. We now state a theorem without proof and interested readers may look into the proof given by Zemanian [109].

Theorem 17. Every distribution on $\mathbb{R}$ has infinitely many primitives defined by

$$\langle f^{(-1)}, \varphi \rangle = K \langle f^{(-1)}, \varphi_0 \rangle + \langle f^{-1}, \chi \rangle$$

where $\varphi_0 \in \mathcal{D}$ such that $\int_{-\infty}^{\infty} \varphi_0(t)\, dt = 1$ and $K = \int_{-\infty}^{\infty} \varphi(t)\, dt$ and $\chi = \varphi - K\varphi_0$. Each primitive is also a distribution. The difference between any two primitives of a distribution is a constant given by

$$C = \langle f_1^{-1}, \varphi_0 \rangle - \langle f_2^{(-1)}, \varphi_0 \rangle$$

See Zemanian [109, pp. 72–78].

Theorem 18. Locally, every distribution is a finite order derivative of a continuous function. More precisely, let I be a fixed finite closed interval in $\mathbb{R}'$, and let f be a distribution defined over a neighborhood of I. There exists a nonnegative integer r and a continuous function $h(t)$ such that

$$\langle f, \varphi \rangle = \langle h^{r+2}, \varphi \rangle \qquad \forall\, \varphi \in \mathcal{D}(I) \text{ [Zemanian 109, p. 92]}$$

Corollary. Let f be a distribution over $\mathbb{R}$ with a bounded support. There exists a nonnegative integer ν and a continuous function $h(t)$ such that $f(t) = h^{(\nu+2)}(t)$ for all t [Zemanian p. 93].

1.6. CHARACTERIZATION OF DISTRIBUTIONS OF COMPACT SUPPORTS

We define the testing function space $E(\mathbb{R}^n)$ as the collection of infinitely differentiable complex-valued functions defined on $\mathbb{R}^n$ equipped with the topology generated by the open neighborhood basis

$$V(m, K) = \{\varphi \mid \varphi \in E(\mathbb{R}^n), \gamma_{m,K}(\varphi) < \epsilon\}$$

where m's are nonnegative integers, ϵ's are arbitrary positive numbers and K's are compact subsets of $\mathbb{R}^n$

$$\gamma_{m,K}(\varphi) = \sup_{\substack{|\alpha| \leq m \\ t \in K}} |\partial^\alpha \varphi|$$

A sequence $\{\varphi_\nu\}_{\nu=1}^{\infty}$ converges to zero as $\nu \to \infty$ if and only if each of the sequences $\{\varphi_\nu^{(k)}(t)\}$ tends to zero uniformly on every compact subset of $\mathbb{R}^n$. Distributions of compact support belong to $E'(\mathbb{R}^n)$, and conversely, any element of $E'(\mathbb{R}^n)$ can be identified as a distribution of compact support. This is because the space $\mathcal{D}(\mathbb{R}^n)$ is a subspace of the space $E(\mathbb{R}^n)$ and because $\mathcal{D}(\mathbb{R}^n)$ is dense in $E(\mathbb{R}^n)$. The identity map $i : \mathcal{D}(\mathbb{R}^n) \rightarrow E(\mathbb{R}^n)$ is continuous. Therefore the restriction of $f \in E'(\mathbb{R}^n)$ to $\mathcal{D}(\mathbb{R}^n)$ is in $\mathcal{D}'(\mathbb{R}^n)$. Any element of $E'(\mathbb{R}^n)$ can be identified by an element of $\mathcal{D}'(\mathbb{R}^n)$ uniquely and hence $E'(\mathbb{R}^n) \subset \mathcal{D}'(\mathbb{R}^n)$. We now give below a theorem that gives a characterization of $E'(\mathbb{R}^n)$.

Theorem 19. Every distribution $f \in E'(\mathbb{R}^n)$ has compact support.

Proof. Let K be the support of f. If K is not compact, then its support must be unbounded. Hence there exists a sequence of points $t_\mu \in K$ that has no finite accumulation point. With each t_μ as center, we draw a ball B_μ that does not intersect any of the balls B_ν for $\mu < \nu$. Thus we are able to draw balls with centers at t_μ such that the balls are all disjoint. Since each of the points t_μ belong to the support of f, we must be able to find φ_μ with support in ball B_μ such that $\langle f, \varphi_\mu \rangle \neq 0$. Now we choose constants c_μ such that

$$c_\mu \langle f, \varphi_\mu \rangle = 1$$

that is, $\langle f, c_\mu \varphi_\mu \rangle = 1$. Set

$$\varphi = \sum_{\mu=1}^{\infty} c_\mu \varphi_\mu \tag{1.29}$$

Since any compact set K intersects only finitely many balls the series in (1.29) is convergent and represents a function in $E(\mathbb{R}^n)$. Now

$$\lim_{N \to \infty} N \left\langle f, \sum_{\mu=1}^{N} c_\mu \varphi_\mu \right\rangle = \langle f, \varphi \rangle$$

$\lim_{N \to \infty} N = \langle f, \varphi \rangle$, a contradiction. $\square$

Therefore the support of any arbitrary $f \in E'(\mathbb{R}^n)$ must be compact.

1.7. CONVOLUTION OF DISTRIBUTIONS

In this section I will assign a meaning to the convolution of two distributions in a way analogous to the convolution of two functions, and the result will be used in the subsequent section on the Fourier transform.

Theorem 20. Let $f \in \mathcal{D}'(\mathbb{R}^n)$ and $\lambda \in E(\mathbb{R}^n)$ such that $\lambda = 1$ in an open set containing the support of f and zero outside a larger set. Then

$$\langle f, \varphi \rangle = \langle f, \lambda \varphi \rangle$$

In other words, the value of $\langle f, \varphi \rangle$ depends upon the value of φ in a neighborhood of the support of f.

Proof. Let λ be an element of $\mathcal{D}(\mathbb{R}^n)$ as stated in the problem. Therefore, for all $\varphi \in \mathcal{D}(\mathbb{R}^n)$ and $f \in \mathcal{D}'(\mathbb{R}^n)$, we have

$$\langle f, \varphi \rangle = \langle f, (1 - \lambda)\varphi + \lambda\varphi \rangle$$
$$= \langle f, (1 - \lambda)\varphi \rangle + \langle f, \lambda\varphi \rangle$$

$(1 - \lambda)\varphi$ is zero in an open set containing the support of f; namely the support of $(1 - \lambda)\varphi$ is contained in the null set of f. Therefore $\langle f, (1 - \lambda)\varphi \rangle = 0$. Hence

$$\langle f, \varphi \rangle = \langle f, \lambda\varphi \rangle \qquad \square$$

1.8. THE DIRECT PRODUCT OF DISTRIBUTIONS

The operation of the tensor product or the direct product of distributions arises in the development of the convolution of distributions. This section presents some very essential features of the direct product to be used in the development of the theory of convolutions.

Let $\varphi(t, \tau) \in \mathcal{D}(\mathbb{R}^2)$ and $f(t) \in \mathcal{D}'_t$ and $g(\tau) \in \mathcal{D}'_\tau$.

Definition. Let $f, g \in \mathcal{D}'$. Their direct product $f \times g$ is defined to be distribution in $\mathcal{D}'_{t,\tau}$ or in $\mathcal{D}'(\mathbb{R}^2)$ by

$$\langle f \times g, \varphi(t, \tau) \rangle = \langle f(t), \langle g(\tau), \varphi(t, \tau) \rangle \rangle \qquad \forall \, \varphi(t, \tau) \in \mathcal{D}(\mathbb{R}^2) \qquad (1.30)$$

Justification of definition (1.30) can be given as follows: Let

$$\psi(t) = \langle g(\tau), \varphi(t, \tau) \rangle$$

Show that $\psi(t) \in \mathcal{D}_t$.

It is evident that $\psi(t)$ has compact support. To show that

$$\psi'(t) = \left\langle g(\tau), \frac{\partial}{\partial t}\varphi(t, \tau) \right\rangle \qquad (1.31)$$

we have to show that

$$\left[\frac{\varphi(t + \Delta t, \tau) - \varphi(t, \tau)}{\Delta t} - \frac{\partial \varphi(t, \tau)}{\partial t} \right] \to 0$$

in D_τ as $\Delta t \to 0$. To prove this, we have to show that

$$\frac{\varphi^{(m)}(t + \Delta t, \tau) - \varphi^{(m)}(t, \tau)}{\Delta t} - \varphi^{(m+1)}(t, \tau)$$

$$= \frac{1}{\Delta t} \int_t^{t+\Delta t} \left[\frac{\partial^{m+1}\varphi(x, \tau)}{\partial x^{m+1}} - \frac{\partial^{(m+1)}\varphi(t, \tau)}{\partial t^{m+1}} \right] dx \to 0$$

uniformly as $\Delta t \to 0$ for all τ lying in any compact subset of $\mathbb{R}$. This proves that

$$\frac{\psi(t + \Delta t) - \psi(t)}{\Delta t} - \left\langle g(\tau), \frac{\partial}{\partial t} \varphi(t, \tau) \right\rangle \to 0 \text{ as } \Delta t \to 0$$

Put differently,

$$\psi'(t) = \left\langle g(\tau), \frac{\partial \varphi}{\partial t}(t, \tau) \right\rangle$$

By using a similar technique and the method of induction, we can show that

$$\psi^{(m)}(t) = \left\langle g(\tau), \frac{\partial^m \varphi}{\partial t^m}(t, \tau) \right\rangle$$

This proves that $\psi(t) \in \mathcal{D}_t$ and that the functional defined by (1.30) is meaningful. We now show that the functional $f \times g$ defined by (1.30) is also continuous on $\mathcal{D}_{t,\tau}$. The linearity is, however, trivial.

Assume that $\varphi_\nu(t, \tau) \to 0$ in $\mathcal{D}_{t,\tau}$ as $\nu \to \infty$. Define

$$\psi_\nu(t) = \langle g(\tau), \varphi_\nu(t, \tau) \rangle$$

Our objective is to show that $\psi_\nu(t) \to 0$ as $\nu \to \infty$ in $\mathcal{D}_t$:

$$\psi_\nu^{(k)}(t) = \langle g(\tau), \varphi_\nu^{(k)}(t, \tau) \rangle$$

Here $\varphi_\nu^{(k)}(t, \tau)$ stands for $\frac{\partial^k \varphi_\nu}{\partial t^k}(t, \tau)$. If $\psi_\nu(t)$ does not go to zero in $\mathcal{D}_t$, then for some fixed k and an $\epsilon > 0$ there exists a sequence $\{t_\nu\}_{\nu=1}^\infty$ such that

$$\left| \psi_\nu^{(k)}(t_\nu) \right| = \left| \left\langle g(t), \frac{\partial^k}{\partial t^k} \varphi_\nu(t, \tau) \right\rangle \right|_{t=t_\nu} \geq \epsilon \tag{1.32}$$

$\forall \nu = 1, 2, 3, 4, \ldots$. If $\varphi_\nu(t, \tau) \to 0$ as $\nu \to \infty$ in $\mathcal{D}_{t,\tau}$, then there exists a compact set K of $\mathbb{R}^2$ containing the supports of all $\varphi_\nu(t, \tau)$ and

$$\sup_{t,\tau} \left| \frac{\partial^k}{\partial t^k} \varphi_\nu(t, \tau) \right| \to 0 \qquad \text{as } \nu \to \infty$$

Therefore $\sup_t \frac{\partial^k}{\partial t^k} \varphi_\nu(t, \tau) \to 0$ as $\nu \to \infty$ uniformly $\forall \tau \in \mathbb{R}$. Likewise $\psi_\nu^{(k)}(t_\nu) \to 0$ as $\nu \to \infty$. This contradicts (1.32).

Therefore $f \times g$ as defined by (1.30) $\in \mathcal{D}_{t,\tau}'$. We now prove that $\delta(t) \times \delta(\tau) = \delta(t, \tau)$:

$$\langle \delta(t) \times \delta(\tau), \varphi(t, \tau) \rangle = \langle \delta(t), \langle \delta(\tau), \varphi(t, \tau) \rangle \rangle$$

$$= \langle \delta(t), \varphi(t, 0) \rangle$$

$$= \varphi(0, 0)$$

$$= \langle \delta(t, \tau), \varphi(t, \tau) \rangle$$

Therefore

$$\delta(t) \times \delta(\tau) = \delta(t, \tau)$$

One can see that $\delta(t) \times \delta(\tau) = \delta(\tau) \times \delta(t)$. If f and g are locally integrable functions and if f, g are the corresponding regular distributions, then

$$\langle f \times g, \varphi(t, \tau) \rangle = \left\langle f(t), \int_{-\infty}^{\infty} \varphi(t, \tau) g(\tau) \, d\tau \right\rangle$$

$$= \int_{-\infty}^{\infty} \int_{-\infty}^{\infty} \varphi(t, \tau) g(\tau) f(t) \, d\tau \, dt$$

$$= \int_{-\infty}^{\infty} \int_{-\infty}^{\infty} \varphi(t, \tau) f(t) g(\tau) \, dt \, d\tau$$

The support of $\varphi(t, \tau) \, g(\tau) \, f(t)$ is bounded, so the switch in the order of integration is justified. This shows that

$$f \times g = g \times f$$

It can be also shown that

$$\delta(t) \times 1_+(\tau) = 1_+(\tau) \times \delta(t)$$

In general, we have $f \times g = g \times f$. This result is true in view of the fact that the space of testing functions of the form

$$\varphi(t, \tau) = \sum_{\nu} \psi_\nu(t) \varphi_\nu(\tau)$$

(where $\psi_\nu(t) \in \mathcal{D}_t$ and $\varphi_\nu(\tau) \in \mathcal{D}_\tau$ and the summation has a finite number of terms) is dense in $\mathcal{D}_{t,\tau}$. See Zemanian [109, pp. 119–120].

1.9. THE CONVOLUTION OF FUNCTIONS

The convolution $h(t)$ of two function $f(t)$ and $g(t)$ defined on the real line is given by

$$h(t) = (f * g)(t) = \int_{-\infty}^{\infty} f(\tau) g(t - \tau) \, d\tau \tag{1.33}$$

provided that the integral exists. We want to extend the definition of convolution to distributions so that the definition (1.33) may be true for the convolution of regular distributions.

If $\varphi \in \mathcal{D}$, then

$$\langle h, \varphi \rangle = \langle f * g, \varphi \rangle$$

$$= \int_{-\infty}^{\infty} \varphi(t) \int_{-\infty}^{\infty} f(\tau) g(t - \tau) \, d\tau \, dt$$

Use $\tau = x$, $t = x + y$,

$$\langle f * g, \varphi \rangle = \int_{-\infty}^{\infty} \int_{-\infty}^{\infty} f(x)g(y)\varphi(x + y)\,dx\,dy \tag{1.34}$$

The form (1.34) resembles the definition of the direct product of regular distributions. Therefore we should define the convolution $f * g$ of two distributions f, g by

$$\langle f * g, \varphi \rangle = \langle f(t) \times g(\tau), \varphi(t + \tau) \rangle$$
$$= \langle f(t), \langle g(\tau), \varphi(t + \tau) \rangle \rangle \tag{1.35}$$

But a problem arises in the definition (1.35) in that the support of $\varphi(t + \tau)$ is not bounded. In fact, if the support of $\varphi(t)$ is contained in the closed interval $[a, b]$, then the support of $\varphi(t + \tau)$ should be in $[(t, \tau) : a \leq t + \tau \leq b]$, which is unbounded (see Figure 1.1). Now let $\Omega = $ [support of $f \times g$] $\cap$ support of $\varphi(t + \tau)$ and let $\lambda(t, \tau)$ be an element of $\mathcal{D}_{t,\tau}$ such that $\lambda \equiv 1$ over an open set containing Ω. We can now give a meaning to the convolution $f * g$ whenever Ω is bounded by

$$\langle f * g, \varphi \rangle = \langle f(t) \times g(\tau), \lambda(t, \tau)\varphi(t + \tau) \rangle \tag{1.36}$$

This replacement is legitimate, since the value of a distribution depends upon the value of φ in a neighborhood of the support of a distribution and is not altered by changing the value of $\varphi(t, \tau)$ outside a neighborhood of the support of $f \times g$. (See Theorem 19.)

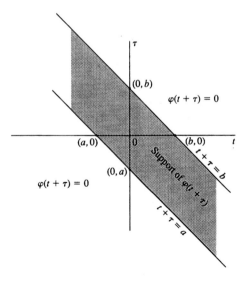

Figure 1.1

Theorem 21. Let $f \in \mathcal{D}'$. A sufficient condition for the validity of the representation (1.36) for the convolution $f * g$ as a distribution on $\mathbb{R}$ of two distributions f and g over $\mathbb{R}$ is that

$$\Omega = [\text{support } f \times g] \cap \text{support } \varphi(t + \tau)$$

is bounded. This is ensured when one of the following conditions hold:

i. Either f or g has a bounded support.
ii. Both f and g have supports bounded to the left; namely there exists a constant T_1 such that $f(t)$, $g(t)$ are both zero for $t < T_1$.
iii. Both f and g have supports bounded to the right; namely there exists a constant T_2 such that $f(t)$ and $g(t)$ both are zero for $t > T_2$.

The proof is easy and is left as an exercise to the readers. The interested reader may consult Zemanian [109, pp. 123–124].

Examples. Show that

$$\delta * f = f \tag{1.37}$$

$$\delta^{(m)} * f = f^{(m)} \tag{1.38}$$

We know that

$$\langle f * \delta, \varphi \rangle = \left\langle f(t), \langle \delta(\tau), \lambda(t, \tau)\varphi(t + \tau) \rangle \right\rangle$$
$$= \langle f(t), \lambda(t, 0)\varphi(t) \rangle = \langle f(t), \varphi(t) \rangle$$
$$\langle f * \delta^{(k)}, \varphi \rangle = \left\langle f(t), \langle \delta^{(k)}(\tau), \lambda(t, \tau)\varphi(t + \tau) \rangle \right\rangle$$
$$= \left\langle f(t), \langle \delta(\tau), (-1)^k D^k(\lambda(t, \tau)\varphi(t + \tau)) \rangle \right\rangle$$
$$= \langle f(t), (-1)^k \varphi^{(k)}(t)\lambda(0, \tau) \rangle$$
$$= \langle f(t), (-1)^k \varphi^{(k)}(t) \rangle$$
$$= \langle D^k f, \varphi \rangle.$$

So if $P(D)$ is a differentiation operator

$$a_0 + a_1 D + a_2 D^2 + \cdots + a_n D^n$$

then

$$P(D)f = (a_0 \delta + a_2 \delta'' + \cdots + a_n \delta^{(n)}) * f$$
$$= [P(D)\delta] * f$$

It can be easily shown that

$$D^m(f * g) = f^{(p)} * g^{(q)}$$

where $p + q = m$, $p, q = 0, 1, 2, \ldots$. In a special case

$$D^m(f * g) = f^{(m)} * g = f * g^{(m)}$$

1.10. REGULARIZATION OF DISTRIBUTIONS

Sometimes in distributional analysis we map a distribution into a C^∞ function by a convolution process that is called regularization of the distribution. We state without proof a theorem on regularization. The proof of this theorem can be found in any book on distribution theory.

Theorem 22. Let $f \in \mathcal{D}'$ and $\varphi \in \mathcal{D}$. Then $h = f * \varphi$ is an ordinary function given by

$$h(t) = \langle f(x), \varphi(t - x) \rangle$$

Also $h(t) \in C^\infty$, and

$$h^{(k)}(t) = \langle f(x), \varphi^{(k)}(t - x) \rangle$$

See Zemanian [109, pp. 132–133].

1.11. THE CONTINUITY OF THE CONVOLUTION PROCESS

Let $\mathcal{D}'_R$ and $\mathcal{D}'_L$ be distribution spaces whose supports are bounded to the left and to the right, respectively. We say that a sequence $\{f_i\}_{i=1}^\infty$ of distributions in $\mathcal{D}'_R$ converges to a distribution $f \in \mathcal{D}'_R$ if each of the elements $f_1, f_2, f_3, \ldots$ are in $\mathcal{D}'_R$ and $\lim_{i \to 0} f_i = f$ in the weak distributional sense. Similarly we define the convergence in $\mathcal{D}'_L$. We now state the following theorem without proof:

Theorem 23. Let g be a given distribution, and let $\{f_i\}$ be a sequence of distributions tending to f weakly. Then the sequence $f_i * g$ converges $f * g$ in $\mathcal{D}'$ if one of the following conditions is satisfied:

   i. $g \in E'$.
   ii. $\{f_\nu\}$ converges to f in E'.
   iii. $\{f_\nu\}$ converges in $\mathcal{D}'_R$ and g is also in $\mathcal{D}'_R$.
   iv. $\{f_\nu\}$ converges in $\mathcal{D}'_L$ and g is also in $\mathcal{D}'_L$.

1.12. FOURIER TRANSFORMS AND TEMPERED DISTRIBUTIONS

1.12.1. The Testing Function Space $S(\mathbb{R}^n)$

A C^∞ function φ defined on $\mathbb{R}^n$ belongs to $S(\mathbb{R}^n)$ if

$$\sup_{\substack{|\alpha| \le k \\ x \in \mathbb{R}^n}} \left| x^m \partial^\alpha \varphi(x) \right| < \infty \tag{1.39}$$

for each $k, |m| = 0, 1, 2, 3, \ldots$. From (1.39) we can derive that

$$\lim_{\substack{|x| \to \infty \\ |m|, |\alpha| = 0,1,2,3,\ldots}} x^m \partial^\alpha \varphi(x) = 0 \tag{1.40}$$

if $\varphi \in S(\mathbb{R}^n)$. Another way of describing conditions (1.39), (1.40) is

$$\sup_{\substack{|\alpha| \le r \\ x \in \mathbb{R}^n}} \lim_{|x| \to \infty} (1 + |x|^2)^r \partial^\alpha \varphi(x) = 0 \tag{1.41}$$

$r, |\alpha| = 0, 1, 2, 3, \ldots$. The set of conditions (1.39), (1.40), and (1.41) are equivalent conditions. As an example, $e^{-(x_1^2 + x_2^2 + \ldots x_n^2)} \in S(\mathbb{R}^n)$.

The topology of the testing function space $S(\mathbb{R}^n)$ is generated by the sequence of seminorms $\{\gamma_{k,|m|}\}, k, |m| = 0, 1, 2, \ldots$, where

$$\gamma_{k,|m|}(\varphi) = \sup_{\substack{x \in \mathbb{R}^n \\ |\alpha| \le k}} |x^m \partial^\alpha \varphi(x)|, \varphi \in S(\mathbb{R}^n) \tag{1.42}$$

$k, |m| = 0, 1, 2, 3, \ldots$.

The topology generated by the separating collection of the sequence of seminorms $\{\gamma_{k,m}\}_{k,|m|=0}^\infty$ as defined by (1.42) can also be generated by the sequence of seminorms $\{\gamma_r\}_{r=0}^\infty$, where

$$\gamma_r(\varphi) = \sup_{\substack{x \in \mathbb{R}^n \\ |\alpha| \le r}} (1 + |x|^2)^r |\partial^\alpha \varphi|, \varphi \in S(\mathbb{R}^n) \tag{1.43}$$

We can also generate this topology by the separating collection of seminorms $\{\gamma_{|\alpha|,|m|}\}$ $|\alpha|, |m| = 0, 1, 2, 3, \ldots$, where

$$\gamma_{|\alpha|,|m|}(\varphi) = \sup_{x \in \mathbb{R}^n} |x^m \partial^\alpha \varphi(x)|, \quad \varphi \in S(\mathbb{R}^n) \tag{1.44}$$

Therefore a sequence $\varphi_\nu(x)$ in $S(\mathbb{R}^n) \to 0$ as $\nu \to \infty$ if and only if $x^m \partial^\alpha \varphi_\nu(x) \to 0$ uniformly on $\mathbb{R}^n$ for each $|m|, |\alpha| = 0, 1, 2, 3, \ldots$. Consider

$$\varphi_m(x) = \frac{1}{m} e^{-|x|^2}$$

$$\varphi_m(x) \in S(\mathbb{R}^n) \quad \text{and}$$

$$\gamma_{|\alpha|,|\beta|}(\varphi) = \sup_x |x^\alpha \partial^\beta \varphi_m| \to 0 \qquad \text{as } |m| \to \infty$$

Clearly $\varphi_m(x) \to 0$, $m \to \infty$, in the topology of $S(\mathbb{R}^n)$.

The function $e^{-(|x_1|+|x_2|...|x_n|)}$ diminishes to zero faster than any polynomial as $|x| \to \infty$, but this function does not belong to $S(\mathbb{R}^n)$, since $e^{-|x|}$ is not differentiable at the origin.

1.13. THE SPACE OF DISTRIBUTIONS OF SLOW GROWTH $S'(\mathbb{R}^n)$

A continuous linear functional over $S(\mathbb{R}^n)$ is called a *distribution of slow growth*. The space $\mathcal{D}(\mathbb{R}^n) \subset S(\mathbb{R}^n)$, the identity map (embedding) $\mathcal{D}(\mathbb{R}^n) \to S(\mathbb{R}^n)$ is continuous, and $\mathcal{D}(\mathbb{R}^n)$ is dense in $S(\mathbb{R}^n)$. A distribution of slow growth can be identified one to one by an element of $\mathcal{D}'(\mathbb{R}^n)$. Therefore

$$\mathcal{D}'(\mathbb{R}^n) \supset S'(\mathbb{R}^n)$$

Every element of $\mathcal{D}(\mathbb{R}^n)$ and of $S(\mathbb{R}^n)$ can be identified as a regular distribution in $S'(\mathbb{R}^n)$. Therefore

$$E'(\mathbb{R}^n) \subset \mathcal{D}(\mathbb{R}^n) \subset S(\mathbb{R}^n) \subset S'(\mathbb{R}^n) \subset \mathcal{D}'(\mathbb{R}^n)$$

A distribution of slow growth is also called as a *tempered distribution*.

$S'(\mathbb{R}^n)$ is a proper subset of $\mathcal{D}'(\mathbb{R}^n)$. The distribution $\sum_{n=-\infty}^{\infty} \delta(t-n)e^{n^2} \in \mathcal{D}'(\mathbb{R})$ does not belong to $S'(\mathbb{R})$. If $\varphi \in \mathcal{D}$,

$$\left\langle \sum \delta(t-n)e^{n^2}, \varphi(t) \right\rangle = \sum_{n=-\infty}^{\infty} e^{n^2} \varphi(n)$$

So $\sum_{n=-\infty}^{\infty} \delta(t-n)e^{n^2}$ exists as an element of D'. It does not belong to S', for

$$\left\langle \sum_{-\infty}^{\infty} \delta(t-n)e^{n^2}, \varphi(t) \right\rangle = \sum_{n=-\infty}^{\infty} \varphi(n)e^{n^2}$$

which diverges for $\varphi(x) = e^{-x^2}$.

1.14. A BOUNDEDNESS PROPERTY OF DISTRIBUTIONS OF SLOW GROWTH AND ITS STRUCTURE FORMULA

Theorem 24. Let $f \in S'(\mathbb{R}^n)$, and let $\varphi \in S(\mathbb{R}^n)$. Then there exists a constant $c > 0$ and a nonnegative integer r such that

$$|\langle f, \varphi \rangle| \le c\gamma_r(\varphi) \tag{1.45}$$

where $\gamma_r(\varphi)$ is the same as defined by (1.43).

Proof. If (1.45) is not true, then for any integer $m > 0$ we can find a function $\varphi_m(x) \in S(\mathbb{R}^n)$ such that

$$|\langle f, \varphi_m \rangle| > m\gamma_m(\varphi_m)$$

$$\left| \left\langle f, \frac{\varphi_m}{m\gamma_m(\varphi_m)} \right\rangle \right| > 1 \tag{1.46}$$

$$\gamma_m\left(\frac{\varphi_m}{m\gamma_m(\varphi_m)} \right) = \frac{1}{m} \to 0 \qquad \text{as } m \to \infty$$

$$\gamma_i\left(\frac{\varphi_m}{m\gamma_m(\varphi_m)} \right) \to 0 \qquad \text{as } m \to \infty$$

for each $i = 0, 1, 2, 3, \ldots$, which contradicts (1.46). Therefore (1.45) must be true. $\square$

Since the sequence of seminorms $\{\gamma_r\}_{r=0}^{\infty}$ is separating, it follows that the space $S(\mathbb{R}^n)$ is metrizable. Further it is a locally convex, sequentially complete, and Hausdorff TVS, and therefore it is also a Fréchet space.

1.15. A CHARACTERIZATION FORMULA FOR TEMPERED DISTRIBUTIONS

Theorem 25. A distribution in $\mathbb{R}^n$ is tempered if and only if it is a finite sum of derivatives of continuous functions growing at infinity slower than some polynomial.

Proof. Sufficiency is obvious. So we need only to prove the necessity.

From Theorem 23 it follows that if $f \in S'(\mathbb{R}^n)$ and $\varphi \in S(\mathbb{R}^n)$, there exists a nonnegative integer r and $C > 0$ such that

$$|\langle f, \varphi \rangle| \le C \sum_{|\alpha| \le r} \sup_{x \in \mathbb{R}^n} |(1 + |x|^2)^r |\varphi^{(\alpha)}(x)|$$

$$|\langle f, \varphi \rangle| \le \sum_{|\alpha| \le n+r} C' \left\| (1 + |x|^2)^r \left(\frac{\partial}{\partial x} \right)^\alpha \varphi(x) \right\|_{L^1(\mathbb{R}^n)}$$

We have used the fact that

$$\int_{-\infty}^{x_n} \cdots \int_{-\infty}^{x_2} \int_{-\infty}^{x_1} \triangleright \varphi(x)\, dx = \varphi(x)$$

where

$$\triangleright = \frac{\partial}{\partial x_n} \frac{\partial}{\partial x_{n-1}} \cdots \frac{\partial}{\partial x_2} \frac{\partial}{\partial x_1}$$

Now let N be the number of n-tuples α such that

$$|\alpha| \leq n + r$$

We consider the product space $L^1 \times \cdots \times L^1 = (L^1)^N$ and the injection

$$J : \varphi \rightarrow \left((1 + |x|^2)^r \left(\frac{\partial}{\partial x} \right)^\alpha \varphi \right)_{|\alpha| \leq n+r}$$

of $S(\mathbb{R}^n)$ into $(L^1)^N$.

Using the Hahn-Banach theorem, we extend it as a continuous linear form in the whole space $(L^1)^N$. But the dual of $(L^1)^N$ is canonically isomorphic with $(L^\infty)^N$. Therefore there exist N L^∞ functions h_α, $|\alpha| \leq n + r$ such that

$$f = \sum_{|\alpha| \leq n+r} \partial^\alpha [(1 + |x|^2)^r (-1)^{|\alpha|} h_\alpha]$$

Set

$$g_\alpha(x) = \int_0^{x_1} \cdots \int_0^{x_n} h_\alpha(t_1, t_2, \ldots t_n) dt_1 \ldots dt_n$$

$$|g_\alpha(x)| \leq |x_1||x_2|\ldots|x_n||h_\alpha|_{L^\infty}$$

Furthermore we have

$$h_\alpha(x_1, x_2, \ldots, x_n) = \triangleright g_\alpha$$

Therefore we have

$$f = \sum_{|\alpha| \leq r+2n} \partial^\alpha \left[(1 + |x|^2)^r (-1)^\alpha \triangleright g_\alpha \right]$$

$$= \sum_{|\alpha| \leq r+3n} \left[(1 + |x|^2)^r P_\alpha \left(x, \frac{1}{1 + |x|^2} \right) \partial^\alpha g_\alpha \right]$$

$P_\alpha(x, t)$ is a polynomial in $x_1, x_2, \ldots, x_n$ and t depending upon α. $\square$

Theorem 26. Let $\mathcal{F}$ be the Fourier transformation operator defined by

$$(\mathcal{F}\varphi)(x) = \int_{\mathbb{R}^n} \varphi(t) e^{it \cdot x} dt$$

Then

$$\mathcal{F} : S(\mathbb{R}^n) \rightarrow S(\mathbb{R}^n)$$

is a homeomorphism and so also the operator $\mathcal{F}^{-1}$ defined by

$$(\mathcal{F}^{-1}\varphi)(x) = \frac{1}{(2\pi)^n} \int_{\mathbb{R}^n} \varphi(t)e^{-it\cdot x} dt$$

Proof. Let $\varphi \in S(\mathbb{R}^n)$. Then

$$(\mathcal{F}\varphi)^{(k)}(x) = \int_{\mathbb{R}^n} \varphi(t)(it)^k e^{it\cdot x} dt$$

Using Theorems 4 and 8, we get

$$(ix)^\alpha (\mathcal{F}\varphi)^{(k)}(x) = \int_{\mathbb{R}^n} (ix)^\alpha \varphi(t)(it)^k e^{it\cdot x} dt$$

$$= \int_{\mathbb{R}^n} (-D)^\alpha [\varphi(t)(it)^k] e^{it\cdot x} dt.$$

Therefore $|x^\alpha \partial^k \mathcal{F}\varphi| < \infty$. From this result it follows that if $\varphi \in S$, then $\mathcal{F}\varphi \in S$, and if $\varphi_\nu \to 0$ in S, then $(\mathcal{F}\varphi_\nu) \to 0$ in S. This indicates that the operator $\mathcal{F} : S \to S$ is continuous. It is also one to one as $\mathcal{F}\varphi = 0 \Rightarrow \varphi \equiv 0$. $\mathcal{F}^{-1}$ exists and is given by

$$(\mathcal{F}^{-1}\varphi)(x) = \frac{1}{(2\pi)^n} \int_{\mathbb{R}^n} \varphi(t)e^{-it\cdot x} dt$$

Now we can show similarly that $\mathcal{F}^{-1} : S \to S$ is continuous. Since $\mathcal{F}, \mathcal{F}^{-1}$ are both continuous, $\mathcal{F} : S$ onto itself is a homeomorphism. $\square$

1.16. FOURIER TRANSFORM OF TEMPERED DISTRIBUTIONS

Let $f \in S'(\mathbb{R}^n)$ then its Fourier transform $\mathcal{F}f$ is defined by

$$\langle \mathcal{F}f, \varphi \rangle = \langle f, \mathcal{F}\varphi \rangle \qquad \forall \, \varphi \in S \tag{1.47}$$

this is an analogue of the classical Parseval's relation, namely

$$\int_{-\infty}^{\infty} (\mathcal{F}f)(x)\varphi(x)\,dx = \int_{-\infty}^{\infty} f(x)(\mathcal{F}\varphi)(x)\,dx \qquad \forall \, \varphi \in S \tag{1.48}$$

if f is absolutely integrable. So the classical Parseval relation (1.48) may be encompassed by the definition (1.47) as a special case. There are other types of Parseval relations such as

$$\int_{-\infty}^{\infty} (\mathcal{F}f)(w)(\mathcal{F}\varphi)(w)\,dw = 2\pi \int_{-\infty}^{\infty} f(t)\varphi(-t)\,dt \tag{1.49}$$

But the majority of mathematicians use (1.49) to extend the Fourier transform to distributions, and we will stay with the majority.

The definition (1.47) says that the Fourier transform of a tempered distribution is a tempered distribution, and it assigns the same number to $\varphi \in S(\mathbb{R}^n)$ as f assigns to $(\mathcal{F}\varphi)(w)$. We can also define the inverse Fourier transformation operator by

$$\langle \mathcal{F}^{-1}f, \varphi \rangle = \langle f, \mathcal{F}^{-1}\varphi \rangle \tag{1.50}$$

for all $f \in S'(\mathbb{R}^n)$. It is easy now to see that

$$\mathcal{F}\mathcal{F}^{-1} = \mathcal{F}^{-1}\mathcal{F} = I$$

so

$$\mathcal{F} : S'(\mathbb{R}^n) \to S'(\mathbb{R}^n)$$

is also a homeomorphism (in the weak topology on $S'(\mathbb{R}^n)$).

Examples. Consider the Fourier transform of the δ function and 1:

$$\langle \mathcal{F}\delta, \varphi \rangle = \langle \delta, \mathcal{F}\varphi \rangle = \left\langle \delta(x), \int_{-\infty}^{\infty} \varphi(t)e^{itx}dt \right\rangle$$

$$= \int_{-\infty}^{\infty} \varphi(t)\, dt$$

$$= \langle 1, \varphi \rangle$$

$$\langle \mathcal{F}^{-1}\delta, \varphi \rangle = \left\langle \delta, \frac{1}{2\pi}\int_{-\infty}^{\infty} \varphi(t)e^{-itx}dt \right\rangle$$

$$= \int_{-\infty}^{\infty} \frac{1}{2\pi}\varphi(t)\, dt$$

$$\mathcal{F}^{-1}\delta = \frac{1}{2\pi}$$

Therefore

$$\mathcal{F}\left(\frac{1}{2\pi}\right) = \delta$$

or

$$\mathcal{F}1 = 2\pi\delta$$

$$\mathcal{F}\delta = 1$$

$$\mathcal{F}\delta' = (-ix)$$

$$\mathcal{F}\delta^{(k)} = (-ix)^k$$

and so on, can be easily verified. Note that the derivative of a tempered distribution f is also defined by

$$\langle \partial^\alpha f, \varphi \rangle = \langle f, (-1)^{|\alpha|}\partial^\alpha \varphi \rangle$$

Theorem 27.* Let $f \in E'$. That is, let f be a distribution of compact support. Then the Fourier transform of f is given by

$$(\mathcal{F}f)(x) = \langle f(t), e^{it \cdot x} \rangle$$

We will need to prove four lemmas before we can prove the main theorem.

Lemma 2. Let $f \in E'$ and $F(x) = \langle f(t), e^{itx} \rangle$. Then $F(x)$ is infinitely differentiable, and $F^{(k)}(x) = \langle f(t), (it)^k e^{itx} \rangle$.

Let us prove this lemma for $k = 1$. The general result can be proved by using similar techniques and the method of induction.

$$\frac{F(x + \Delta x) - F(x)}{\Delta x} - \langle f(t), ite^{itx} \rangle = \left\langle f(t), \frac{it}{\Delta x} \int_x^{x+\Delta x} [e^{itz} - e^{itx}] \, dz \right\rangle \quad (1.51)$$

$$\theta(\Delta x) = \frac{it}{\Delta x} \int_x^{x+\Delta x} [e^{itz} - e^{itx}] \, dz$$

$$D_t^k \theta = \frac{it}{\Delta x} \int_x^{x+\Delta x} [e^{itz}(iz)^k - e^{itx}(ix)^k] \, dz$$

$$+ \frac{ik}{\Delta x} \int_x^{x+\Delta x} [e^{itz}(iz)^{k-1} - (ix)^{k-1}e^{itx}] \, dz$$

$\to 0$ or $\Delta x \to 0$ uniformly for t lying in any compact subset of $\mathbb{R}$. Therefore $\theta(\Delta x) \to 0$ in E as $\Delta x \to 0$. Hence letting $\Delta x \to 0$ in (1.51), we get

$$F'(x) = \langle f(t), ite^{itx} \rangle$$

$$= \left\langle f(t), \frac{\partial}{\partial x} e^{itx} \right\rangle$$

Similarly, by induction, we can show that

$$F^{(k)}(x) = \langle f(t), (it)^k e^{itx} \rangle$$

Lemma 3. The integral

$$\int_{-N}^N \langle f(x), e^{ixt} \rangle \varphi(t) \, dt = \left\langle f(x), \int_{-N}^N \varphi(t)e^{ixt} \, dt \right\rangle \qquad \forall \, \varphi \in S$$

We can partition the interval $[-N, N]$ to define the integral over $[-N, N]$. Then

$$\int_{-N}^N \langle f(x), e^{ixt} \rangle \varphi(t) \, dt = \lim_{\|\Delta\| \to 0} \sum_{i=1}^n \langle f(x), e^{ixt_i} \rangle \varphi(t_i) \Delta t_i$$

---

*Theorem 27 is also true when $f \in E'(\mathbb{R}^n)$. The proof is similar.

$$= \lim_{\|\Delta\| \to 0} \left\langle f(x), \sum e^{ixt_i} \varphi(t_i) \Delta t_i \right\rangle$$

$$\int_{-N}^{N} \varphi(t) e^{ixt} dt - \sum_{i=1}^{n} e^{ixt_i} \varphi(t_i) \Delta t_i \to 0$$

uniformly over any compact subset of the real line (for x) due to uniform continuity of functions involved. The derivative with respect to x of the foregoing summation will also tend to zero uniformly with respect to x, lying on any compact subset of the real line. Therefore

$$\int_{-N}^{N} \langle f(x), e^{ixt} \rangle \varphi(t) \, dt = \left\langle f(x) \int_{-N}^{N} \varphi(t) e^{ixt} dt \right\rangle$$

Lemma 4. Let $\varphi \in S$ and $N \in \mathbb{R}$. Then

$$\left\langle f(x), \int_{N}^{\infty} \varphi(t) e^{ixt} dt \right\rangle \to 0 \qquad \text{as } N \to \infty$$

also

$$\left\langle f(x), \int_{-\infty}^{-N} \varphi(t) e^{ixt} dt \right\rangle \to 0 \qquad \text{as } N \to \infty$$

We will sketch the proof of only one of these two, since the other one can be proved similarly.

Let

$$\psi(x) = \int_{N}^{\infty} \varphi(t) e^{ixt} dt$$

$$\psi^{(k)}(x) = \int_{N}^{\infty} \varphi(t)(it)^k e^{ixt} dt$$

$$|\psi^{(k)}(x)| \le \int_{N}^{\infty} |t|^k |\varphi(t)| dt \to 0 \qquad \text{as } N \to \infty$$

uniformly for all $x \in \mathbb{R}$. This completes the proof of Lemma 3.

Lemma 5. Let $f \in S'$ and $\varphi \in S$. Then

$$\int_{-\infty}^{\infty} \langle f(t), e^{ixt} \rangle \varphi(t) \, dt = \left\langle f(x), \int_{-\infty}^{\infty} e^{ixt} \varphi(t) \, dt \right\rangle$$

Proof. By Lemma 3

$$\int_{-N}^{N} \langle f(x), e^{ixt} \rangle \varphi(t) \, dt = \left\langle f(x), \int_{-N}^{N} e^{ixt} \varphi(t) \, dt \right\rangle$$

$$= \left\langle f(x), \int_{-N}^{N} e^{ixt}\varphi(t)\,dt + \int_{N}^{\infty} e^{ixt}\varphi(t)\,dt + \int_{-\infty}^{-N} e^{ixt}\varphi(t)\,dt \right\rangle$$

$$- \left\langle f(x), \int_{N}^{\infty} e^{ixt}\varphi(t)\,dt \right\rangle - \left\langle f(x), \int_{-\infty}^{-N} e^{ixt}\varphi(t)\,dt \right\rangle$$

Now let $N \to \infty$:

$$\int_{-\infty}^{\infty} \langle f(x), e^{ixt}\rangle\varphi(t)\,dt = \left\langle f(x), \int_{-\infty}^{\infty} e^{ixt}\varphi(t)\,dt \right\rangle$$

Proof of Theorem 27. Let

$$\langle \mathcal{F}f, \varphi \rangle = \langle f, \mathcal{F}\varphi \rangle$$

$$= \left\langle f(x), \int_{-\infty}^{\infty} \varphi(t)e^{itx}\,dt \right\rangle$$

$$= \int_{-\infty}^{\infty} \langle f(x), e^{itx}\rangle\varphi(t)\,dt$$

$$= \langle \langle f(x), e^{itx}\rangle, \varphi(t)\rangle$$

Therefore

$$(\mathcal{F}f)(t) = \langle f(x), e^{itx}\rangle$$

$$(\mathcal{F}f)(x) = \langle f(t), e^{itx}\rangle \qquad \square$$

Applications.

Case 1.

$$\mathcal{F}\delta^{(k)}(t) = \langle \delta^{(k)}(t), e^{itx}\rangle$$

$$= \langle \delta(t), (-1)^k(ix)^k e^{itx}\rangle$$

$$(\mathcal{F}\delta^{(k)})(x) = (-ix)^k$$

Case 2. If f and g are two distributions with compact supports, then their convolution is also of compact support and their Fourier transform is given by

$$\mathcal{F}(f * g)(w) = \langle (f * g)(t), e^{iwt}\rangle$$

$$= \langle f(t), \langle g(\tau), e^{iw(t+\tau)}\rangle\rangle$$

$$= \langle f(t), e^{iwt}\rangle\langle g(\tau), e^{iw\tau}\rangle$$

$$= (\mathcal{F}f)(w)(\mathcal{F}g)(w)$$

1.17. FOURIER TRANSFORM OF DISTRIBUTIONS IN $\mathcal{D}'(\mathbb{R}^n)$

We will prove our theorem for $\mathcal{D}' \equiv \mathcal{D}'(\mathbb{R}^1)$ (a method of Leon Ehrenpreis [36]) and the general results for $\mathcal{D}'(\mathbb{R}^n)$ are similar. Our main objective will be to prove that the space Z of testing function space, which is entire and satisfies the condition

$$|z^k \Phi(z)| \leq c_k e^{a|y|}, \qquad k = 0, 1, 2, 3, \ldots, \ \forall \ \Phi \in Z \tag{1.52}$$

is the class of those entire functions that are the Fourier transforms of $\varphi \in \mathcal{D}$ whose support is contained in the interval $[-a, a]$.

Theorem 28. The Fourier transform of an element $\varphi \in \mathcal{D}$ with support in $[-a, a]$ can be extended as an entire function $\Phi(z)$ such that

$$|z^k \Phi(z)| \leq c_k e^{a|y|} \qquad \text{for each } k = 0, 1, 2, \ldots$$

Conversely, if there is class of entire functions Φ such that

$$|z^k \Phi(z)| \leq c_k e^{a|y|} \tag{1.53}$$

then the inverse Fourier transform $\varphi(t)$ of $\Phi(z)$ defined by

$$\varphi(x) = \frac{1}{2\pi} \int_{-\infty}^{\infty} \Phi(x) e^{-ixt} dx$$

is an element of $\mathcal{D}$ with support contained in the interval $[-a, a]$.

Proof. Let $\varphi(t) \in \mathcal{D}$ such that the support of $\varphi(t)$ is contained in the interval $[-a, a]$. Define $\Phi(z)$,

$$\Phi(z) = \int_{-\infty}^{\infty} \varphi(t) e^{itz} dt$$

Clearly $\Phi(z)$ is the extension of the Fourier transform of $\varphi(t) \in \mathcal{D}$ as an entire function. Now

$$z^k \Phi(z) = \int_{-\infty}^{\infty} \varphi(t) z^k e^{itz} dt = i^k \int_{-\infty}^{\infty} \varphi^{(k)}(t) e^{itz} dt$$

$$|z^k \Phi(z)| \leq e^{a|y|} \int_{-a}^{a} |\varphi^{(k)}(t)| \, dt, \ z = x + iy \tag{1.54}$$

$$|z^k \Phi(z)| \leq c_k e^{a|y|}$$

where

$$c_k = \int_{-a}^{a} |\varphi^{(k)}(t)| \, dt$$

Conversely, let $\Phi(z) \in Z$ such that (1.53) is satisfied. Denote the inverse Fourier transform of $\Phi(z)$ by $\varphi(t)$. Then we have

$$\varphi(t) = \frac{1}{2\pi} \int_{-\infty}^{\infty} \Phi(w)e^{-iwt}\,dw \qquad \text{and } \varphi(t) \in c^{\infty}(\mathbb{R})$$

$$= \frac{1}{\pi} \int_{-\infty}^{\infty} \Phi(w + iy)e^{-i(w+iy)t}\,dw \qquad \text{(by contour integration)}$$

$$= \frac{1}{2\pi} \int_{-\infty}^{\infty} \frac{\Phi(w + iy)(w + iy)^2}{(w + iy)^2}e^{-i(w+iy)t}\,dw$$

$$|\varphi(t)| \leq \frac{c_2 e^{yt}}{2\pi} \int_{-\infty}^{\infty} \frac{dw}{w^2 + y^2}e^{a|y|}$$

$$\leq \frac{c_2}{2|y|}e^{yt+a|y|}$$

For $y > 0$,

$$|\varphi(t)| \leq \frac{c_2}{2y}e^{y(a+t)} \to 0$$

as $y \to \infty$ when $t < -a$.

For $y < 0$,

$$|\varphi(t)| \leq \frac{c_2}{2|y|}e^{y(t-a)} \to 0$$

as $y \to -\infty$ when $t > a$. Therefore $\varphi(t) = 0$ for $|t| > a$. $\square$

Our mission is not fulfilled as yet. We want the Fourier transform to be a homeomorphism from $\mathcal{D}$ onto Z. To achieve this objective, we have to transport the topology of the space $\mathcal{D}$ onto Z by means of Fourier transform operator F and define convergence in the space Z accordingly.

We look into the inequality (1.53) and note the fact that, when $\varphi_\nu \to 0$ in $\mathcal{D}$, there exists a compact subset, say, the interval $[-a, a]$ containing the support of each $\varphi_\nu(t)$. Therefore we define that a sequence $\{\Phi_\nu(z)\}_{\nu=1}^{\infty}$ in Z converges to zero in Z as $\nu \to \infty$ if and only if

1. Each $\Phi_\nu(z) \in Z$.
2. There exist constants $c_k > 0$ independent of ν such that

$$|z^k \Phi_\nu(z)| \leq c_k e^{a|y|}, \qquad k = 0, 1, 2, \ldots$$

3. $\{\Phi_\nu(z)\}_{\nu=1}^{\infty}$ converges uniformly on every bounded domain of the z-plane.

It is now a simple exercise to show that

$$\varphi_\nu(t) \to 0 \text{ in } \mathcal{D} \Leftrightarrow \Phi_\nu(z) \to 0 \text{ in } Z; \quad \varphi_\nu(t) = (\mathcal{F}^{-1}\Phi)(t)$$

With this definition of convergence in Z, the Fourier transform becomes a homeomorphism for $\mathcal{D}$ onto Z. The space Z' is called the *space of ultradistributions*.

We have proved the one-dimensional case only. The general case can be cited and proved analogously. It is left as an exercise. So we define the Fourier transform of $\varphi \in \mathcal{D}'(\mathbb{R}^n)$ by the relation

$$\langle \mathcal{F}f, \varphi \rangle = \langle f, \mathcal{F}\varphi \rangle \qquad \forall \, \varphi \in Z(\mathbb{R}^n)$$

Note that $\mathcal{F}f$ is an ultradistribution that assigns the same number to $\varphi \in Z$ as f assigns to $\mathcal{F}\varphi$. The generalized function

$$\sum_{0}^{\infty} e^{n^2} \delta(t - n) \in \mathcal{D}'$$

but it does not belong to S'. We can find its Fourier transform as follows: Let

$$f = \sum_{n=0}^{\infty} e^{n^2} \delta(t - n)$$

$$\langle \mathcal{F}f, \varphi \rangle = \left\langle f, \int_{-\infty}^{\infty} \varphi(x)e^{itx}dx \right\rangle$$

$$= \sum_{n=1}^{\infty} e^{n^2} \int_{-\infty}^{\infty} \varphi(x)e^{inx}dx$$

$$= \left\langle \sum_{n=0}^{\infty} e^{n^2} e^{inx}, \varphi(x) \right\rangle$$

Therefore

$$\mathcal{F}f = \sum_{n=0}^{\infty} e^{n^2} e^{inx}$$

is an ultradistribution.

EXERCISES

1. Prove that in a topological vector space E over the field of complex numbers, a set different from $\varnothing$ and E cannot be both open and closed.

2. Let E be a vector space, and let U be a subset of E that is convex, balanced, and absorbing. Prove that the sets $\frac{1}{n}U$ $(n = 1, 2, \ldots)$ form a neighborhood base at zero in a topology of E that is compatible with the linear structure of E. What happens when you drop the assumption that U is convex?

3. Prove that the product of a family of topological vector spaces E_α, $\alpha \in A$ is Hausdorff if and only if every E_α is Hausdorff.

4. Prove that a TVS E is compact if and only if it consists of a single element 0.

5. The convex balanced hull of a subset A of a vector space E is the smallest balanced convex set containing A. Give an example of a closed subset of the plane $\mathbb{R}^2$ whose convex hull is not closed.

6. Prove that the Schwartz testing function space $\mathcal{D}(\mathbb{R}^n)$ is not metrizable.

7. Let $\{E_n\}_{n=1}^\infty$ be a sequence of Fréchet spaces. Prove that the product space $E = \prod_{i=1}^\infty E_i$ is a Fréchet space.

8. Let K be a compact subset of $\mathbb{R}^n$, and let $C_C^\infty(K)$ be the space of complex valued functions defined on $\mathbb{R}^n$ having supports in K. Define the topology on $C_C^\infty(K)$ whose neighborhood basis about the zero element is

$$\gamma(m, \epsilon) = \left\{ \varphi : C_C^\infty(K), \sup_{x \in K} \sum_{|\alpha| \le m} |\partial^\alpha \varphi| \le \epsilon \right\}$$

as $m = 0, 1, 2, 3, \ldots$ and $\epsilon > 0$ vary in all possible ways. Prove that $C_C^\infty(K)$ is a Fréchet space.

9. Let $C_\infty(\mathbb{R}^n)$ be the space of continuous function in $\mathbb{R}^n$ that converges to zero at infinity, equipped with the topology of uniform convergence on $\mathbb{R}^n$:

$$\varphi \to \sup_{x \in \mathbb{R}^n} |\varphi(x)|$$

Prove that $C_\infty(\mathbb{R}^n)$ is a Banach space.

10. Give an example of a continuous function f in $\mathbb{R}^n$ with the following two properties: (a) there is no polynomial P in $\mathbb{R}^n$ such that

$$|f(x)| \le |P(x)| \qquad \forall\, x \in \mathbb{R}^n$$

(b) The distribution $\varphi \to \int \varphi(x) f(x)\, dx$ is tempered.

11. Is the function $e^{|x|}$ Fourier transformable in the distributional sense? (Hint: Determine if $e^{|x|}$ is an ultradistribution. Find the Fourier transform of $e^{i|x|}$.)

12. Compute the Fourier transform of the Heavyside unit function

$$u(t) = \begin{cases} 1, & t > 0 \\ 0, & t < 0 \end{cases}$$

13. If $f, g \in S(\mathbb{R}^n)$, define

$$(f * g)(x) = \int_{\mathbb{R}^n} f(t - x) g(x)\, dt$$

Find the Fourier transform of $(f * g)(x)$.

14. By using the Paley-Wiener theorem, or otherwise, show that if $P\left(\frac{\partial}{\partial x}\right)$ is a differential operator with constant coefficients (not all zero) in $\mathbb{R}^n$. The equation $P\left(\frac{\partial}{\partial x}\right)u = 0$ has only one solution in the space E' that is $u \equiv 0$.

Note. Readers who have a strong background in functional analysis may want to use Treves [101] and [103] to solve these problems. These are the prime sources.

2

THE RIEMANN-HILBERT PROBLEM

2.1. SOME COROLLARIES ON CAUCHY INTEGRALS

The first four sections of this chapter briefly present some of very useful results given by Muskhelishvilli [65] and Carrier, Krook, and Pearson [17] for Cauchy integrals. Let $F(z)$ be defined by

$$F(z) = \frac{1}{2\pi i} \int_C \frac{f(t)\,dt}{t - z} \tag{2.1}$$

where C is some curve in the complex t-plane, $f(t)$ is a complex-valued function prescribed on C, and z is a point not on C. The curve C may be an arc or a closed contour, or more generally a collection of such arcs and closed contours. For suitable curves C and functions f, $F(z)$ will be analytic function of z. For example, if $f(t)$ is continuous on C and C is a smooth curve, then $F(z)$ will be analytic. However, the conditions on C and $f(t)$ can be further relaxed and still yield the analyticity of $F(z)$. If C is a closed contour, the positive direction will be counterclockwise and $F_+(t_0)$ and $F_-(t_0)$ where $t_0 \in C$ are limits of $F(z)$ as $z \to t_0$ from inside and outside of C, respectively. If C is a curve extending from point A to point B and if we orient ourselves at point t_0 on C so as to be facing in the positive direction of integration from point A to point B, then the limits of $F(z)$ as $z \to t_0$ from the left and from the right, if limits exist, will be denoted by $F_+(t_0)$ and $F_-(t_0)$, respectively. We define the principal value $F_p(t_0)$ of the integral (2.1) by

$$F_p(t_0) \overset{\Delta}{=} \frac{1}{2\pi i}(P) \int_C \frac{f(t)\,dt}{t - t_0}$$

$$= \frac{1}{2\pi i} \lim_{\epsilon \to 0} \int_{C - C_\epsilon} \frac{f(t)\,dt}{t - t_0}$$

where C_ϵ is the portion of the curve C contained within a small circle of radius ϵ, centered on t_0. Our objective is to find a relation between $F_p(t_0)$, $F_+(t_0)$, and $F_-(t_0)$. To derive such a relation, let us first assume that $f(t)$ is analytic at point t_0. That is to say, there exists a circle ξ with center t_0 and radius $r > 0$ such that $f(t)$ is analytic in the disc $\{t : |t - t_0| < r\}$.

Let z be a point in the circle situated to the left of the curve C. Note that

$$\int_{C_\epsilon} \frac{f(t)}{t - z}\, dt = \int_{C_1} \frac{f(t)}{t - z}\, dt \qquad \text{(Cauchy's theorem)}$$

where C_1 is the arc of a circle with center t_0 and radius $\epsilon < r$, as shown in Figure 2.1.

$$F(z) = \frac{1}{2\pi i} \int_{C-C_\epsilon} \frac{f(t)\, dt}{t - z} + \frac{1}{2\pi i} \int_{C_1} \frac{f(t)\, dt}{t - z} \qquad (2.2)$$

Therefore letting $z \to t_{0+}$ in (2.2), we have

$$F_+(t_0) = \frac{1}{2\pi i} \int_{C-C_\epsilon} \frac{f(t)\, dt}{t - t_0} + \frac{1}{2\pi i} \int_{C_1} \frac{f(t)\, dt}{t - t_0} \qquad (2.3)$$

Now let $\epsilon \to 0$, and (2.3) becomes

$$F_+(t_0) = F_p(t_0) + \frac{1}{2}f(t_0) \qquad (2.4)$$

Similarly, by changing the sides of z and C_1, we get

$$F_-(t_0) = F_p(t_0) - \frac{1}{2}f(t_0) \qquad (2.5)$$

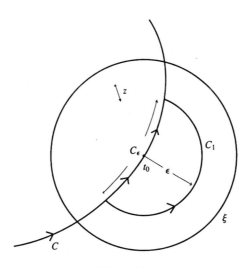

Figure 2.1

The relations (2.4) and (2.5) are known as *Plemelj formulas*. It is a simple exercise to show relations (2.4) and (2.5) by assuming only that $f(t)$ is continuous on C and satisfies the Lipschitz condition

$$|f(t_1) - f(t_0)| < A|t_1 - t_0|^\alpha \tag{2.6}$$

for all points t_1 on C and outside C in some neighborhood of t_0, where A and α are constants, with $0 < \alpha \leq 1$. We are now assuming that $f(t)$ satisfies (2.6) in a neighborhood of t_0, containing the curves C_ϵ and C_1.

Let

$$F(z) = \frac{1}{2\pi i} \int_C \frac{f(t)}{t - z} dt \tag{2.7}$$

and C_ϵ be an arc of C lying in a neighborhood of the point t_0 satisfying (2.6). Then

$$
\begin{aligned}
F(z) &= \frac{1}{2\pi i} \int_{C-C_\epsilon} \frac{f(t)}{t - z} dt + \frac{1}{2\pi i} \int_{C_\epsilon} \frac{f(t)}{t - z} dt \\
&= \frac{1}{2\pi i} \int_{C-C_\epsilon} \frac{f(t)\, dt}{t - z} + \frac{1}{2\pi i} \int_{C_\epsilon} \frac{f(t) - f(z)}{t - z} dt + \frac{f(z)}{2\pi i} \int_{C_\epsilon} \frac{1}{t - z} dt \\
&= \frac{1}{2\pi i} \int_{C-C_\epsilon} \frac{f(t)}{t - z} dt + \frac{1}{2\pi i} \int_{C_\epsilon} \frac{f(t) - f(z)}{t - z} dt + \frac{f(z)}{2\pi i} \int_{C_1} \frac{1}{t - z} dt
\end{aligned}
\tag{2.8}
$$

where C_1 is an arc of a circle lying on the side of the curve C opposite to the side point z lies on, such that the end points of C_1 lie in the region of C where the Lipschitz condition holds true.

Letting $z \to t_0$ from the left and then $\epsilon \to 0$, in (2.8) we get

$$F_+(t_0) = F_p(t_0) + \frac{1}{2} f(t_0)$$

Note that by Cauchy's theorem $\int_{C_\epsilon} \frac{1}{t-z} dt = \int_{C_1} \frac{1}{t-z} dt \to \pi i$ and that the second integral in (2.8) $\to 0$ as $z \to t_0$ from the left and then $\epsilon \to 0$ (see Example 1).

Similarly we can prove that

$$F_-(t_0) = F_p(t_0) - \frac{1}{2} f(t_0)$$

This shows that to prove (2.4) and (2.5) the condition of analyticity at t_0 can be replaced by a Lipschitz condition in a neighborhood of the point t_0. A stronger result is given in Section 2.2.

2.2. RIEMANN'S PROBLEM

In his Ph.D. dissertation Riemann considered the problem of determining a function $W_+(z) = u(x, y) + iv(x, y)$ that is analytic inside a closed contour C such that the boundary values of its real and imaginary parts on the contour C satisfy the linear relation

$$\alpha(t)u(t) + \beta(t)v(t) = \gamma(t) \qquad \forall\, t \text{ on } C \tag{2.9}$$

where $\alpha(t), \beta(t), \gamma(t)$ are given real functions. In view of the Riemann mapping theorem concerning conformal mapping the closed contour C can be taken as the unit circle. We define an analytic function $W_-(z)$ outside the circle C by the relation

$$W_-(z) = \overline{W}_+\left(\frac{1}{\bar{z}}\right) \tag{2.10}$$

For z on C we have

$$z\bar{z} = |z|^2 = 1$$

That is, $z = \frac{1}{\bar{z}}\ \forall\, z$ on C.

Therefore $z \rightarrow t$ implies that $\frac{1}{\bar{z}} \rightarrow t$ and

$$W_-(t) = \overline{W}_+(t)$$

or

$$\overline{W}_-(t) = W_+(t) = u(t) + iv(t)$$

or

$$W_-(t) = u(t) - iv(t)$$

Since

$$W_+(t) = u(t) + iv(t)$$

we have

$$u(t) = \frac{W_+(t) + W_-(t)}{2} \tag{2.11}$$

and

$$v(t) = \frac{W_+(t) - W_-(t)}{2i} \tag{2.12}$$

Substituting for $u(t), v(t)$ from (2.11) and (2.12) in (2.9), we get

$$\frac{\alpha(t) - i\beta(t)}{2}W_+(t) + \frac{\alpha(t) + i\beta(t)}{2}W_-(t) = \gamma(t) \tag{2.13}$$

The Riemann problem is now reduced to finding functions $W_+(z), W_-(z)$ analytic inside and outside the circle C, respectively, such that their boundary values on the circle satisfies the linear relation (2.13). The behavior of $W_-(z)$ at ∞ must also be prescribed for completeness. From the relation (2.10) it follows that $W_-(z) \rightarrow \overline{W}_+(0)$ as $z \rightarrow \infty$.

2.2.1. The Hilbert Problem

Hilbert generalized the Riemann problem posed above by determining a function $W(z)$ analytic for all values of z not lying on the curve C such that for t on C

$$W_+(t) = g(t)W_-(t) + f(t) \tag{2.14}$$

where $f(t)$ and $g(t)$ are given complex-valued functions (or even generalized functions as considered in this book) and $W_+(t)$ and $W_-(t)$ are limits of $W(z)$ as $z \to t$ from inside C and outside C, respectively.

2.2.2. Riemann-Hilbert Problem

Hilbert in his problem described above considered only the case of a closed curve or circle C. But a more general case where C is an arc, a closed contour, or a collection of arcs and contours has come to be called the *Riemann-Hilbert problem*. The behavior of $W(z)$ at ∞ must be specified, as must also be the behavior of $W(z)$ near the ends of C if C is an arc. If C is an arc extending from point A to point B and the positive direction on the arc C is considered as the direction pointed by an arrow sign pointed toward B, then $W_+(t)$ is interpreted as the limit of $W(z)$ as z approaches t from the left of C and $W_-(t)$ is interpreted as the limit of $W(z)$ as z approaches t from the right of C.

2.3. CARLEMAN'S APPROACH TO SOLVING THE RIEMANN-HILBERT PROBLEM

Carleman encountered the Riemann-Hilbert problem in his work on singular integral equations and devised a very effective means of solving it. The method is first to find a nonzero holomorphic function $L(z)$ that is analytic everywhere on the z plane except possibly on the curve C, satisfying

$$L_+(t) = g(t)L_-(t) \tag{2.15}$$

and such that $L_+(t), L_-(t)$ are nonzero. Substituting (2.15) in (2.14), we get

$$\frac{W_+(t)}{L_+(t)} - \frac{W_-(t)}{L_-(t)} = \frac{f(t)}{L_+(t)} \tag{2.16}$$

Since $L(z) \neq 0$, the function $\frac{W(z)}{L(z)}$ is analytic for z not on C such that (2.15) is satisfied. Since $L(z)$ is known, $W(z)$ is known. If we take

$$M(z) = \frac{1}{2\pi i} \int_C \frac{f(t)/L_+(t)}{t - z} dt$$

then using the discontinuity theorem we have

$$\frac{f(t)}{L_+(t)} = M_+(t) - M_-(t)$$

Therefore the equation (2.16) becomes

$$\frac{W_+(t)}{L_+(t)} - M_+(t) = \frac{W_-(t)}{L_-(t)} - M_-(t)$$

and so the function $\frac{W(z)}{L(z)} - M(z)$ is continuous on C and in view of Morera's theorem is entire. Now the behavior of $W(z)$ at ∞ is prescribed to be of polynomial order if we can select a polynomial $p_m(z)$ such that the solution $W(z)$ of the equation $\frac{W(z)}{L(z)} - M(z) = p_m(z)$ satisfies the order condition $W(z) = o(z^n)$, $z \to \infty$.

We now proceed to find $L(z)$. From (2.15) we have

$$\ln L_+(t) - \ln L_-(t) = \ln\big(g(t)\big) \tag{2.17}$$

We can solve (2.17) at least for the case where C is an arc, since we do not have to worry about the multiplicities of values of $\ln g(t)$. We also assume that $g(z)$ is continuous at end points z_1 and z_2 of the arc having the values $g(z_1)$ and $g(z_2)$ at their end points. Using the discontinuity theorem, we then get

$$\ln L(z) = \frac{1}{2\pi i} \int_C \frac{\ln\big(g(t)\big)}{t - z} dt$$

which, for example, is $Q(z)$. Thus

$$L(z) = e^{Q(z)} \neq 0$$

Therefore

$$L_+(t) = e^{Q_+(t)}$$

$$L_-(t) = e^{Q_-(t)}$$

and

$$\frac{L_+(t)}{L_-(t)} = e^{[Q_+(t)-Q_-(t)]} = e^{\ln(g(t))}$$

$$= g(t)$$

We have now solved (2.15).

We know that $t^\alpha, 0 > \alpha > -1$, and that $\ln t$ are integrable in the neighborhood of the origin, even though these functions blow up at the origin. Such functions are said to have integrable singularities at the origin. Now, if C is an arc, we permit $f(t)$ to have integrable singularities at the end points. If $f(t)$ has an integrable singularity at an end point t_e, then $F(z)$ can grow no faster than some power $\beta > -1$ of $|z - t_e|$. In solving some singular integral equations of physical problems, such types of functions do arise.

Now there are two important cases to be considered: (1) $L(z)$ is discontinuous at the end points z_1 and z_2 of the arc C and (2) when C is a circle, $\ln g(t)$ may have multiplicities of values on C. These two cases are resolved quite easily as follows:

Case 1. C is an arc with beginning end point z_1 and the final end point z_2 with a singularity at the point z_1. The case where $L(z)$ has a point of discontinuity at z_2 can be resolved similarly. Assuming z_1 to be a point of discontinuity, we get

$$Q(z) = \frac{1}{2\pi i} \int_C \frac{\ln(g(t))\, dt}{t - z}$$

$$\sim -\frac{1}{2\pi i} \ln(g(z_1)) \ln(z_1 - z)$$

or

$$L(z) \sim (z - z_1)^{-1/2\pi i} \ln(g(z_1))$$

$$\sim (z - z_1)^{a+bi} \qquad (a, b \text{ being real})$$

$$L(z)(z - z_1)^{k_1} \sim (z - z_1)^{a+k_1+ib}$$

We choose k_1 such that $-1 < a + k_1 < 0$. Thus the singularities at end point are taken care of in previous section, since $L(z)(z - z_1)^{k_1}$ has an integrable singularity at $z = z_1$. All that we have to do is to replace $L(z)$ by $\tilde{L}(z) = L(z)(z - z_1)^{k_1}$.

Case 2. Now we resolve the case where $g(t)$ has singularities at $t = z_1$ and C is a circle of unit radius. So $\ln g(t)$ will change values by a multiple of $2\pi i$. We now can avoid the multiplicity of values as follows: Define

$$g_0(t) = (t - z_0)^{-n} g(t)$$

where z_0 is a point within C_0. Now define

$$N(z) = \begin{cases} L(z) & \text{for } z \text{ inside } C \\ (z - z_0)^n L(z), & \text{for } z \text{ outside } C \end{cases}$$

Our problem is now resolved by finding the solution of

$$N_+(t) = g_0(t) N_-(t)$$

where $g_0(t)$ is a single-valued function on C. The same procedure as was used for the arc may now be applied. Having determined $N(z)$, we get $L(z)$ as follows:

$$L(z) = \begin{cases} N(z) & \text{for } z \text{ inside } C \\ \dfrac{N(z)}{(z-z_0)^n} & \text{for } z \text{ outside } C \end{cases}$$

We now solve a few related problems:

Example 1. Let $f(t)$ be continuous on a smooth and rectifiable curve C and satisfy

$$|f(t_1) - f(t_0)| < A|t_1 - t_0|^\alpha$$

for all points t_1 on C and outside C in some neighborhood of $t_0 \in C$ where A and α are constants with $0 < \alpha \leq 1$. Show that $F_p(t_0)$ exists and the Plemelj formulas hold.

Solution. Refer to Figure 2.1, and set

$$F(z) = \frac{1}{2\pi i} \int_C \frac{f(t)\, dt}{t - z}$$

$$= \frac{1}{2\pi i} \int_{C-C_\epsilon} \frac{f(t)\, dt}{t - z} + \frac{1}{2\pi i} \int_{C_\epsilon} \frac{f(t) - f(z)}{t - z} dt + \frac{1}{2\pi i} \int_{C_\epsilon} \frac{f(z)}{t - z} dt$$

In view of Cauchy's theorem, we have

$$\frac{1}{2\pi i} \int_{C_\epsilon} \frac{f(z)}{t - z} dt = \frac{1}{2\pi i} \int_{C_1} \frac{f(z)}{t - z} dt$$

Now let $z \to t_0+$. Therefore

$$F_+(t_0) = \frac{1}{2\pi i} \int_{C-C_\epsilon} \frac{f(t)}{t - t_0} dt + \frac{1}{2\pi i} \lim_{z \to t_0} \int_{C_\epsilon} \frac{f(t) - f(z)}{t - z} dt$$

$$+ \frac{1}{2\pi i} f(t_0) \int_{C_1} \frac{dt}{t - t_0}$$

The proof for the existence of $F_p(t_0)$ is sketched in the next example. Letting $\epsilon \to 0$, we get

$$F_+(t_0) = F_p(t_0) + \frac{f(t_0)}{2} + \lim_{\epsilon \to 0} \lim_{z \to t_0} \frac{1}{2\pi i} \int_{C_\epsilon} \frac{f(t) - f(z)}{t - z} dt$$

Now we show that the last double limit in the above equation is zero.

$$\left| \frac{1}{2\pi i} \int_{C_\epsilon} \frac{f(t) - f(z)}{t - z} dt \right| \le \frac{1}{2\pi i} A \int_{C_\epsilon} \frac{|t - z|^\alpha |dt|}{|t - z|}$$

Denote $\int_{C_\epsilon} \frac{|t - z|^\alpha}{|t - z|} |dt|$ by I and put $t - z = pe^{\varphi i}$, where $p = |t - z|$ and $\varphi = \arg(t - z) = $ the angle between line zt and the line zt_0. Therefore

$$\frac{dt}{t - z} = \frac{dp}{p} + i d\varphi$$

$$\frac{|dt|}{|t - z|} = \frac{|dt|}{p} \le \frac{|dp|}{p} + |d\varphi|$$

Hence

$$I \le \int_{C_\epsilon} |t - z|^\alpha \frac{|dt|}{|t - z|}$$

$$\le \int_{C_\epsilon} p^\alpha \left[\frac{|dp|}{p} + d\varphi \right]$$

$$= \int_{C_\epsilon} [p^{\alpha-1} |dp| + p^\alpha |d\varphi|]$$

With z being fixed, we can split up the arc C_ϵ into subarcs (t_0, t_1), (t_1, t_2), $(t_2, t_3), \ldots, (t_{n-1}, t_n)$ such that in each of these subintervals p is monotonic on C. For the sake of definiteness, we assume that p is monotonic* increasing for t between t_{i-1} and t_i then

$$\int_{t_{i-1}}^{t_i} \left[p^{\alpha-1} |dp| \right] = \int_{t_{i-1}}^{t_i} p^{\alpha-1} dp = \left. \frac{p^\alpha}{\alpha} \right|_{t_{i-1}}^{t_i}$$

$$= \frac{1}{\alpha} \left[p_i^\alpha - p_{i-1}^\alpha \right]$$

Thus we see that

$$I \le \frac{1}{\alpha} \left[p_1^\alpha + 2p_2^\alpha + 2p_3^\alpha + \cdots + p_n^\alpha \right] + p_j^\alpha \Phi$$

where Φ is the total variation of the angle φ and p_j is the maximum of all p_i's, as discussed above. Clearly $\Phi \le 2\pi$. Now $z - t = z - t_0 + t_0 - t$. Therefore $|z - t| \le |z - t_0| + \epsilon$, and

$$\overline{\lim_{z \to t_0}} I \le \frac{1}{2} \left[2(n-1)\epsilon^\alpha + \epsilon^\alpha 2\pi \right]$$

Letting $\epsilon \to 0$, we have

$$\lim_{\epsilon \to 0} \lim_{z \to t_0} I = 0$$

Note that a geometrical interpretation of Φ can be that it represents the angle subtended by the arc C_ϵ at z. We can assume that the curve C has a corner at point t_0, making an angle θ at t_0 as shown in Figure 2.2. Assume also that the function $f(t)$ satisfies the H condition in a small neighborhood of the point t_0 as stated in (2.6).

Our Plemelj formula under the same set of conditions takes the form

$$F_+(t_0) = \left(1 - \frac{\theta}{2\pi} \right) f(t_0) + F_p(t_0)$$

$$F_-(t_0) = -\frac{\theta}{2\pi} f(t_0) + F_p(t_0) \tag{2.18}$$

From (2.18) it follows that

$$f(t_0) = F_+(t_0) - F_-(t_0)$$

$$F_p(t_0) = \frac{1}{2} \left[F_+(t_0) + F_-(t_0) \right] - \left(\frac{1}{2} - \frac{\theta}{2\pi} \right) f(t_0) \tag{2.19}$$

The proofs for these formulas become simpler if the H condition around t_0 is replaced by analyticity of f at t_0. In that case we can also say that $f(t)$ is analytic in a

---

*It is assumed that the arc C_ϵ has at the most n oscillations, n being finite and ≥ 0. We say that the curve C oscillates at point t if its slope changes sign at t.

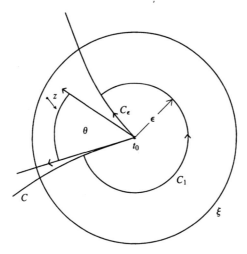

Figure 2.2

neighborhood of t_0, and we can replace $\frac{1}{2\pi i} \int_{C_\epsilon} \frac{f(t)}{t-z} \, dt$ by $\frac{1}{2\pi i} \int_{C_1} \frac{f(t)\, dt}{t-z}$ with appropriate choice of C_ϵ and C_1.

We now give a much stronger result proved by Muskhelishvili [65, p. 38].

Example 2. Let $f(t)$ be a complex-valued function continuous on a smooth curve C satisfying H condition:

$$|f(t) - f(t_0)| \leq A|t - t_0|^\alpha$$

for all points t on C in some neighborhood of $t_0 \in C$ where A and α are constants with $0 < \alpha \leq 1$. Show that $F_p(t_0)$ exists and the Plemelj formula holds.

Solution. By writing $f(t) = f(t) - f(t_0) + f(t_0)$ and performing calculation as in the foregoing theorem, we get

$$F_+(t_0) = \frac{1}{2}f(t_0) + F_p(t_0) + \lim_{\epsilon \to 0} \lim_{z \to t_0} \frac{1}{2\pi i} \int_{C_\epsilon} \frac{f(t) - f(t_0)}{t - z} dt$$

We can assume the existence of $F_p(t_0)$ for the time being:

$$F_+(t_0) = \frac{1}{2}f(t_0) + F_p(t_0) + \lim_{\epsilon \to 0} \frac{1}{2\pi i} \int_{C_\epsilon} \frac{f(t) - f(t_0)}{t - t_0} dt$$

See Muskhelishvili [65, p. 38].

Let $|t - t_0| = r$, $\varphi = \arg(t - t_0) =$ the angle that the chord $t_0 t$ makes with the tangent vector to the curve C at t_0. Then

$$t - t_0 = re^{i\varphi}$$

$$\ln(t - t_0) = \ln r + i\varphi$$

$$\frac{dt}{t - t_0} = \frac{dr}{r} + id\varphi$$

Let

$$I = \frac{1}{2\pi i} \int_{C_\epsilon} \frac{f(t) - f(t_0)}{t - t_0} dt$$

Therefore

$$|I| \le \frac{1}{2\pi} A \int_{C_\epsilon} |t - t_0|^\alpha \frac{|dt|}{|t - t_0|}$$

or

$$|I| \le \frac{A}{2\pi} \int_{C_\epsilon} r^\alpha \left[\frac{|dr|}{r} + |d\varphi|\right]$$

$$\le \frac{A}{2\pi} \int_{C_\epsilon} r^\alpha \frac{|dr|}{r} + \frac{A}{2\pi} \int_{C_\epsilon} r^\alpha |d\varphi|$$

$$|I| \le \frac{A}{2\pi\alpha} \left[r_1^\alpha + r_2^\alpha\right] + \frac{A}{2\pi} \epsilon^\alpha m$$

as $\int_{C_\epsilon} |d\varphi| \le m$. This is because for a smooth curve we have

$$\frac{dr}{ds} = \pm \cos \alpha'$$

$$0 \le \alpha' < \alpha_0 < \frac{\pi}{2} \qquad [65, \text{p. } 10]$$

where α' is the acute angle that the tangent to C at t makes [65, pp. 10, 425] with the chord $t_0 t$. The upper sign refers to the upper arc, and the lower sign refers to the lower arc. Therefore r is monotonic with s.

Let $r_1 = |t_0 - t_f|$ and $r_2 = |t_0 - t_e|$ where t_e, t_f are two end points of the curve C_ϵ. By construction (see Figure 2.1), $r_1 < \epsilon, r_2 < \epsilon$. Therefore

$$|I| \le \frac{A}{2\pi\alpha} 2\epsilon^\alpha + \frac{A}{2\pi} m\epsilon \to 0 \qquad \text{as } \epsilon \to 0$$

One of the Plemelj formulas follows. The other Plemelj formula can be proved similarly. The existence of $F_p(t_0) = \frac{1}{2\pi i}(P) \int_C \frac{f(t)}{t - t_0} dt$ can be proved by doing similar manipulations.

Example 3 (Inversion of the Cauchy's integral). Let C be a closed contour, and let $f(t)$ be a function that is continuous on C and satisfies H condition on C in a neighborhood of point $t_0 \in C$. Let

$$g(t_0) = \frac{1}{\pi i}(P) \int_C \frac{f(t)}{t - t_0} dt$$

To prove that

$$f(t_0) = \frac{1}{\pi i}(P) \int_C \frac{g(t)}{t - t_0} dt$$

let $F(z)$ be defined as

$$F(z) = \frac{1}{\pi i} \int_C \frac{f(t)}{t - z} dt, \qquad z \notin C$$

Then

$$F_+(t_0) - F_-(t_0) = f(t_0)$$

$$F_+(t_0) + F_-(t_0) = \frac{1}{\pi i}(P) \int_C \frac{f(t)}{t - t_0} dt$$

Now define a function $G(z)$ as

$$G(z) = \begin{cases} F(z), & z \text{ inside } C \\ -F(z), & z \text{ outside } C \end{cases}$$

Then

$$G_+(t_0) + G_-(t_0) = F_+(t_0) - F_-(t_0) = f(t_0)$$

$$G_+(t_0) - G_-(t_0) = F_+(t_0) + F_-(t_0) = \frac{1}{\pi i} \int_C \frac{f(t)}{t - t_0} dt = g(t_0)$$

$$G(z) = \frac{1}{2\pi i} \int_C \frac{g(t)}{t - z} dt$$

Now we can write

$$G_+(t_0) + G_-(t_0) = \frac{1}{\pi i}(P) \int_C \frac{g(t)}{t - t_0} dt$$

Therefore

$$f(t_0) = \frac{1}{\pi i}(P) \int_C \frac{g(t)}{t - t_0} dt$$

Note that if $f(t)$ satisfies the H condition in a neighborhood of point t_0 belonging to C, then so does $g(t)$. For proof, see Muskhelishvili [65].

The foregoing formulas are also true when C is the union of contours L_0, L_1, $L_2, L_3, \ldots, L_p$ where $L_1, L_2, \ldots, L_p$ are nonintersecting contours contained in L_0 and S_+ stands for connected region bounded by one or several smooth nonintersecting contours $L_0, L_1, \ldots, L_p$. The region S_- is the complement of $S_+ \cup L$. Here $L = L_0 \cup L_1 \cup L_2 \cup \cdots \cup L_p$. See Figure 2.3.

If L_0 exists, then $S_- = S_1^- \cup S_2^- \cup \cdots \cup S_p^- \cup S_0^-$ where $S_1^-, S_2^-, \ldots, S_p^-$ are the interiors of contours $L_1, L_2, \ldots, L_p$, respectively, and S_0^- is the exterior of L_0. If L_0 does not exist, then $S_- = S_1^- \cup S_2^- \cup \cdots \cup S_p^-$ and $S_+ = $ complement of $S_- \cup L_1 \cup L_2 \cup \cdots \cup L_p$. Obviously S_+ in this case is unbounded. The above

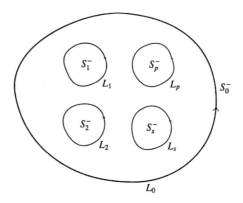

Figure 2.3

inversion formula can be put in the form

$$f(t_0) = \frac{1}{\pi i}(P) \int_C \left[\frac{1}{\pi i}(P) \int_C \frac{f(\tau)\,d\tau}{\tau - t} \right] \frac{dt}{t - t_0}$$

which is called the *Poincaré-Bertrand formula*.

2.4. THE HILBERT INVERSION FORMULA FOR PERIODIC FUNCTIONS

The Hilbert transform $(Hf)(\theta)$ of a periodic function $f(t)$ with period 2π is defined by

$$(Hf)(\theta) = \frac{1}{2\pi}(P) \int_{-\pi}^{\pi} f(\theta - t) \cot \frac{t}{2}\, dt$$

provided the integral exists. With a little careful calculation it can be shown that

$$(Hf)(\theta) = \frac{1}{2\pi}(P) \int_{-\pi}^{\pi} f(t) \cot \left(\frac{\theta - t}{2} \right) dt$$

$$= \frac{1}{2\pi}(P) \int_{0}^{2\pi} f(t) \cot \left(\frac{\theta - t}{2} \right) dt$$

When the periodic function f is L^p integrable for $p > 1$ over the interval $[-\pi, \pi]$, it can be shown that

$$H^2 f = -f + \frac{1}{2\pi} \int_{-\pi}^{\pi} f(t)\, dt \qquad \text{a.e.}$$

We give the most elementary proof of this theorem by using the Poincaré-Bertrand formula as follows:

Let $f(t)$ be a complex-valued function, satisfying H condition on the unit circle or be analytic on C. Then in view of the Poincaré-Bertrand formula, we have

$$\frac{1}{\pi i}(P)\int_C \frac{f(t)}{x-t}\,dt = g(x) \tag{2.20}$$

$$\frac{1}{\pi i}(P)\int_C \frac{g(t)}{x-t}\,dt = f(x) \tag{2.21}$$

Let $t = e^{i\theta}$, $x = e^{i\varphi}$,

$$\frac{dt}{x-t} = \frac{ie^{i\theta}\,d\theta}{e^{i\varphi} - e^{i\theta}} = \frac{1}{2}\cot\frac{\theta-\varphi}{2}\,d\theta + \frac{i}{2}\,d\theta$$

$$\frac{1}{\pi i}(P)\int_C f(e^{i\theta})\left[\frac{1}{2}\cot\frac{\theta-\varphi}{2}\,d\theta + \frac{i}{2}\,d\theta\right] = g(e^{i\theta})$$

$$\frac{1}{\pi i}(P)\int_C g(e^{i\theta})\left[\frac{1}{2}\cot\frac{\theta-\varphi}{2}\,d\theta + \frac{i}{2}\,d\theta\right] = f(e^{i\theta})$$

Now put

$$f(e^{\theta i}) = \Phi(\theta)$$
$$g(e^{i\theta}) = \psi(\theta)$$

Therefore

$$-\frac{1}{2\pi i}(P)\int_0^{2\pi}\Phi(\theta)\cot\left(\frac{\varphi-\theta}{2}\right)d\theta$$

$$+\frac{1}{2\pi}\int_0^{2\pi}\Phi(\theta)\,d\theta = \psi(\theta) \tag{2.22}$$

$$-\frac{1}{2\pi i}(P)\int_0^{2\pi}\psi(\theta)\cot\left(\frac{\varphi-\theta}{2}\right)d\theta$$

$$+\frac{1}{2\pi}\int_0^{2\pi}\psi(\theta)\,d\theta = \Phi(\theta) \tag{2.23}$$

or

$$i(H\Phi)(\theta) + (I\Phi)(\theta) = \psi(\theta) \tag{2.24}$$

$$i(H\psi)(\theta) + (I\psi)(\theta) = \Phi(\theta) \tag{2.25}$$

$$(I\Phi)(\theta) = \frac{1}{2\pi}\int_0^{2\pi}\Phi(\theta)\,d\theta$$

Eliminating ψ between (2.24) and (2.25), we get

$$iH[iH\varphi + I\varphi]\theta + I[i(H\Phi) + I\Phi](\theta) = \Phi(\theta)$$

ASTON UNIVERSITY
LIBRARY & INFORMATION SERVICES

Since $IH\Phi = 0$ and $HI\varphi = 0$, we have

$$-H^2\varphi + I^2\Phi(\theta) = \Phi(\theta)$$

or

$$H^2\Phi - (I\Phi)\theta = -\Phi(\theta)$$

$$(H^2\Phi)(\theta) = -\Phi(\theta) + \frac{1}{2\pi}\int_0^{2\pi}\Phi(\theta)\,d\theta$$

If $\int_0^{2\pi}\Phi(\theta)\,d\theta = 0$, we get

$$(H^2\Phi)(\theta) = -\Phi(\theta)$$

The Hilbert transform of periodic functions f with period 2τ, as defined in Chapter 7, is given by

$$(Hf)(x) = \frac{1}{2\tau}(P)\int_{-\tau}^{\tau} f(x-t)\cot\left(\frac{t\pi}{2\tau}\right) \tag{2.26}$$

The inversion formula for this transform is

$$(H^2f)(x) = -f(x) + \frac{1}{2\tau}\int_{-\tau}^{\tau} f(x)\,dx \tag{2.27}$$

The inversion formula (2.27) at point x for the Hilbert transform of periodic function f with period 2τ is valid under the assumption that f is continuous over the period $[-\pi, \pi]$ and satisfies the H conditions in a neighborhood of the point x (see Exercise 1). Butzer and Nessel [12] show that this inversion formula is true for any function that is periodic with period 2π and is L^p integrable over the interval $[-\pi, \pi]$ for $p > 1$. But their inversion formula can be easily generalized for arbitrary period 2τ. By a careful calculation it can be shown that the definition (2.26) is also equivalent to the following two forms:

$$(Hf)(x) = \frac{1}{2\tau}(P)\int_{-\tau}^{\tau} f(t)\cot\left(\frac{(x-t)\pi}{2\tau}\right)\,dt \tag{2.28}$$

or

$$(Hf)(x) = \frac{1}{2\tau}(P)\int_0^{2\tau} f(x-t)\cot\left(\frac{t\pi}{2\tau}\right)\,dt \tag{2.29}$$

Example 4. Show that there exists an analytic functions $F(z)$ defined inside the unit circle whose limit as $z \to \theta$ on the boundary of the circle is of the form $f(\theta) + ig(\theta)$ where $g(\theta) = (Hf)(\theta)$. In this sense f and g are conjugate to each other. Finally, show that if $f(\theta) = \sum_{1}^{\infty} [a_n \cos n\theta + b_n \sin n\theta]$, then

$$g(\theta) = \sum_{n=1}^{\infty} (a_n \sin n\theta - b_n \cos n\theta)$$

Note that the constant term in the Fourier series of $f(\theta)$ is missing. This is because the Hilbert transform of a constant (periodic function) is zero.

Solution. The required analytic function $F(z)$ should satisfy the following properties:

1. $F(z)$ is analytic on the unit disk $z : |z| \leq 1$.
2. $F(0) = 0$.

Let us denote the unit circle $|z| = 1$ by C. Then

$$F(z) = \frac{1}{2\pi i} \int_C \frac{F(t)\, dt}{t - z} \tag{2.30}$$

Again

$$\int_C \frac{F(z)\, dz}{z} = 2\pi i F(0) = 0$$

Therefore

$$\int_0^{2\pi} F(e^{i\varphi})\, d\varphi = 0$$

$$\int_C [f(e^{i\varphi})\, d\varphi + ig(e^{i\varphi})\, d\varphi] = 0$$

$$\int_C f(\cos \varphi, \sin \varphi)\, d\varphi = \int_C g(\cos \varphi, \sin \varphi)\, d\varphi = 0$$

Letting $z \to t_0 = e^{i\theta}$ in (2.30) and using (2.4), we get

$$F_+(t_0) = \frac{1}{2} F(t_0) + \frac{1}{2\pi i}(P) \int_C \frac{F(t)\, dt}{t - t_0} \tag{2.31}$$

$F_+(t_0) = F(t_0)$ as F is analytic on C, the unit circle. Therefore from (2.30) we get

$$F(t_0) = \frac{1}{2} F(t_0) + \frac{1}{2\pi i}(P) \int_C \frac{F(t)\, dt}{t - t_0}$$

or

$$F(t_0) = \frac{1}{\pi i}(P) \int_C \frac{F(t)\, dt}{t - t_0}$$

Let

$$t = e^{i\varphi}$$

$$t_0 = e^{i\theta}$$

Therefore

$$F(e^{i\theta}) = \frac{1}{2\pi i} \int_C F(e^{i\varphi}) \left[\cot \frac{\varphi - \theta}{2} + i \right] d\varphi$$

$$= \frac{1}{2\pi i} \int_C F(e^{i\varphi}) \cot \frac{\varphi - \theta}{2} d\varphi$$

$$F(e^{i\theta}) = i(HF)(\theta)$$

$$f(e^{\theta i}) + ig(e^{i\theta}) = i(Hf)(\theta) - (Hg)(\theta)$$

Equating the real and imaginary parts, we set

$$g = Hf$$

$$f = -Hg$$

Again, if $z = e^{\theta i} = \cos \theta + i \sin \theta$, then $z^n = \cos n\theta + i \sin n\theta$. By the foregoing result, we have

$$H(\cos n\theta) = \sin n\theta$$

$$H(\sin n\theta) = -\cos n\theta$$

If

$$f(\theta) = \sum_{n=1}^{\infty} (a_n \cos n\theta + b_n \sin n\theta)$$

then formally

$$g(\theta) = \sum_{n=1}^{\infty} a_n H(\cos n\theta) + b_n H(\sin n\theta)$$

H is the operator of the Hilbert transformation:

$$g(\theta) = \sum_{n=1}^{\infty} a_n \sin n\theta - b_n \cos n\theta$$

It will be shown in Chapter 7 that

$$He^{itn} = \frac{1}{\pi} \lim_{N \to \infty} P \int_{-N}^{N} \frac{e^{itn}}{x - t} dt$$

$$= -ie^{ixn}$$

$$= \sin nx - i \cos nx$$

Therefore, equating the real and imaginary parts, we get

$$H(\cos nt) = \sin nx$$

$$H(\sin nt) = -\cos nx$$

Example 5. Solve the following integral equation for f:

$$(P) \int_{-1}^{1} \frac{f(t) \, dt}{t - x} = h(x)$$

the function h being continuous in the interval $(-1, 1)$. The function f is to be continuous in the interval $(-1, 1)$ and may have integrable singularities at $t = \pm 1$.

Solution. Take

$$F(z) = \frac{1}{2\pi i} \int_{-1}^{1} \frac{f(t)}{t - z} \, dt, \qquad \operatorname{Im} z \neq 0$$

$$F_+(x) - F_-(x) = f(x)$$

$$F_+(x) + F_-(x) = \frac{1}{\pi i}(P) \int_{-1}^{1} \frac{f(t) \, dt}{t - x}$$

The given integral equation reduces to the form

$$\pi i \big(F_+(x) + F_-(x) \big) = h(x)$$

$$F_+(x) = (-1)F_-(x) + \frac{h(x)}{\pi i}, \qquad g(x) = -1$$

We find a nonzero function $L(z)$, satisfying

$$\frac{L_+(x)}{L_-(x)} = -1$$

Take

$$\ln L(z) = \frac{1}{2\pi i} \int_{-1}^{1} \frac{\ln \big(g(t) \big) \, dt}{t - z}$$

$$= \frac{1}{2\pi i} \int_{-1}^{1} \frac{\ln(-1)}{t - z} \, dt = \frac{1}{2} \ln \left(\frac{1 - z}{-(1 + z)} \right) = \ln \sqrt{\frac{z - 1}{z + 1}}$$

$$L(z) = \sqrt{\frac{z - 1}{z + 1}}$$

We now make an adjustment for $L(z)$ by choosing

$$L(z) = \frac{1}{z-1}\sqrt{\frac{z-1}{z+1}} = \frac{1}{\sqrt{(z-1)(z+1)}}$$

This adjustment is done in order to let $L(z)$ grow algebraically with index between -1 and 0 as $z \to \pm 1$. The given integral equation can be written

$$\frac{F_+(x)}{L_+(x)} - \frac{F_-(x)}{L_-(x)} = \frac{h(x)}{L_+(x)\pi i}$$

$$\frac{F(z)}{L(z)} = \frac{1}{2\pi i}\int_{-1}^{1}\frac{h(t)}{L_+(t)\pi i(t-z)}dt + k_1$$

where k_1 is an arbitrary constant. Thus

$$F(z) = 0(1), \; z \to \infty$$

$$= 0(z \pm 1)^{-1/2} \qquad \text{as } z \to \pm 1$$

which is what we wish to have.

$$F(z) = \frac{L(z)}{2\pi i}\int_{-1}^{1}\frac{h(t)}{L_+(t)\pi i(t-z)}dt + k_1 L(z)$$

Now

$$f(x) = F_+(x) - F_-(x)$$

$$= k_1\left[L_+(x) - L_-(x)\right] + \left[L_+(x) + L_-(x)\right]\frac{h(x)}{L_+(x)2\pi i}$$

$$+ \left[L_+(x) - L_-(x)\right]\frac{1}{2\pi i}(P)\int_{-1}^{1}\frac{h(t)}{L_+(t)(t-x)\pi i}$$

$$L(z) = \frac{1}{\sqrt{z-1}\sqrt{z+1}}$$

$$L_+(t) = \frac{-i}{\sqrt{1-t^2}}, \quad L_-(t) = \frac{i}{\sqrt{1-t^2}}$$

Therefore

$$f(x) = -\frac{k_1 2i}{\sqrt{1-t^2}} - \frac{2i}{\sqrt{1-t^2}}\frac{1}{2\pi i}(P)\int_{-1}^{1}\frac{h(t)}{L_+(t)(t-x)\pi i}dt$$

$$= -\frac{2ik_1}{\sqrt{1-t^2}} - \frac{1}{\pi\sqrt{1-t^2}}(P)\int_{-1}^{1}\frac{h(t)}{L_+(t)(t-x)\pi i}dt$$

$$= \frac{2ik_1}{\sqrt{1-t^2}} - \frac{1}{\pi\sqrt{1+t^2}} \int_{-1}^{1} \frac{\sqrt{1-t^2}\, h(t)}{(t-x)\pi i^2}\, dt$$

$$= \frac{2ik_1}{\sqrt{1-t^2}} + \frac{1}{\pi^2} \frac{1}{\sqrt{1-t^2}}(P) \int_{-1}^{1} \frac{\sqrt{1-t^2}\, h(t)}{(t-x)}\, dt$$

$$= \frac{A}{\sqrt{1-t^2}} + \frac{1}{\pi^2\sqrt{1-t^2}}(P) \int_{-1}^{1} \frac{\sqrt{1-t^2}\, h(t)\, dt}{t-x}$$

where A is an arbitrary constant. Note that

$$(P) \int_{-1}^{1} \frac{1}{\sqrt{1-t^2}} \frac{1}{t-x}\, dt = 0$$

for all x lying in the interval $(-1, 1)$.

Example 6. Solve

$$kf(x) = (P) \int_{-1}^{1} \frac{f(t)}{t-x}\, dt + g(x)$$

where k is an arbitrary constant, and $f(x)$, $g(x)$ are continuous real-valued functions defined on the interval $(-1, 1)$. An integrable singularity for $f(x)$ at $x = \pm 1$ is also permitted.

Solution. Let

$$F(z) = \frac{1}{2\pi i} \int_{-1}^{1} \frac{f(t)}{t-z}\, dt$$

Then

$$F_+(x) - F_-(x) = f(x)$$

$$F_+(x) + F_-(x) = \frac{1}{\pi i}(P) \int_{-1}^{1} \frac{f(t)}{t-x}\, dt$$

Therefore the given integral equation reduces to

$$k[F_+(x) - F_-(x)] = \pi i [F_+(x) + F_-(x)] + g(x)$$

$$F_+(x) = \frac{k+\pi i}{k-\pi i} F_-(x) + \frac{g(x)}{k-\pi i}$$

We first find a nonzero function $L(z)$, satisfying

$$\frac{L_+(x)}{L_-(x)} = \frac{k+\pi i}{k-\pi i}$$

such that $L(z)$ does not vanish on the complex z-plane. A suitable $L(z)$ is

$$L(z) = \frac{1}{z-1} e^{Q(z)}$$

where

$$Q(z) = \frac{1}{2\pi i} \int_{-1}^{1} \frac{dt}{t-z} \ln\left(\frac{k+\pi i}{k-\pi i}\right) dt$$

Now, if

$$k + \pi i = \sqrt{k^2 + \pi^2} e^{\beta i}, \qquad 0 < \beta < \pi$$

then

$$k - \pi i = \sqrt{k^2 + \pi^2} e^{-\beta i}$$

$$\frac{1}{2\pi i} \ln \frac{k+\pi i}{k-\pi i} = \frac{2\beta i}{2\pi i} = \frac{\beta}{\pi}$$

That is,

$$0 < \frac{1}{2\pi i} \ln\left(\frac{k+\pi i}{k-\pi i}\right) < 1$$

$$L_+(x) = \frac{1}{x-1} e^{Q_+(x)}$$

$$= \frac{1}{x-1} e \left\{ \frac{1}{2} \ln \frac{k+\pi i}{k-\pi i} + \frac{1}{2\pi i}(P) \int_{-1}^{1} \frac{\ln\left[(k+\pi i)/(k-\pi i)\right]}{t-x} dt \right\}$$

$$= \frac{1}{x-1} \sqrt{\frac{k+\pi i}{k-\pi i}} e^{\omega(x)}$$

$$= -\frac{1}{(1-x)^{1-(\beta/\pi)}(1+x)^{\beta/\pi}} \sqrt{\frac{k+\pi i}{k-\pi i}}$$

where

$$\omega(x) = \frac{\ln\left[(k+\pi i)/(k-\pi i)\right]}{2\pi i} \ln\left(\frac{x-1}{x+1}\right) = \ln\left(\frac{1-x}{1+x}\right)^{\beta/\pi}$$

Similarly

$$L_-(x) = \frac{1}{x-1} e^{Q_-(x)}$$

$$= \frac{1}{x-1} e^{-1/2} \ln\left(\frac{k+\pi i}{k-\pi i}\right) + (P) \int_{-1}^{1} \frac{\ln\left[(k+\pi i)/(k-\pi i)\right]}{2\pi i(t-x)} dt$$

or

$$L_-(x) = \frac{1}{x-1}\sqrt{\frac{k-\pi i}{k+\pi i}}e^{\omega(x)}$$

$$= -\frac{1}{(1-x)^{1-(\beta/\pi)}(1+x)^{\beta/\pi}}\sqrt{\frac{k-\pi i}{k+\pi i}}$$

Clearly an appropriate $L(z)$ has been found.

Now the given equation can be written

$$F_+(x) = \frac{L_+(x)}{L_-(x)}F_-(x) + \frac{g(x)}{L_+(x)}$$

$$\frac{F_+(x)}{L_+(x)} - \frac{F_-(x)}{L_-(x)} = \frac{g(x)}{L_+(x)(k-\pi i)}$$

Under the assumption that $\frac{F(z)}{L(z)} = 0(1)$ as $z \to \infty$, the most general solution is

$$\frac{F(z)}{L(z)} = \frac{1}{2\pi i}\int_{-1}^{1}\frac{g(t)\,dt}{(t-z)L_+(t)(k-\pi i)} + k_1$$

where k_1 is an arbitrary constant. We make the simplification

$$f(x) = F_+(x) - F_-(x) = k_1[L_+(x) - L_-(x)]$$

$$+ L_+(x)\left[\frac{1}{2}\frac{g(x)}{L_+(x)(k-\pi i)} + \frac{1}{2\pi i}(P)\int_{-1}^{1}\frac{g(t)dt}{(t-x)L_+(t)(k-\pi i)}\right]$$

$$- L_-(x)\left[-\frac{1}{2}\frac{g(x)}{L_+(x)(k-\pi i)} + \frac{1}{2\pi i}(P)\int_{-1}^{1}\frac{g(t)dt}{(t-x)L_+(t)(k-\pi i)}\right]$$

$$= k_1[L_+(x) - L_-(x)] + \frac{1}{2}[L_+(x) + L_-(x)]\frac{g(x)}{L_+(x)(k-\pi i)}$$

$$+ [L_+(x) - L_-(x)]\frac{1}{2\pi i}(P)\int_{-1}^{1}\frac{g(t)\,dt}{(t-x)L_+(t)(k-\pi i)}$$

$$= \frac{kg(x)}{k^2+\pi^2} + \frac{1}{(k^2+\pi^2)}\frac{e^{\omega(x)}}{(x-1)}(P)\int_{-1}^{1}(t-1)\frac{g(t)e^{-\omega(t)}dt}{t-x}$$

$$+ \frac{2\pi i k_1}{x-1}\frac{e^{\omega(x)}}{\sqrt{k^2+\pi^2}}$$

2.5. THE HILBERT TRANSFORM ON THE REAL LINE

Let f be a function defined on the real line, then its Hilbert transform $(Hf)(x)$ is defined by

$$(Hf)(x) = \frac{1}{\pi}(P) \int_{-\infty}^{\infty} \frac{f(t)}{x-t} dt$$

$$= \lim_{\epsilon \to 0^+} \frac{1}{\pi} \left(\int_{-\infty}^{x-\epsilon} + \int_{x+\epsilon}^{\infty} \right) \frac{f(t)}{x-t} dt \qquad (2.32)$$

provided that the limit exists. Our definition of the Hilbert transform differs from that of Titchmarsh [99] in sign only, but this does not effect our basic results. Titchmarsh has discussed the properties of this Hilbert transform in great length and has proved, among many others, the following theorems (the statement of these results in Tricomi [103] is especially clear).

Theorem 1. If the function $f(x) \in L^p(\mathbb{R})$, $p > 1$, then the formula (2.32) defines almost everywhere a function $g(x) = (Hf)(x)$ that also belongs to L^p and whose Hilbert transform coincides almost everywhere with $-f(x)$. In other words, we have the inversion formula for the Hilbert transformation H in the space $L^p(\mathbb{R})$ as follows:

$$H^2 f = -f \qquad \text{a.e.}$$

The following relation between f and g hold

$$\left. \begin{array}{ll} (Hf)(x) = g(x) & \text{a.e.} \\ (Hg)(x) = -f(x) & \text{a.e.} \end{array} \right\} \qquad (2.33)$$

Relations (2.33) are also called *reciprocity relations* between f and g.

We now elaborate the concept of an analytic function in a region being regular. Consider the function

$$f(z) = \frac{z^2 + 1}{z - i}$$

The function $f(z)$ is analytic everywhere on the complex z-plane except at $z = i$, where it is not defined. But we can prescribe the value $2i$ to f at $z = i$ to make f analytic. Since the only singularity of f in the complex plane is a removable singularity the function, $f(z)$ is said to be regular in the complex plane. From now on we can say that $f(z)$ is regular in a region Ω if the only singularities of $f(z)$ in Ω (if any) are the removable singularities.

Theorem 2. Generalized Parseval's Identity. Let the functions $f_1(x)$ and $f_2(x)$ belong to the classes L^{p_1} and L^{p_2}, respectively, $p_1, p_2 > 1$. If

$$\frac{1}{p_1} + \frac{1}{p_2} = 1$$

we have

$$\int_{-\infty}^{\infty} f_1(x) f_2(x) \, dx = \int_{-\infty}^{\infty} [(Hf_1)(x)][(Hf_2)(x)] \, dx \qquad (2.34)$$

Let

$$f_1(x) = f(x) \in L^p(\mathbb{R})$$

and let

$$f_2(x) = (Hg)(x)$$

where $g \in L^p(\mathbb{R})$. Then we can derive from (2.34) and Theorem 1 the following formula:

$$\int_{-\infty}^{\infty} f(x)(Hg)(x)\, dx = \int_{-\infty}^{\infty} [(Hf)(x)][(H(Hg))(x)\, dx$$

$$= \int_{-\infty}^{\infty} (-Hf)(x)g(x)\, dx \qquad (2.35)$$

In duality notation we can put this result into the form

$$\langle f, Hg \rangle = \langle -Hf, g \rangle$$

or $\qquad\qquad\qquad\qquad\qquad\qquad\qquad\qquad\qquad\qquad\qquad\qquad (2.36)$

$$\langle Hf, g \rangle = \langle f, -Hg \rangle$$

Relations (2.35) and (2.36) are also called *Parseval's identities*.

Theorem 3. Let $F(z)$ be an analytic function regular in the region $Im\, z > 0$ such that

$$\int_{-\infty}^{\infty} |F(x + iy)|^p\, dx < k \qquad (p > 1)$$

where k is a constant independent of y. Then as $y \to 0^+$, $F(x + iy)$ converges for almost all x to a limit function $u(x) + iv(x)$, which we denote by $F(x + i0)$. The functions $u(x)$ and $v(x)$ are both L^p functions satisfying the reciprocity relations

$$v(x) = \frac{1}{\pi}(P) \int_{-\infty}^{\infty} \frac{u(t)}{x - t}\, dt$$

$$u(x) = -\frac{1}{\pi} \int_{-\infty}^{\infty} \frac{v(t)}{x - t}\, dt \qquad (2.37)$$

Let us now apply this theorem to a function $F(z) = \frac{1}{z+i}$. Clearly $F(z)$ has no singularity in the region $Im\, z > 0$. It will be treated as an analytic function that is regular in the region $Im\, z > 0$.

$$F(x + iy) = \frac{1}{x + i(y + 1)}$$

$$\int_{-\infty}^{\infty} |F(x + iy)|^p\, dx = \int_{-\infty}^{\infty} \frac{1}{[x^2 + (y + 1)^2]^{p/2}}\, dx$$

Put $x = (y + 1) \tan \theta$, $-\frac{\pi}{2} \le \theta \le \frac{\pi}{2}$:

$$\int_{-\infty}^{\infty} |F(x + iy)|^p \, dx = \int_{-\pi/2}^{\pi/2} \frac{(y + 1) \sec^2 \theta}{(y + 1)^p \sec^p \theta} \, d\theta$$

$$\le 2 \int_0^{\pi/2} \cos^{p-2} \theta \, d\theta$$

$$\le 2 \frac{\sqrt{(p - 1)/2} \sqrt{1/2}}{2 \sqrt{p/2}}$$

Therefore the function $F(z) = \frac{1}{z+i}$ satisfies the requirement for the validity of Theorem 3,

$$F(z) = \frac{1}{z + i}$$

$$\lim_{y \to 0^+} F(z) = \frac{1}{x + i} = \frac{x - i}{x^2 + 1} = u(x) + iv(x)$$

$$u(x) = \frac{x}{x^2 + 1}$$

$$v(x) = \frac{-1}{x^2 + 1}$$

Using the method of contour integration, it is easy to show that

$$\frac{-1}{x^2 + 1} = (P) \int_{-\infty}^{\infty} \frac{1}{\pi} \frac{t}{t^2 + 1} \frac{1}{x - t} \, dt$$

which proves that $v(x) = (Hu)(x)$.

To these results of Titchmarsh [99] we add the following theorem:

Theorem 4. Let $f(z)$ be analytic in the region $\operatorname{Im} z = y \ge 0$ such that

$$f(z) = o(1), \qquad |z| \to \infty \text{ uniformly } \forall \, \theta \in [0, \pi]$$

Then

$$f(x) = i(Hf)(x)$$

that is,

$$v(x) = (Hu)(x) \tag{2.38}$$

$$u(x) = -(Hv)(x) \qquad \forall \, x \in \mathbb{R}$$

where $f(x) = u(x) + iv(x)$.

Proof. By Cauchy's theorem it can be shown that

$$f(z) = \frac{1}{2\pi i} \int_{-\infty}^{\infty} \frac{f(t)}{t - z} dt \qquad (2.39)$$

This result is obtained by performing the integration of $\frac{1}{2\pi i} \frac{f(t)}{t-z}$ over the contour consisting of the real line segment $-R \le x \le R$ and the upper semicircle with center at 0 and radius R and finally letting $R \to \infty$. Denoting the semicircular arc by C_R, we can see that

$$\int_{C_R} \frac{f(t)}{t - z} dt \to 0 \qquad \text{as } R \to \infty$$

For R sufficiently large and $\epsilon > 0$, we get

$$\left| \int_{C_R} \frac{f(t)}{t - z} dt \right| \le \epsilon \int \frac{R\, d\theta}{R - |z|}$$

$$\overline{\lim}_{R \to \infty} \left| \int_{C_R} \frac{f(t)}{t - z} dt \right| \le \epsilon$$

Since ϵ is arbitrary small number > 0,

$$\lim_{R \to \infty} \int_{C_R} \frac{f(t)}{t - z} dt = 0$$

Result (2.39) follows. Letting $z \to x$ in (2.39), we get

$$f(x) = \frac{1}{2\pi i}(P) \int_{-\infty}^{\infty} \frac{f(t)}{t - x} dt + \frac{f(x)}{2}$$

or

$$f(x) = \frac{1}{\pi i}(P) \int_{-\infty}^{\infty} \frac{f(t)}{t - x} dt$$

$$f(x) = i(Hf)(x)$$

After equating the real and imaginary parts, we see the Hilbert receprocity relation (2.38).

To illustrate an application of Theorem 4, let us take

$$f(z) = \frac{1}{z + i} = \frac{1}{x + i(y + 1)}$$

$$= \frac{x - i(y + 1)}{x^2 + (y + 1)^2}$$

So

$$f(x + i0) = f(x) = \frac{x - i}{x^2 + 1}$$

$$u(x) = \frac{x}{x^2 + 1}$$

$$v(x) = \frac{-1}{x^2 + 1}$$

In this case it is easy to show by using the technique of contour integration that

$$v(x) = (Hu)(x) \qquad \square$$

Tricomi [103] proves the next theorem by using the results proved by Titchmarsh [99] in Theorems 1 through 3.

Theorem 5. Let the functions $\varphi_1(x)$ and $\varphi_2(x)$ belong to classes $L^{p_1}[-1, 1]$, and $L^{p_2}[-1, 1]$, respectively. Then, if $\frac{1}{p_1} + \frac{1}{p_2} < 1$, we have

$$(H)[\varphi_1(H\varphi_2) + \varphi_2(H\varphi_1)] = (H\varphi_1)(H\varphi_2) - \varphi_1\varphi_2$$

almost everywhere. Here $(H)\varphi$ stands for the finite Hilbert transform of φ:

$$(H)(\varphi)(x) = \frac{1}{\pi}(P) \int_{-1}^{1} \frac{\varphi(t)}{x - t} dt.$$

We will now demonstrate the method used by Tricomi to find an inversion formula for the finite Hilbert transform.

The finite Hilbert transformation does not possess the uniqueness property. For $(H)\left(\frac{1}{\sqrt{1-t^2}}\right) = 0$;

$$(H)\left(\frac{1}{\sqrt{1 - t^2}}\right) = \frac{1}{\pi}(P) \int_{-1}^{1} \frac{(1 - t^2)^{-1/2}}{x - t} dt$$

Put $t = \frac{1-y^2}{1+y^2}, y > 0$:

$$(H)\left(\frac{1}{\sqrt{1 - t^2}}\right) = -\frac{2}{\pi} \int_{0}^{\infty} \frac{dy}{(1 - x) - (1 + x)y^2}, \qquad -1 < x < 1$$

$$= \frac{2}{\pi(1 + x)} \int_{0}^{\infty} \frac{dy}{y^2 - \left(\frac{1-x}{1+x}\right)}$$

$$= \frac{1}{\pi} \frac{1}{\sqrt{1 - x^2}} \ln \left| \left(\frac{\sqrt{1 + xy} - \sqrt{1 - x}}{\sqrt{1 + xy} + \sqrt{1 - x}} \right) \right|_{0}^{\infty} = 0$$

When $|x| > 1$, $\frac{1-x}{1+x} < 0$, we have

$$(H)\left(\frac{1}{\sqrt{1-t^2}}\right) = \frac{2}{\pi(1+x)}\int_0^\infty \frac{dy}{y^2 + (x-1)/(x+1)}$$

$$= \frac{2}{\pi(1+x)}\sqrt{\frac{x+1}{x-1}}\tan^{-1}\left[\frac{y}{\sqrt{(x-1)/(x+1)}}\right]\Bigg|_0^\infty$$

$$= \frac{1}{\sqrt{x^2-1}}$$

$$(H)(\sqrt{1-t^2}) = -\frac{1}{\pi}(P)\int_{-1}^1 \frac{1-t^2}{\sqrt{1-t^2}}\frac{dt}{t-x}, \qquad |x| < 1$$

$$= -\frac{1}{\pi}(P)\int_{-1}^1 \frac{-t^2}{\sqrt{1-t^2}}\frac{dt}{t-x}$$

$$= -\frac{1}{\pi}(P)\int_{-1}^1 \frac{x^2-t^2}{\sqrt{1-t^2}}\frac{dt}{t-x}$$

$$= \frac{2}{\pi}(P)\int_{-1}^1 \frac{x+t}{\sqrt{1-t^2}}dt$$

$$= \frac{1}{\pi}(P)\int_{-1}^1 \frac{x}{\sqrt{1-t^2}}dt = x$$

We now can solve the aerofoil equation

$$f(x) = \frac{1}{\pi}(P)\int_{-1}^1 \frac{\varphi(t)}{x-t}dt = (H)(\varphi) \tag{2.40}$$

We want to find the solutions of this equation in the space L^p, $1 < p < 2$. This class of function is referred to as the class 2–0. The reason to choose $1 < p < 2$ is that the Hilbert transform theory is very convenient to deal with in the space L^p, $p > 1$, and that the function $\frac{1}{\sqrt{1-t^2}}$ in the process of finding the solution must also belong to the space L^p, which is possible when $p < 2$. We take $\varphi_1(x) = \varphi(x)$, $\varphi_2(x) = \sqrt{1-x^2}$, $\varphi_2(x) \in L^{p_2}[-1,1]$, for any $p_2 > 1$. By Theorem 5,

$$(H)\left[t\varphi(t) + \sqrt{1-t^2}f(t)\right] = (H)\left[t\varphi_1(t) + \varphi_2 f(t)\right]$$

$$= (H)\left[\varphi_1(H)\varphi_2 + \varphi_2(H)\varphi_1\right]$$

$$= \left[(H)\varphi_1(H)(\varphi_2) - \varphi_1\varphi_2\right]$$

$$= \left[xf(x) - \sqrt{1-x^2}\,\varphi(x)\right] \tag{2.41}$$

We obtain

$$(H)[t\varphi(t)] = \frac{1}{\pi}(P)\int_{-1}^{1} \frac{t - x + x}{x - t}\varphi(t)\,dt$$

$$= -\frac{1}{\pi}\int_{-1}^{1}\varphi(t)\,dt + xf(x)$$

$$= -C + xf(x)$$

Therefore from (2.41),

$$-C + xf(x) + (H)\left[\sqrt{1 - t^2}\,f(t)\right] = xf(x) - \sqrt{1 - x^2}\,\varphi(x)$$

$$\varphi(x) = \frac{C}{\sqrt{1 - x^2}} - \frac{1}{\pi}(P)\int_{-1}^{1}\frac{\sqrt{1 - t^2}\,f(t)}{\sqrt{1 - x^2}\,(x - t)}\,dt \qquad (2.42)$$

To verify that (2.42) is a solution of (2.41) is a difficult task. For this we refer the readers to Tricomi [103] on the integral equation. By virtue of the fact that $(H)\frac{1}{\sqrt{1-x^2}} = 0$, C in (2.42) can be taken to be an arbitrary constant.

This solution was also obtained using the technique of solving the Hilbert problem. In fact, by using the Hilbert problem technique, we can obtain many more solutions. This is because in the Hilbert problem technique our classes of functions are associated with analytic functions and not necessarily with the 2–0 class of functions.

2.6. FINITE HILBERT TRANSFORM AS APPLIED TO AEROFOIL THEORIES

We give below a brief account as to how Hilbert transform is applied to wing theory as discussed by Robinson and Laurmann [83].

Figure 2.4 shows a pair of thin parallel, unstaggered aerofoils of equal chord. For thin aerofoils whose thickness effects can be neglected, the horizontal and the vertical

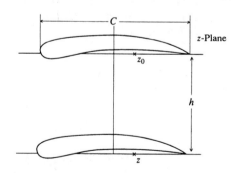

Figure 2.4

induced velocities on the aerofoil surface become

$$u' = -U \left(\frac{\alpha_0}{2} \tan \frac{\theta}{2} + \sum_{n=1}^{\infty} \alpha_n \sin n\theta \right)$$

$$v' = U \left(\frac{\alpha_0}{2} + \sum_{n=1}^{\infty} \alpha_n \cos n\theta \right)$$

$$v' = v - U \sin \alpha$$

$$u' = u - U \cos \alpha$$

See Robinson and Laurmann [83, p. 142]. In the two-dimensional case the vorticity vector is always normal to the z-plane.

The complex velocity potential due to an isolated vortex of strength σ is given by

$$\pi(z) = \frac{i\sigma}{2\pi} \log(z - z_0) \tag{2.43}$$

The corresponding velocity is

$$w(z) = \frac{i\sigma}{2\pi} \frac{1}{z - z_0} \tag{2.44}$$

We will assume that (2.43) and (2.44) for the case of two aerofoils forming a bi-plane are also possible for the self-induced velocity due to the vorticity distribution of each foil. For the vorticity distribution on the upper aerofoil, we put

$$\gamma_1 = -2u_1' = 2U \left[\frac{\alpha_0}{2} \tan \left(\frac{\theta_1}{2} \right) + \sum_{n=1}^{\infty} \alpha_n \sin n\theta_1 \right] \qquad \pi \geq \theta_1 > 0$$

The suffix 1 stands for the result concerning the upper aerofoil. A similar expression is taken for the lower aerofoil with suffix 2. The induced velocity at a point z_0 of the upper aerofoil is found by integration of (2.44) over the vorticity distribution on the two aerofoils

$$w_1'(z_0) = \frac{1}{2\pi i} \int_{-(e/2)+(ih/2)}^{(e/2)+(ih/2)} \frac{\gamma_1(z)}{z - z_0} |dz| + \frac{1}{2\pi i} \int_{-(e/2)-(ih/2)}^{(e/2)-(ih/2)} \frac{\gamma_2(z)}{z - z_0} |dz|$$

where h is the distance apart of the aerofoil and e is their length.

Separating the real and imaginary parts in the foregoing result, we get the components of the total velocity at the upper wing

$$u_1(z_0) = U \cos \alpha + \frac{1}{2\pi} \int_{-e/2}^{e/2} \frac{\gamma_1(x)(y - y_0)}{(x_0 - x)^2 + (y_0 - y)^2} dx$$

$$+ \frac{1}{2\pi} \int_{-e/2}^{e/2} \frac{\gamma_1(x)(y - y_0)}{(x_0 - x)^2 + (y_0 - y)^2} dx$$

and

$$v_1(z_0) = U \sin \alpha + \frac{1}{2\pi} \int_{-e/2}^{e/2} \frac{\gamma_1(x)(x_0 - x)}{(x_0 - x)^2 + (y_0 - y)^2} dx$$

$$+ \frac{1}{2\pi} \int_{-e/2}^{e/2} \frac{\gamma_2(x)(x_0 - x)}{(x_0 - x)^2 + (y_0 - y)^2} dx$$

or

$$u_1(z) = U \cos \alpha - \frac{1}{2\pi} \int_{-e/2}^{e/2} \frac{\gamma_2(x)(x_0 - x)}{(x_0 - x)^2 + (y_0 - y)^2} dx - \frac{\gamma_1(z_0)}{2}$$

and

$$v_1(z_0) = U \sin \alpha + \frac{1}{2\pi} \int_{-e/2}^{e/2} \frac{\gamma_2(x)(x_0 - x)}{(x_0 - x)^2 + (y_0 - y)^2} dx$$

$$+ \frac{1}{2\pi}(P) \int_{-e/2}^{e/2} \frac{\gamma_1(x)}{x_0 - x} dx$$

Put

$$t = \frac{2x}{3}$$

The last integral in the last equation becomes

$$\frac{1}{\pi}(P) \int_{-e/2}^{e/2} \frac{\gamma_1(x)}{x_0 - x} dx = \frac{1}{\pi}(P) \int_{-1}^{1} \frac{\gamma_1(te/2)}{2x_0/e - t} dt$$

which is the finite Hilbert transform of $\gamma_1(\frac{te}{2})$.

By working out the total force required to uplift the plane, we obtain the finite Hilbert transform. Robinson and Laurmann [83, p. 187] have shown that the z component of the velocity induced by the bailing vortices at the point $(0, y, z)$ in a three-dimensional aerofoil is

$$w(0, y, z) = -\frac{1}{4\pi}(P) \int_{-s_0}^{s_0} \frac{\gamma(y_1)}{y_1 - y} dy_1$$

It is assumed that the wing extends from $y = -s_0$ to $y = s_0$.

2.7. THE RIEMANN-HILBERT PROBLEM APPLIED TO CRACK PROBLEMS

Inglis [92] in 1913 studied the distribution of stress in the neighborhood of an elliptic crack with semi-major axis c and semi-minor axis b contained in an infinite thin plate. Griffith considered the case where $b = 0$ when the elliptic crack degenerates into a straight line of length $2c$. A straight-line crack can be considered as a degenerate case of an elliptic crack when b the semi-minor axis is zero. For this reason a straight-line crack occupying the segment

$$y = 0, \qquad -c \le x \le c$$

is called a *Griffith crack*. Sneddon and Lowengrub [92] in their discussion on the crack problem in the classical theory of elasticity give a derivation by G. B. Kolosov [1909] to solve the two-dimensional equation arising in the field of two-dimensional elasticity and show how such equations were solved by Kolosov by introducing complex analytic functions. They state that "the major development of the present century in the field of two-dimensional elasticity has been Muskhelishvili's work on the complex form of the two-dimensional equations due to G. B. Kolosov [1909]."

2.8. REDUCTION OF A GRIFFITH CRACK PROBLEM TO THE HILBERT PROBLEM

Sneddon and Lowengrub [92] give a very good account of how Muskhelishvili used his technique of solving a Hilbert problem to solve the complex form of the two-dimensional equations due to Kolosov [1909] arising in connection with finding the stress distribution in the neighborhood of a Griffith crack or a series of Griffith cracks. For details see [92, pp. 39–40]. The stress around a Griffith crack is related with function $\varphi(z)$, satisfying the condition

$$[\varphi'(x)]^+ + [\varphi'(x)]^- = -p(x), \qquad x \in L \qquad (2.45)$$

$$\varphi^+(x) - \varphi^-(x) = 0 \qquad x \in X - L \qquad (2.46)$$

$$\varphi(z) = 0(z^{-1}), \qquad \text{as } |z| \to \infty$$

where L is the union of all the Griffith cracks lying along the x-axis or it may also be a single Griffith crack along the x-axis. It follows that $\varphi(z)$ is holomorphic in the complement of L, and therefore $\varphi(z)$ is also analytic there and is $0(z^{-2})$ as $z \to \infty$. Therefore the form of $\varphi(z)$ is obtained by solving for the Hilbert problem (2.45).

Let us now restrict to the case of one crack $-c \le x \le c$. By the technique discussed in the previous sections of this chapter, it follows that the solution to (2.45) is

$$\varphi'(z) = \frac{-1}{2\pi i (c^2 - z^2)^{1/2}} \int_{-c}^{c} \frac{p(t)\sqrt{c^2 - t^2}}{t - z} dt \qquad (2.47)$$

We can see that $\varphi'(z) = 0\left(\frac{1}{z^2}\right)$ as $|z| \to \infty$. When we integrate (2.47) to get $\varphi(z)$, we take as zero the constant of integration in view of the asymptotic order of $\varphi(z)$ as $z \to \infty$. The condition (2.46) is automatically met. That is, $\varphi(z)$ is analytic outside the line segment

$$-c \le x \le c, \qquad y = 0$$

2.9. FURTHER APPLICATIONS OF THE HILBERT TRANSFORM

2.9.1. The Hilbert Transform

In 1987 R. K. Donnelly [35] gave a solution to the differential equation (singular differential equation)

$$P(\mathcal{D}, H)u = f \qquad (2.48)$$

in the space $\mathcal{D}' \cap H'(\mathcal{D})$ where $\mathcal{D} \equiv \frac{d}{dt}$ and H is the operator of one-dimensional Hilbert transform and $P(x, y)$ is a polynomial of finite degree in x and y. The space $H'(\mathcal{D})$ and $\mathcal{D}'$ are defined in Chapters 4 and 1, respectively. It turns out that the space $\mathcal{D}' \cap H'(\mathcal{D})$ is closed with respect to operator $\mathcal{D}$ of differentiation with respect to t and the operator H of the Hilbert transforms. Donnelly also gave a solution to the operator equation

$$P(\mathcal{D}, H)U = f$$

where

$$f \in \mathcal{D}'(\mathbb{R}^n) \cap H'\big(\mathcal{D}(\mathbb{R}^n)\big)$$

and $P(x, y) = P(x_1, x_2, \ldots, x_n, y)$ is a polynomial of finite degree in $n + 1$ variables $x_1, x_2, \ldots, x_n, y$. $\mathcal{D}$ stands for the operator

$$\left(\frac{\partial}{\partial x}, \frac{\partial}{\partial x_2}, \ldots, \frac{\partial}{\partial x_n} \right) \equiv (\mathcal{D}_1, \mathcal{D}_2, \ldots, \mathcal{D}_n)$$

H is understood to be the n-dimensional operator of the Hilbert transformation.

2.9.2. The Hibert Transform and the Dispersion Relations

The term "dispersion relations" comes from the observation of Kronig and Kramers in the theory of the dispersion of light by gaseous atoms or molecules. They observed that a relation exists between the real and the imaginary parts of the complex index of refraction, which is called a *dispersion relation*. This dispersion relation is closely associated with the Hilbert transform discussed in this chapter. Dispersion relations have been used to connect the real and imaginary parts of the scattering amplitude through Hilbert transform in high-energy physics. A good account of the role played by the Hilbert transform in the dispersion relation is given by Wu and Omura [107, Ch. 7]. Readers interested in the applications of the Hilbert transform to dispersion relations may look into [107]. Another good reference is the book by J. E. Marsden, *Basic Complex Analysis*, 1972, pp. 411–433.

EXERCISES

1. Solve the integral equation

$$a(x)f(x) = \lambda (P) \int_{-1}^{1} \frac{f(t)\,dt}{t - x} + g(x)$$

for x in $(-1, 1)$ where λ is real and > 0 and where $a(x)$, $g(x)$ are prescribed real functions. You may assume that f and g are both continuous functions in the interval $(-1, 1)$ and that $f(t)$ is allowed to have integrable, singularity in the neighborhood of $t = \pm 1$. *Hint:* Start with

$$F(z) = \frac{1}{2\pi i} \int_{-1}^{1} \frac{f(t)}{t - z}\,dt$$

$F(z)$ may have algebraic singularity of degree > -1 at the point ± 1, in which case

$$\left(a(x) - \lambda \pi i \right) F_+(x) = \left(a(x) + \lambda \pi i \right) F_-(x) + g(x)$$

and so on. The answer takes the form

$$f(x) = \frac{a(x)g(x)}{a^2(x) + \lambda^2 \pi^2}$$

$$+ \frac{\lambda e^{\omega(x)}}{\sqrt{a^2(x) + \lambda^2 \pi^2}}(P) \int_{-1}^{1} \frac{g(t)e^{-\omega(x)}\,dt}{(t - x)\sqrt{a^2(x) + \lambda^2 \pi^2}}$$

$$+ \frac{k e^{\omega(x)}}{(1 - x)\sqrt{a^2(x) + \lambda^2 \pi^2}}$$

where k is an arbitrary constant and

$$\omega(x) = \frac{1}{2\pi i}(P) \int_{-1}^{1} \frac{dt}{t - x} \ln \frac{a(t) + \lambda \pi i}{a(t) - \lambda \pi i}.$$

2. Prove that if $f(t)$ satisfies H condition on the unit circle $|t| = 1$, then $g(\theta) = f(e^{\theta i})$ also satisfies H condition in the interval $(-\infty, \infty)$. Note that $g(\theta)$ is a periodic function with period 2π.

3. Solve the integral equation

$$(P) \int_{0}^{1} \frac{f(t)\,dt}{t - x} + (P) \int_{2}^{3} \frac{f(t)\,dt}{t - x} = x$$

for x either in the interval $(0, 1)$ or $(2, 3)$.

4. (a) Prove that

$$(P) \int_{-1}^{1} \frac{1}{\sqrt{1-x^2}} \frac{dx}{(x-y)} = 0 \qquad \forall\, y \in (-1,1).$$

(b) Define

$$(Hf)(x) = \frac{1}{\pi}(P) \int_{-1}^{1} \frac{f(t)}{t-x} dt \qquad \forall\, x \in (-1,1)$$

$$= 0 \qquad x \notin (-1,1)$$

Prove that $(H^2 f)(x) = -f(x)$. *Hint:* Use the Titchmarsh inversion formula on the real line.

(c) If $(Hf)(x) = g(x)$, prove that $f(x) = -(Hg)(x) + \frac{c}{\sqrt{1-x^2}}$, where c is an arbitrary constant, $-1 < x < 1$.

(d) Generalize the class of functions for which this result is true.

5. Show that the general solution of

$$(P) \int_{C} \left[\frac{1}{t-t_0} + P(t-t_0) \right] f(t)\, dt = g(t_0)$$

where C is a simple closed curve and $P(t)$ is a given entire function of t, is

$$f(t) = -\frac{1}{\pi^2}(P) \int_{C} \frac{g(\tau)\, d\tau}{\tau - t} - \frac{1}{\pi^2} \int_{C} g(\tau)P(\tau - t)\, d\tau$$

6. Derive the solution to Example 4 from that of Example 5.

7. Let C, C_ϵ, and C_1 be the curves as discussed in Figure 2.1. Assume that $f(t)$ satisfies H condition on C. Prove that $F_p(t_0)$ exists. We define $F_p(t_0)$ as

$$\lim_{\epsilon \to 0} \lim_{z \to t_0} \frac{1}{2\pi i} \int_{C-C_\epsilon} \frac{f(t)\, dt}{t-z} = \lim_{\epsilon \to 0} \frac{1}{2\pi i} \int_{C-C_\epsilon} \frac{f(t)}{t-t_0} dt.$$

Hint: Write $f(t) = f(t) - f(t_0) + f(t_0)$.

3

THE HILBERT TRANSFORM OF DISTRIBUTIONS IN $\mathcal{D}'_{L^p}$, $1 < p < \infty$

3.1. INTRODUCTION

Laurent Schwartz [87] developed the theory of Fourier transform of tempered distributions by using a Parseval-type relation as follows: Let f be a tempered distribution $f \in S'$. Then the Fourier transform $\mathcal{F}f$ as a tempered distribution is given by the relation

$$\langle \mathcal{F}f, \varphi \rangle = \langle f, \mathcal{F}\varphi \rangle \qquad \forall \, \varphi \in S \tag{3.1}$$

where the expression $\mathcal{F}\varphi$ in (3.1) is understood to be the classical Fourier transform of φ. It is a well-known fact that the classical Fourier transform operator $\mathcal{F}$ is a homeomorphism from S onto itself. As a consequence the operator $\mathcal{F}$ defined by (3.1) is a homeomorphism from S' onto itself with respect to its weak (or strong) topology. One can use relation (3.1) to calculate the Fourier transform of tempered distributions.

Since the classical inverse Fourier transform operator $\mathcal{F}^{-1}$ is also a homeomorphism from S onto itself, Schwartz defines the inverse Fourier transform operator on S' by using an analogous relation

$$\langle \mathcal{F}^{-1}f, \varphi \rangle = \langle f, \mathcal{F}^{-1}\varphi \rangle \qquad \forall \, \varphi \in S \tag{3.2}$$

One can easily check that operators $\mathcal{F}$ and $\mathcal{F}^{-1}$, as defined by (3.1) and (3.2), are inverse transforms of each other.

The main objective of this chapter is to develop an analogous theory for the Hilbert transform of distributions in $\mathcal{D}'_{L^p}$, $1 < p < \infty$. Shown will be many applications of this theory.

89

In problems of physics we sometimes need to find harmonic functions $U(x, y)$ in the region $y > 0$ whose limit as $y \to 0^+$ does not exist in the pointwise sense but does exist in the distributional sense. The theory of Hilbert transform of distributions that we are going to develop will provide an answer to the existence and the uniqueness of the solution to such problems.

The Hilbert transform of distributions in various subspaces of $\mathcal{D}'$ were investigated by a number of authors such as Schwartz [87], Beltrami and Wohlers [5, 6], Gelfand and Shilov [43], Lauwerier [58], Mitrovic [61, 62, 63], Newcomb [68], Orton [72, 73], and Horvath [50]. Notable among the literature are the techniques followed by Orton [72, 73] and Schwartz [87]. Methods followed by Orton are dependent upon the analytic representation of distributions. This is not quite constructive for distributions not having compact support, and as such, the methods used by her cannot be applied with sufficient ease to applied problems dealing with computation of Hilbert transforms of distributions having noncompact supports.

I now give a brief summary of the methods of Orton. She uses the fact that every $f \in \mathcal{D}'$ has an analytic representation. In other words, for $f \in \mathcal{D}'$ there exists a function $F(z)$ defined and analytic on the complement of the support of f such that

$$\lim_{y \to 0} \int_{-\infty}^{\infty} [F(x + iy) - F(x - iy)] \varphi \, dx = \langle f(x), \varphi(x) \rangle \qquad \forall \varphi \in \mathcal{D} \qquad (3.3)$$

This deep result was proved by Tillman [98]. Orton then defines the Hilbert transform $Hf(x)$ of a Schwartz distribution $f \in \mathcal{D}'$, relative to its analytic representation $F(z)$, as the distributional limit $Hf(x)$ defined by

$$Hf(x) = \lim_{y \to 0} i [F(x + iy) + F(x - iy)] \qquad (3.4)$$

She easily justifies the nomenclature considering a distribution f of compact support; for in that case the analytic representation $F(z)$ of f will be given by

$$F(z) = \frac{1}{2\pi i} \left\langle f(t), \frac{1}{t - z} \right\rangle, \qquad \text{Im } z \neq 0$$

and in this case the Hilbert transform $Hf(x)$ of f will be given by

$$Hf(x) = \lim_{y \to 0} i [F(x + iy) + F(x - iy)]$$

$$= \lim_{y \to 0} -\frac{1}{\pi i} \left\langle f(x), \frac{x - t}{(x - t)^2 + y^2} \right\rangle$$

$$= \lim_{y \to 0} \frac{1}{\pi i} f(x) * \frac{x}{x^2 + y^2}$$

$$= \frac{1}{\pi i} f(x) * \text{p.v.} \left(\frac{1}{x}\right) \qquad (3.5)$$

As the above demonstration indicates, Orton's method is not very useful in solving some applied problems involving the Hilbert transform of distributions. Be it as it may, the theory developed by her is of great theoretical significance.

Laurent Schwartz [87] defines the Hilbert transform of a distribution $f \in \mathcal{D}'_{L^p}$ as $-\left(p.v.\frac{1}{x}\right) * f$ where $*$ stands for operation of convolution. His method involves very complex mathematical theories and cannot be used with sufficient ease to applied problems. Nevertheless his definition is of great theoretical significance.

The work done by Dragisa Mitrovic [61, 62, 63] in this direction is more oriented toward analytic representation of distributions, distributional representations of analytic functions, and distributional boundary-value problems, for example, and not toward the Hilbert transform of distributions as such.

In this chapter I will develop a theory of the Hilbert transform for the Schwartz distribution space $\mathcal{D}'_{L^p}$, $1 < p < \infty$, that will be fairly simple and easily accessible to applied scientists. I will also develop the calculus of the Hilbert transform on $\mathcal{D}'_{L^p}$, $1 < p < \infty$, and demonstrate its application to solve some singular equations and integro-differential equations. Applications of this technique to solve some related boundary value problems will be also shown. I will use the theory to solve problems on analytic representation of distributions, distributional representation of analytic functions, and distributional Hilbert problem. Related theoretical problems will also be discussed.

I will be using notation and terminology commonly used in analysis. The letter $\mathbb{N}$ will be used to indicate the set of nonnegative integers and the pairing between a testing function space and its dual is denoted by $\langle f, \varphi \rangle$. If f is a distribution then the notation $f(t)$ is used to indicate that the testing functions on which f operates have t as their variable, but whenever no confusion arises we will tend to drop the argument of a distribution. The space of C^∞ functions defined on $\mathbb{R}$ and having compact supports will be denoted by $\mathcal{D}$ and not by $\mathcal{D}(\mathbb{R})$, but the space of C^∞ functions defined on $\mathbb{R}^n$, $n > 1$, and having compact supports will be denoted by $\mathcal{D}(\mathbb{R}^n)$. Throughout this book p is a real number greater than one, unless otherwise stated.

3.2. CLASSICAL HILBERT TRANSFORM

The Hilbert transform $(Hf)(x)$ at the point $x \in \mathbb{R}$ of a function $f(t)$ defined a.e. on $\mathbb{R}$ is given by the following integral limit:

$$(Hf)(x) = \lim_{\epsilon \to 0^+} \frac{1}{\pi} \int_{|x-t|>\epsilon} \frac{f(t)}{x - t} \, dt \tag{3.6}$$

provided that the limit in (3.6) exists. The expression $(Hf)(x)$ as defined by (3.6) is sometimes denoted by

$$\frac{1}{\pi} \, p.v. \int_{-\infty}^{\infty} \frac{f(t)}{x - t} \, dt$$

or by

$$\frac{1}{\pi} (P) \int_{-\infty}^{\infty} \frac{f(t)}{x - t} dt$$

The symbols P and p.v. are put to indicate that the above integrals are taken in Cauchy principal value sense. Some authors omit the factor $1/\pi$ in the definition (3.6) and replace it by a "$-$" or a "$+$" sign. I will stick to the notation used in the definition (3.6) throughout this book.

Let us now consider briefly some of the important properties of the classical Hilbert transform operator H defined by (3.6), for which we will briefly recall some results of Chapter 2.

1. If $f \in L^p$, $1 < p < \infty$, then $(Hf)(x)$ exists a.e. on $\mathbb{R}$, and it belongs to L^p. As a matter of fact the limit in (3.6) exists in L^p sense too [99]. There exists $C_p > 0$ independent of f such that

$$\|Hf\|_p \le C_p \|f\|_p$$

2. If $f(t)$ and $g(t)$ belong to L^p, and $L^{p'}$ respectively, where $1 < p < \infty$ and $p' = p/(p - 1)$, then

$$\int_{-\infty}^{\infty} (Hf)(x)g(x) \, dx = - \int_{-\infty}^{\infty} f(x)(Hg)(x) \, dx \qquad (3.7)$$

In duality notation, relation (3.7) can be written

$$\langle Hf, g \rangle = \langle f, -Hg \rangle \qquad (3.8)$$

Note that the existence of the integrals in (3.7) follows from property 1.

3. Let H be the operator defined by (3.6). Then

$$(H^2 f)(x) = -f(x), \qquad \text{a.e.} \qquad (3.9)$$

and in L^p sense as well $\forall\, f \in L^p$, $1 < p < \infty$. Relation (3.9) can be expressed by saying that on L^p, $1 < p < \infty$,

$$H^2 = -I \qquad (3.10)$$

4. If $f \in L^p$, $1 < p < \infty$, then $u(x, y)$, $v(x, y)$ defined by

$$u(x, y) = \frac{1}{\pi} \int_{-\infty}^{\infty} \frac{f(t)y}{(t - x)^2 + y^2} dt$$

$$v(x, y) = \frac{1}{\pi} \int_{-\infty}^{\infty} \frac{f(t)(t - x)}{(t - x)^2 + y^2} dt$$

Both functions exist for $y \ne 0$ and belong to L^p when treated as functions of x for each fixed $y \ne 0$. Functions u and v are also conjugate harmonic functions in the region $y \neq 0$. Furthermore

$$\lim_{y \to 0^+} u(x, y) = f(x) \tag{3.11}$$

$$\lim_{y \to 0^+} v(x, y) = -Hf(x) \qquad , \tag{3.12}$$

where the limits hold in L^p sense and also a.e. [88].

5. If $f \in L^2$, then the integral $\lim_{N \to \infty} \int_{-N}^{N} f(t)e^{ixt}\,dt$ exists. That is, there exists a function in L^2 to which the limit as $\lim_{N \to \infty} \int_{-N}^{N} f(t)e^{ixt}\,dt$ converges in L^2 sense.

We denote the limit of the integral in property 5 by $(\mathcal{F}f)(x)$ and call it the Fourier transform of f. Thus

$$(\mathcal{F}f)(x) = \operatorname*{l.i.m.}_{N \to \infty} \int_{-N}^{N} f(t)e^{ixt}\,dt \tag{3.13}$$

Let $\mathcal{F}$ be the operator defined by (3.13) and H be the operator defined by (3.6), then [75]

$$(\mathcal{F}Hf)(x) = i\operatorname{sgn}(x)(\mathcal{F}f)(x) \qquad \forall f \in L^2 \tag{3.14}$$

In this chapter our theory of distributional Hilbert transform will be centred upon the Parseval-type relation (3.8).

3.3. SCHWARTZ TESTING FUNCTION SPACE, $\mathcal{D}_{L^p}$, $1 < p < \infty$

A function $\varphi(t)$ defined on $\mathbb{R}$ belongs to the space $\mathcal{D}_{L^p}$ iff

1. $\varphi(t) \in C^\infty$.
2. $\varphi^{(k)}(t) \in L^p \; \forall k \in \mathbb{N}$; $\mathbb{N}$ is the set of nonnegative integers.

Note that $\mathcal{D}$ and S are both subspaces of $\mathcal{D}_{L^p}$, $1 < p < \infty$.

3.3.1. The Topology on the Space $\mathcal{D}_{L^p}$

The topology on the space $\mathcal{D}_{L^p}$ is generated by the separating collection of seminorms $\{\gamma_k\}_{k \in \mathbb{N}}$ [1, 108, 110], where

$$\gamma_k(\varphi) = \left(\int_{-\infty}^{\infty} |\varphi^{(k)}(t)|^p \, dt \right)^{1/p}$$

Therefore sequence $\{\varphi_\mu\}_{\mu=1}^\infty$ converges to an element φ in $\mathcal{D}_{L^p}$ as $\mu \to \infty$ iff

$$\gamma_{(k)}(\varphi_\mu - \varphi) \to 0 \qquad \text{as } \mu \to \infty \; \forall k \in \mathbb{N}$$

A sequence $\{\varphi_\mu\}_{\mu=1}^\infty$ is said to be a Cauchy sequence in $\mathcal{D}_{L^p}$ iff for all $k \in \mathbb{N}$,

$$\gamma_k(\varphi_m - \varphi_n) \to 0 \qquad \text{as } m, n \to \infty$$

independently of each other. It is well known that $\mathcal{D}_{L^p}$, $1 < p < \infty$, is a locally convex, Hausdorff and sequentially complete topological vector space [87, 67]. It is also known that if $\varphi \in \mathcal{D}_{L^p}$, $1 < p < \infty$, then $\lim_{|x| \to \infty} \varphi^{(k)}(x) = 0$ for each $k \in \mathbb{N}$ [87, 42, 67].

If $\{\varphi_\mu\}_{\mu=1}^\infty$ is a sequence converging to zero in $\mathcal{D}_{L^p}$, then it is well known that the sequence $\{\varphi_\mu^{(k)}(t)\}_{\mu=1}^\infty$ converges to zero uniformly on $\mathbb{R}$ for each $k \in \mathbb{N}$ [87, 42, 67]. Therefore, for each $\epsilon > 0$, $\exists n \in \mathbb{N}$ independent of t such that

$$\left| \varphi_\mu^{(k)}(t) \right| < \epsilon \qquad \forall \mu > n \tag{3.15}$$

Note that n may depend on k and φ.

In the sequel for $1 < p < \infty$ we will write $p' = p/(p-1)$ and denote by $\mathcal{D}'_{L^p}$ the dual of the space $\mathcal{D}_{L^{p'}}$; that is, $\mathcal{D}'_{L^p} = (\mathcal{D}_{L^{p'}})'$ and so $(\mathcal{D}_{L^p})' = \mathcal{D}'_{L^{p'}}$

It can now be seen quite easily, in view of the above-mentioned properties, that a distribution with compact support is an element of $\mathcal{D}'_{L^p}$. We have $\mathcal{D} \subset \mathcal{D}_{L^p}$ and convergence of any sequence in $\mathcal{D}$ implies its convergence in $\mathcal{D}_{L^p}$. Therefore the restriction of $f \in \mathcal{D}'_{L^p}$ to $\mathcal{D}$ is in $\mathcal{D}'$. In view of the fact that $\mathcal{D}$ is dense in $\mathcal{D}_{L^p}$ [87, 67] it follows that there is one to one correspondence between $f \in \mathcal{D}'_{L^p}$ and its restriction to $\mathcal{D}$. Consequently

$$\mathcal{D}' \supset \mathcal{D}'_{L^p}, \qquad 1 < p < \infty \tag{3.16}$$

We now prove the following:

Theorem 1. Let H be the operator defined by (3.6). Then, for $1 < p < \infty$, H is a linear homeomorphism from $\mathcal{D}_{L^p}$ onto itself and $H^{-1} = -H$.

Proof. Let $\varphi \in \mathcal{D}_{L^p}$, then

$$(H\varphi)(x) = \frac{1}{\pi} \lim_{\epsilon \to 0^+} \int_{|t-x|>\epsilon} \frac{\varphi(t)}{x-t} dt \tag{3.17}$$

$$= -\frac{1}{\pi} \lim_{\epsilon \to 0^+} \int_{|t|>\epsilon} \frac{\varphi(t+x)}{t} dt \tag{3.18}$$

$$= -\frac{1}{\pi} \int_{|t|>N} \frac{\varphi(t+x)}{t} dt - \frac{1}{\pi} \lim_{\epsilon \to 0^+} \int_{N \geq |t|>\epsilon} \frac{\varphi(t+x)}{t} dt \tag{3.19}$$

$$= -\frac{1}{\pi} \int_{|t|>N} \frac{\varphi(t+x)}{t} dt - \frac{1}{\pi} \lim_{\epsilon \to 0^+} \int_{N \geq |t|>\epsilon} \frac{\varphi(t+x) - \varphi(x)}{t} dt \tag{3.20}$$

$$= -\frac{1}{\pi} \int_{|t|>N} \frac{\varphi(t+x)}{t} dt - \frac{1}{\pi} \int_{N \geq |t|} \psi(x,t) dt \tag{3.21}$$

where

$$\psi(x, t) = \begin{cases} \dfrac{\varphi(t + x) - \varphi(x)}{t}, & t \neq 0 \\ \varphi'(x), & t = 0 \end{cases} \tag{3.22}$$

Clearly $\psi(x, t)$ defined by (3.22) belongs to $C^\infty(\mathbb{R}^2)$, so (3.21) is justified. It is easy to see by Holder's inequality that

$$I_N(x) = \left| \int_{|t|>N} \frac{\varphi'(t + x)}{t} dt \right| \leq \|\varphi'\|_p \frac{N^{-1/p}}{(1 - p')^{1/p'}} \tag{3.23}$$

Therefore the sequence in $I_N(x)$ converges uniformly on $\mathbb{R}$ as $N \to \infty$. Hence invoking standard theorems on the uniform convergence of of integrals [48], we get from (3.21),

$$
\begin{aligned}
(H\varphi)'(x) &= -\frac{1}{\pi} \int_{|t|>N} \frac{\varphi'(t + x)}{t} dt - \frac{1}{\pi} \int_{N \geq |t|} \frac{\partial}{\partial x} \psi(x, t) \, dt \\
&= -\frac{1}{\pi} \int_{|t|>N} \frac{\varphi'(t + x)}{t} dt - \frac{1}{\pi} \lim_{\epsilon \to 0^+} \int_{N \geq |t| > \epsilon} \frac{\partial}{\partial x} \psi(x, t) \, dt \\
&= -\frac{1}{\pi} \int_{|t|>N} \frac{\varphi'(t + x)}{t} dt - \frac{1}{\pi} \lim_{\epsilon \to 0^+} \int_{N \geq |t| > \epsilon} \frac{\varphi(t + x) - \varphi(x)}{t} dt \\
&= -\frac{1}{\pi} \int_{|t|>N} \frac{\varphi'(t + x)}{t} dt - \frac{1}{\pi} \lim_{\epsilon \to 0^+} \int_{N \geq |t| > \epsilon} \frac{\varphi'(t + x)}{t} dt \\
&= -\frac{1}{\pi} \lim_{\epsilon \to 0^+} \int_{|t|>\epsilon} \frac{\varphi'(t + x)}{t} dt \\
&= \frac{1}{\pi} \lim_{\epsilon \to 0^+} \int_{|t|>\epsilon} \frac{\varphi'(t)}{x - t} dt \\
&= (H\varphi')(x)
\end{aligned}
$$

Using a similar technique, we can show by induction that

$$(H\varphi)^{(k)}(x) = (H\varphi^{(k)})(x) \qquad \forall k \in \mathbb{N} \tag{3.24}$$

By the properties of H quoted in Section 3.1, we get from (3.24)

$$\left\| (H\varphi)^{(k)} \right\|_p \leq C_p \left\| \varphi^{(k)} \right\|_p \qquad \forall k \in \mathbb{N} \tag{3.25}$$

Therefore $(H\varphi)(x) \in \mathcal{D}_{L^p}$, $1 < p < \infty$. Linearity of H is trivial to prove. H is a linear mapping from $\mathcal{D}_{L^p}$ onto itself.

To prove that H is one to one, assume that for $\varphi \in \mathcal{D}_{L^p}$, $(H\varphi)(x) = 0$. Operating on both sides of the above equation by H, we get $H^2\varphi = 0$. Thus $-\varphi = 0$; that is, $\varphi = 0$. So $H\varphi = 0 \Rightarrow \varphi = 0$. Now we show that H is onto.

Let $\varphi \in \mathcal{D}_{L^p}$. Then $(-H\varphi) \in \mathcal{D}_{L^p}$. But $H(-H\varphi) = -H^2\varphi = \varphi$. Therefore for every $\varphi \in \mathcal{D}_{L^p}$ there exists $-H\varphi \in \mathcal{D}_{L^p}$ that is mapped by H to φ. Consequently H is also onto.

Clearly H is a one to one and onto mapping from $\mathcal{D}_{L^p}$ onto itself. Therefore H^{-1} is defined on $\mathcal{D}_{L^p}$. Since $-H^2\varphi = \varphi$ for all $\varphi \in \mathcal{D}_{L^p}$, it follows that

$$H^{-1} = -H \tag{3.26}$$

The continuity of H and H^{-1} follow from (3.25) and (3.26), respectively. $\square$

3.4. THE HILBERT TRANSFORM OF DISTRIBUTIONS IN $\mathcal{D}'_{L^p}$, $1 < p < \infty$

In analogy to Parseval's relation (3.7) for the classical Hilbert transform operator H, we define the Hilbert transform Hf of $f \in \mathcal{D}'_{L^p}$ by the relation

$$\langle Hf, \varphi \rangle = \langle f, -H\varphi \rangle \qquad \forall \varphi \in \mathcal{D}_{L^{p'}}, \; p' = \frac{p}{p-1} \tag{3.27}$$

Clearly the Hilbert transform Hf of $f \in \mathcal{D}'_{L^p}$ is a functional that assigns the same number to φ as f assigns to $-H\varphi$, where $H\varphi$ is the classical Hilbert transform of φ as defined by (3.6). The linearity of H on $\mathcal{D}'_{L^p}$ is trivial and the continuity follows by the result proved in Theorem 1.

3.4.1. Regular Distribution in $\mathcal{D}'_{L^p}$

Let $f \in L^p$, $1 < p < \infty$. We can define

$$\langle f, \varphi \rangle = \int_{-\infty}^{\infty} f(x)\varphi(x)\,dx \qquad \forall \varphi \in \mathcal{D}_{L^{p'}} \tag{3.28}$$

Linearity and continuity of f defined by (3.28) can be easily proved. Let us now find its distributional Hilbert transform Hf. By definition,

$$\langle Hf, \varphi \rangle = \langle f, -H\varphi \rangle$$

$$= -\int_{-\infty}^{\infty} f(x)(H\varphi)(x)\,dx$$

$$= \int_{-\infty}^{\infty} (Hf)(x)\varphi(x)\,dx \qquad \text{(Parseval's relation)}$$

Hence

$$Hf = \frac{1}{\pi}(P)\int_{-\infty}^{\infty} \frac{f(t)}{x-t}\,dt$$

3.5. THE INVERSION THEOREM

Theorem 2. Let $f \in \mathcal{D}'_{L^p}$, $1 < p < \infty$. Then

$$- H^2 f = f \qquad \forall f \in \mathcal{D}'_{L^p} \qquad (3.29)$$

The generalized Hilbert transform H defined by (3.27) is an isomorphism from $\mathcal{D}'_{L^p}$ onto itself, and $H^{-1} = -H$ is valid on $\mathcal{D}'_{L^p}$.

Proof.

$$\begin{aligned}
\langle H^2 f, \varphi \rangle &= \langle H(Hf), \varphi \rangle \\
&= \langle Hf, -H\varphi \rangle \qquad \forall \varphi \in \mathcal{D}_{L^{p'}}, \; p' = p/(p-1) \\
&= \langle f, (-H)(-H)\varphi \rangle \\
&= \langle -f, \varphi \rangle \qquad \forall \varphi \in \mathcal{D}_{L^{p'}} \qquad \text{(by inversion formula)}
\end{aligned}$$

Therefore $H^2 f = -f$:

$$H^2 = -I \qquad (3.30)$$

By definition, if $f \in \mathcal{D}'_{L^p}$, then $Hf \in \mathcal{D}'_{L^p}$. Now, if $Hf = 0$, then $H^2 f = 0$ and by (3.30), $f = 0$. Put differently, $Hf = 0 \Rightarrow f = 0$, so H is one to one.

The fact that H is onto follows by (3.30). Indeed for every $f \in \mathcal{D}'_{L^p}$ there exists $(-Hf) \in \mathcal{D}'_{L^p}$ such that $H(-Hf) = f$.

The linearity of H is trivial. Therefore H is a linear isomorphism from $\mathcal{D}'_{L^p}$ onto itself. H^{-1} exists and in view of (3.29)

$$H^{-1} = -H \qquad (3.31)$$

Corollary. For $1 < p < \infty$ the generalized Hilbert transform H defined by (3.27) is a homeomorphism from $\mathcal{D}'_{L^p}$ onto itself with respect to the weak as well as the strong topology on $\mathcal{D}'_{L^p}$.

Proof. In view of Theorem 2, we need only show that H is continuous. We will prove the continuity of H with respect to the weak topology on $\mathcal{D}'_{L^p}$. The corresponding results for the strong topology on $\mathcal{D}'_{L^p}$ will be discussed in Chapter 5 where H is defined as a distributional convolution with $\frac{1}{\pi}$ p.v. $\frac{1}{x}$. In fact it will be proved in Chapter 5 that the latter convolution coincides with the Hilbert transform considered here. So the proof given in Chapter 5, for the continuity of H in the strong topology of $\mathcal{D}'_{L^p}$, will be applicable here too.

Assume that $f_\nu \to 0$ in $\mathcal{D}'_{L^p}$. Then

$$\begin{aligned}
\langle Hf_\nu, \varphi \rangle &= \langle f_\nu, -H\varphi \rangle \qquad \forall \varphi \in \mathcal{D}_{L^{p'}} \\
&\to 0 \qquad\qquad\qquad \text{as } \nu \to \infty
\end{aligned}$$

That is, $Hf_\nu \to 0$ weakly as $\nu \to \infty$. This completes the proof of Theorem 2. $\square$

Theorem 3. Let $f \in \mathcal{D}'_{L^p}$, $1 < p < \infty$, and assume that Df is the operator of distributional differentiation defined on $\mathcal{D}'_{L^p}$ by

$$\langle Df, \varphi \rangle = \langle f, -D\varphi \rangle \qquad \forall \, \varphi \in \mathcal{D}_{L^{p'}} \tag{3.32}$$

Then

$$(HD^k f) = D^k Hf, \qquad \forall \, k \in \mathbb{N} \tag{3.33}$$

Proof. We prove (3.33) for the case $k = 1$, and the general case can then be proved by induction.

$$\langle HDf, \varphi \rangle = \langle Df, -H\varphi \rangle \qquad \forall \, \varphi \in \mathcal{D}_{L^{p'}}, \ p' = p/(p-1), \ 1 < p < \infty$$

$$= \langle f, -D(-H\varphi) \rangle$$

$$= \langle f, DH\varphi \rangle$$

$$= \langle f, HD\varphi \rangle \qquad \text{(by [24])}$$

$$= \langle (-D)(-Hf), \varphi \rangle$$

$$= \langle D(Hf), \varphi \rangle \qquad \forall \, \varphi \in \mathcal{D}_{L^p}$$

Thus we have proved that $HDf = DHf$. $\square$

3.5.1. Some Examples and Applications

Example 1. Show that $H\delta = \frac{1}{\pi} \text{p.v.} \frac{1}{t}$.

Solution.

$$\langle H\delta, \varphi \rangle = \langle \delta, -H\varphi \rangle \qquad \forall \, \varphi \in \mathcal{D}_{L^{p'}}, \ p' = p/(p-1), \ 1 < p < \infty$$

$$= \left\langle \delta, -\frac{1}{\pi} \lim_{\epsilon \to 0^+} \int_{|x-t|>\epsilon} \frac{\varphi(t)}{x-t} dt \right\rangle$$

$$= -\frac{1}{\pi} \lim_{\epsilon \to 0^+} \int_{|t|>\epsilon} \frac{\varphi(t)}{-t} dt$$

$$= \left\langle \frac{1}{\pi} \text{p.v.} \frac{1}{t}, \varphi(t) \right\rangle$$

Therefore

$$H\delta = \frac{1}{\pi} \text{p.v.} \frac{1}{t} \tag{3.34}$$

Example 2. Find $H(\text{p.v.}\frac{1}{t})$.

Solution. Operating on both sides of (3.34) by H, we get

$$H^2\delta = \frac{1}{\pi}H\left(\text{p.v.}\frac{1}{t}\right)$$

$$-\delta\pi = H\left(\text{p.v.}\frac{1}{t}\right)$$

Therefore

$$H\left(\text{p.v.}\frac{1}{t}\right) = -\pi\delta$$

Example 3. For $f \in \mathcal{D}'_{L^p}$, $1 < p < \infty$, find a solution to the operator equation

$$y(x) = Hy + f(x) \tag{3.35}$$

where H in (3.35) is the generalized Hilbert transform operator. Show that this is the only solution to (3.35).

Solution. Operating on both sides of (3.35) by H, we get

$$Hy = H^2y + Hf$$

$$y(x) - f(x) = -y + Hf$$

$$y = \frac{f(x) + Hf}{2}$$

Uniqueness: If y_1 and y_2 are two solutions to (3.35), then it follows that $H(y_1 - y_2) = (y_1 - y_2)$ or $H(y_1 - y_2) = -(y_1 - y_2)$. By addition we get $H(y_1 - y_2) = 0 \Rightarrow y_1 - y_2 = 0$.

Example 4. Solve the following integral equation if $f \in L^p$, $1 < p < \infty$:

$$Hy = \lim_{\epsilon \to 0+} \int_{|t-x|>\epsilon} \frac{f(t)}{t-x}dt + \delta(x) + \text{p.v.}\frac{1}{x} + \frac{1}{x^2+1} \tag{3.36}$$

Solution. We can rewrite (3.36) as

$$Hy = -\pi Hf + \delta(x) + \text{p.v.}\frac{1}{x} + \frac{1}{x^2+1} \tag{3.37}$$

Operating on both sides of (3.37) by H and using Examples 1 and 2, we get

$$H^2y = -\pi H^2f + \frac{1}{\pi}\text{p.v.}\frac{1}{x} - \pi\delta + H\left(\frac{1}{x^2+1}\right)$$

or

$$y = -\pi f - \frac{1}{\pi} \text{p.v.} \frac{1}{t} + \pi \delta - H\left(\frac{1}{t^2 + 1}\right) \qquad (3.38)$$

Using the technique of contour integration or otherwise, we get

$$H\left(\frac{1}{x^2 + 1}\right) = \frac{t}{t^2 + 1}$$

Therefore

$$y(t) = -\pi f - \frac{1}{\pi} \text{p.v.} \frac{1}{t} + \pi \delta - \frac{t}{t^2 + 1}$$

Example 5. Solve in $\mathcal{D}'_{L^p}$, $1 < p < \infty$, the operator equation

$$\frac{dy}{dx} + H\delta'(x) = 2xe^{-x^2} \qquad (3.39)$$

Solution. Using the properties of H, we can write (3.39) as

$$\frac{d}{dx}(y + H\delta) = -\frac{d}{dx}\left(e^{-x^2}\right)$$

so

$$y + H\delta + e^{-x^2} = k$$

The only constant distribution that belongs to $\mathcal{D}'_{L^p}$ is the zero distribution and therefore $k = 0$. Hence

$$y = -\left(\frac{1}{\pi} \text{p.v.} \frac{1}{x} + e^{-x^2}\right)$$

3.6. APPROXIMATE HILBERT TRANSFORM OF DISTRIBUTIONS

For $1 < p < \infty$ and $f \in \mathcal{D}'_{L^p}$ and for each fixed $\eta > 0$, let us define the numerical valued function F_η by the relation

$$F_\eta \equiv \left\langle f(t), \frac{1}{\pi} \frac{x - t}{(x - t)^2 + \eta^2} \right\rangle \equiv (H_\eta f)(x) \qquad (3.40)$$

It is a simple exercise to show that for any fixed real x and $\eta > 0$, $(x-t)/((x-t)^2 + \eta^2)$ as a function of t belongs to $\mathcal{D}'_{L^{p'}}$, $p' = p/(p - 1) > 1$.

Therefore the function F_η as defined by (3.40) exists (i.e., is meaningful). Since the space $\mathcal{D}'_{L^p}$ is closed with respect to differentiation, it follows that, for any $\eta \in R$

and $\eta > 0$,

$$\frac{\partial^k}{\partial t^k}\left[\frac{x - t}{(x - t)^2 + \eta^2}\right] \in \mathcal{D}_{L^{p'}} \qquad \forall k \in \mathbb{N}$$

as a function of t. Now by using the technique used in [42] or the structure formula for f [47, p. 201], we can show that

$$\frac{\partial^k}{\partial x^k}F_\eta = \left\langle f(t), \frac{1}{\pi}\frac{\partial^k}{\partial x^k}\frac{x - t}{(x - t)^2 + \eta^2}\right\rangle \tag{3.41}$$

We now wish to derive a duality relation for F_η and show that

$$\lim_{\eta \to 0^+} F_\eta = Hf \qquad \text{in } \mathcal{D}'_{L^p}$$

in the weak topology of $\mathcal{D}'_{L^p}$.

We now prove a lemma that will be useful in the sequel.

Lemma 1. Let $f \in \mathcal{D}'_{L^p}$, $1 < p < \infty$, and let F_η be the approximate Hilbert transform of f as defined by (3.40). Then for each $k \in \mathbb{N}\ F_\eta^{(k)}(x) \in L^p$.

Proof. We will sketch the proof for $k = 0$. In other words, we will show that $F_\eta \in L^p$, and the proof for the case $k > 0$ can be given in a similar fashion by induction.

In view of the structure formula for f [87, p. 201], there exists a nonnegative integer r and functions $f_i \in L^p$ such that

$$F_\eta = \sum_{i=1}^{r}\int_{-\infty}^{\infty} f_i(t)\left(-\frac{\partial}{\partial t}\right)^{i-1}\left[\frac{x - t}{(x - t)^2 + \eta^2}\right] dt \tag{3.42}$$

It is a fact proved by Titchmarsh that if $f_i(t) \in L^p$, then the expressions

$$\int_{-\infty}^{\infty} f_i(t)\left[\frac{x - t}{(x - t)^2 + \eta^2}\right] dt$$

and

$$\int_{-\infty}^{\infty} f_i(t)\left[\frac{1}{(t - x)^2 + \eta^2}\right] dt$$

belong to L^p [99, p. 132]. Therefore, using the fact that

$$\left|\frac{x - t}{(x - t)^2 + \eta^2}\right| \leq \frac{1}{2|\eta|}$$

(η real $\neq 0$) in majorising derivatives in (3.42), it follows that each of the terms in the summation in (3.42) is in L^p. $\square$

Theorem 4. Let $f \in \mathcal{D}'_{L^p}$, $1 < p < \infty$, and let F_η be the approximate Hilbert transform of f defined for fixed $\eta > 0$ by (3.40). Then

$$\lim_{\eta \to 0^+} F_\eta = Hf(x)$$

in the weak topology of $\mathcal{D}'_{L^p}$.

Proof. In view of Lemma 1, we know that

$$\langle F_\eta, \varphi(x) \rangle \equiv \int_{-\infty}^{\infty} F_\eta \varphi(x) \, dx$$

exists for all $\varphi \in \mathcal{D}_{L^{p'}}$ and that the regular distribution generated by F_η belongs to $\mathcal{D}'_{L^p}$. Using the relation (3.42), we have

$$\int_{-\infty}^{\infty} F_\eta \varphi(x) \, dx = \sum_{i=1}^{r} \frac{1}{\pi} \int_{-\infty}^{\infty} \varphi(x) \, dx$$

$$\times \left(\int_{-\infty}^{\infty} f_i(t) \left(-\frac{\partial}{\partial t} \right)^{i-1} \left[\frac{x-t}{(x-t)^2 + \eta^2} \right] dt \right) \quad (3.43)$$

In view of Fubini's theorem the right-hand side expression in (3.43) reduces to

$$\sum_{i=1}^{r} \frac{1}{\pi} \int_{-\infty}^{\infty} f_i(t) \left[\int_{-\infty}^{\infty} \left(-\frac{\partial}{\partial t} \right)^{i-1} \left[\frac{x-t}{(x-t)^2 + \eta^2} \right] \varphi(x) \, dx \right] dt$$

$$= \sum_{i=1}^{r} \frac{1}{\pi} \int_{-\infty}^{\infty} f_i(t) \left(-\frac{\partial}{\partial t} \right)^{i-1} \left(\int_{-\infty}^{\infty} \frac{\varphi(x)(x-t)}{(x-t)^2 + \eta^2} \, dx \right) dt$$

$$= \left\langle f(t), -\frac{1}{\pi} \int_{-\infty}^{\infty} \frac{\varphi(x)(x-t)}{(x-t)^2 + \eta^2} \, dx \right\rangle$$

$$= \langle f(t), -(H_\eta \varphi)(t) \rangle \quad \forall \, \varphi \in \mathcal{D}_{L^{p'}} \quad (3.44)$$

Since $\varphi \in \mathcal{D}_{L^{p'}}$, in view of Lemma 1 it follows that $H_\eta \varphi \in \mathcal{D}_{L^{p'}}$. This justifies the existence of the expression in (3.44). Now

$$\left(\frac{d}{dt} \right)^k (H_\eta \varphi)(t) = \left(\frac{d}{dt} \right)^k \int_{-\infty}^{\infty} \frac{\varphi(x)(x-t)}{(x-t)^2 + \eta^2} \, dx$$

$$= \int_{-\infty}^{\infty} \varphi(x) \left(\frac{\partial}{\partial t} \right)^k \left[\frac{x-t}{(x-t)^2 + \eta^2} \right] dx$$

$$= \int_{-\infty}^{\infty} \varphi(x) \left(-\frac{\partial}{\partial x} \right)^k \left[\frac{x-t}{(x-t)^2 + \eta^2} \right] dx$$

$$= \int_{-\infty}^{\infty} \varphi^{(k)}(x) \frac{x-t}{(x-t)^2 + \eta^2} \, dx \quad \text{(by integration by parts)}$$

for $\lim_{|x| \to \infty} \varphi(x) = 0$ whenever $\varphi \in \mathcal{D}_{L^{p'}}$, $1 < p < \infty$. Hence

$$(H_\eta \varphi)^{(k)}(t) = (H_\eta \varphi^{(k)})(t)$$

By Lemma 4, $(H_\eta \varphi^{(k)}) \in L^{p'}$ for all $k \in \mathbb{N}$. Therefore

$$(H_\eta \varphi)(t) \in \mathcal{D}_{L^{p'}}$$

and

$$(H_\eta \varphi)(t) = \frac{1}{\pi} \int_{-\infty}^{\infty} \frac{\varphi(x)(x - t)}{(x - t)^2 + \eta^2} \, dx \to H\varphi(t)$$

in $\mathcal{D}_{L^{p'}}$ as $\eta \to 0^+$ [70]. Hence letting $\eta \to 0^+$ in (3.44), we get

$$\lim_{\eta \to 0^+} \langle H_\eta f, \varphi \rangle = \lim_{\eta \to 0^+} \langle f, -H_\eta \varphi \rangle = \langle f, -H\varphi \rangle = \langle Hf, \varphi \rangle$$

so $\lim_{\eta \to 0^+} H_\eta f = Hf$ for all $f \in \mathcal{D}'_{L^p}$. $\square$

The following theorem can be proved in a way similar to that followed in proving Theorem 4.

Theorem 5. Let $f \in \mathcal{D}'_{L^p}$, $1 < p < \infty$. Then for each $\varphi \in \mathcal{D}_{L^{p'}}$, $p' = p/(p-1)$, we have

$$\lim_{y \to 0^+} \left\langle \left\langle f(t), \frac{1}{\pi} \frac{y}{(t - x)^2 + y^2} \right\rangle, \varphi(x) \right\rangle = \langle f, \varphi \rangle \qquad \forall \, \varphi \in \mathcal{D}_{L^{p'}}$$

that is,

$$\lim_{y \to 0^+} \left\langle f(t), \frac{1}{\pi} \frac{y}{(t - x)^2 + y^2} \right\rangle = f \qquad \text{in } \mathcal{D}'_{L^p} \text{ (weakly)} \qquad (3.45)$$

3.6.1. Analytic Representation

Let $f \in \mathcal{D}'_{L^p}$, $1 < p < \infty$, and let $F(z)$ be the complex-valued function of z defined for $\operatorname{Im} z \neq 0$ by the relation

$$F(z) = \frac{1}{2\pi i} \left\langle f(t), \frac{1}{t - z} \right\rangle \qquad (3.46)$$

Using the structure formula for f and a technique of Lemma 1, we can show that

$$F^{(k)}(z) = \frac{1}{2\pi i} \left\langle f(t), \frac{k!}{(t - z)^{k+1}} \right\rangle, \qquad \operatorname{Im} z \neq 0, \, k \in \mathbb{N}$$

We see that $F(z)$ is analytic in the region $\mathrm{Im}\, z \neq 0$. Using Theorem 4, we can show that

$$\lim_{\epsilon \to 0^+} \langle F(x + i\epsilon) - F(x - i\epsilon), \varphi(x) \rangle = \langle f, \varphi \rangle \qquad \forall\, \varphi \in \mathcal{D}_{L^{p'}}, \; p' = \frac{p}{p-1}$$

Therefore $F(z)$ defined by (3.46) is the analytic representation for $f \in \mathcal{D}'_{L^p}$, $1 < p < \infty$.

3.6.2. Distributional Representation of Analytic Functions

Using the structure formula for $f \in \mathcal{D}'_{L^p}$, $p' = p/(p-1)$, $1 < p < \infty$, we can prove that the function $F(z)$ defined by (3.46) in the open half-plane $\mathrm{Im}\, z = y > 0$ satisfies the following relations:

1. For any $\delta > 0$,

$$\sup_{\substack{-\infty < x < \infty \\ y \geq \delta > 0}} |F(x + iy)| = A_\delta < \infty \tag{3.47}$$

and

$$|F(x + iy)| = O\left(\frac{1}{y^{1/p}}\right), \qquad y \to \infty, \; \frac{1}{p} = \frac{p'-1}{p'} \tag{3.48}$$

Let us now reverse the problem. Let $F(z)$ be analytic on the open upper half-plane and satisfy the relations (3.47) and (3.48) such that $F(x + i\epsilon) \in L^2$, for any $\epsilon > 0$ and $\lim_{\epsilon \to 0^+} F(t + i\epsilon) = f^+(t)$ in $\mathcal{D}'_{L^p}$, $1 < p < \infty$. Can we find $f \in \mathcal{D}'_{L^p}$ such that

$$F(z) = \left\langle f(t), \frac{1}{t-z} \right\rangle, \qquad \mathrm{Im}\, z > 0 \tag{3.49}$$

The answer to this question is affirmative. It is provided by the following theorem:

Theorem 6. Let $F(z)$ be an analytic complex-valued function of the complex variable $z = x + iy$ in the open upper half-plane ($\mathrm{Im}\, z > 0$), satisfying

i. for fixed $y > 0$, $p' = p/(p-1)$, $1 < p < \infty$,

$$F(x + iy) \in L^p \tag{3.50}$$

ii. $\lim_{y \to 0^+} F(x + iy) = f^+(x)$ in $\mathcal{D}'_{L^p}$ (weakly),

$$\sup_{x \in \mathbb{R}} |F(x + iy)| \to 0 \qquad \text{as } y \to \infty \tag{3.51}$$

$$\sup_{\substack{-\infty < x < \infty \\ y \geq \delta > 0}} |F(z)| = A_\delta < \infty \tag{3.52}$$

Then

$$F(z) = \left\langle \frac{1}{2\pi i} f^+(t), \frac{1}{t-z} \right\rangle \qquad \forall\ \mathrm{Re}\, z > 0 \qquad (3.53)$$

Proof. Let us do the complex integration of the function

$$\frac{1}{2\pi i} \frac{F(t + i\epsilon)}{(t - z)}$$

with respect to the variable t along the semicircle of radius R, center at the origin lying on the upper half-plane with the diameter of the semicircle lying on the x-axis. Let A and B be the points at the left and right ends of the diameter on the x-axis. Choose points A' and B' on the circumference of the semicircle lying in the second and the first quadrant, respectively, such that $\angle BOB' = \angle AOA' = \eta > 0$. By using Cauchy integration, we have

$$\frac{1}{2\pi i} \int_{-R}^{R} \frac{F(t + i\epsilon)}{t - z} dt + \frac{1}{2\pi i} \int_{\mathrm{Arc}\ BB'} \frac{F(t + i\epsilon)}{t - z} dt$$

$$+ \frac{1}{2\pi i} \int_{\mathrm{Arc}\ B'A'} \frac{F(t + i\epsilon)}{t - z} dt$$

$$+ \frac{1}{2\pi i} \int_{\mathrm{Arc}\ A'A} \frac{F(t + i\epsilon)}{t - z} dt$$

$$= F(z + i\epsilon) \qquad \mathrm{Im}\, z > 0 \qquad (3.54)$$

Taking $R > |z|$,

$$\left| \frac{1}{2\pi i} \int_{\mathrm{Arc}\ BB'} \frac{F(t + i\epsilon)}{t - z} dt \right| \leq \left| \frac{1}{2\pi i} \right| \frac{\eta R A_\epsilon}{(R - |z|)} \qquad (3.55)$$

Similarly

$$\left| \frac{1}{2\pi i} \int_{\mathrm{Arc}\ A'A} \frac{F(t + i\epsilon)}{t - z} dt \right| \leq \left| \frac{1}{2\pi i} \right| \frac{\eta R A_\epsilon}{(R - |z|)} \qquad (3.56)$$

$$\lim_{R \to \infty} \left| \frac{1}{2\pi i} \int_{\mathrm{Arc}\ B'A'} \frac{F(t + i\epsilon)}{t - z} dt \right|$$

$$\leq \lim_{R \to \infty} \left| \frac{1}{2\pi i} \right| \frac{(\pi - 2\eta)R}{R - |z|} \sup_{\substack{-\infty < x < \infty \\ y \to \infty}} |F(z)| \qquad (3.57)$$

Letting $R \to \infty$ in (3.54), and using (3.55), (3.56), and (3.57) and then in turn letting $\eta \to 0^+$, we get

$$\frac{1}{2\pi i} \int_{-\infty}^{\infty} \frac{F(t + i\epsilon)}{t - z} dt = \begin{cases} F(z + i\epsilon), & \text{Im } z > 0 \\ 0, & \text{Im } z < 0 \end{cases} \tag{3.58}$$

That is,

$$\left\langle \frac{1}{2\pi i} F(t + i\epsilon), \frac{1}{t - z} \right\rangle = \begin{cases} F(z + i\epsilon), & \text{Im } z > 0 \\ 0, & \text{Im } z < 0 \end{cases} \tag{3.59}$$

Letting $\epsilon \to 0+$ in (3.59), we get

$$\left\langle \frac{1}{2\pi i} f^+(t), \frac{1}{t - z} \right\rangle = \begin{cases} F(z), & \text{Im } z > 0 \\ 0, & \text{Im } z < 0 \end{cases} \tag{3.60}$$

The uniqueness of $F(z)$ is evident in view of the representation formula (3.60). $\square$

We will now establish a further result pertaining to the case $\text{Im } z \neq 0$.

Theorem 7. Let $F(z)$ be an analytic complex-valued function of the complex variable $z = x + iy$ in the complex plane cut along the real line satisfying, for any fixed $y \neq 0$,

$$F(x + iy) \in L^p, \qquad 1 < p < \infty \tag{3.61}$$

$$\lim_{y \to 0\pm} F(z) = f^{\pm}(x) \qquad \text{in } \mathcal{D}'_{L^p} \tag{3.62}$$

$$\sup_{-\infty < x < \infty} |F(z)| \to 0 \qquad \text{as } |y| \to \infty \tag{3.63}$$

$$\sup_{\substack{-\infty < x < \infty \\ y \geq \delta > 0}} |F(z)| = A_\delta < \infty \tag{3.64}$$

Then,

$$F(z) = \left\langle \frac{1}{2\pi i} \left(f^+(t) - f^-(t) \right), \frac{1}{t - z} \right\rangle, \qquad \text{Im } z \neq 0 \tag{3.65}$$

Proof. As in Theorem 6 we prove that for $\text{Im } z > 0$,

$$\left\langle \frac{1}{2\pi i} f^+(t), \frac{1}{t - z} \right\rangle = \begin{cases} F(z), & \text{Im } z > 0 \\ 0, & \text{Im } z < 0 \end{cases} \tag{3.66}$$

Similarly, performing the integration of the function

$$\frac{1}{2\pi i} \frac{F(t - i\epsilon)}{t - z}, \qquad \epsilon > 0$$

in the semicircle in the lower half-plane and taking appropriate limits, we get

$$\left\langle \frac{1}{2\pi i} f^-(t), \frac{1}{t-z} \right\rangle = \begin{cases} 0, & \text{Im } z > 0 \\ -F(z), & \text{Im } z < 0 \end{cases} \tag{3.67}$$

We arrive at (3.65) by combining (3.66) and (3.67). The uniqueness of $F(z)$ follows in view of the representation formula as given by (3.65). $\square$

Further results dealing with Hardy spaces and dealing with questions similar to those considered in Theorems 6 and 7 will be discussed in Chapter 6.

3.7. EXISTENCE AND UNIQUENESS OF THE SOLUTION TO A DIRICHLET BOUNDARY-VALUE PROBLEM

This section gives the solutions to some Dirichlet boundary-value problems related to the Hilbert transform of Schwartz distributions. I first present some definitions, examples, and a theorem on these problems.

Definition. A harmonic function $w(x, y)$ defined on the open upper half-plane (Im $z > 0$) is said to belong to the space $\mathcal{H}_p$ if and only if for fixed $y > 0$,

$$w(x, y) \in L^p, \qquad 1 < p < \infty \tag{3.68}$$

when treated as a function of x,

$$\lim_{y \to 0^+} w(x, y) \tag{3.69}$$

exists in the weak topology of $\mathcal{D}'_{L^p}$

$$\sup_{\substack{-\infty < x < \infty \\ y \geq \delta > 0}} |w(x, y)| = A_\delta < \infty \tag{3.70}$$

$$\sup_{-\infty < x < \infty} |w(x, y)| \to 0 \qquad \text{as } y \to \infty \tag{3.71}$$

We now prove the following:

Theorem 8. Let $u(x, y)$, $v(x, y)$ be the conjugate harmonic functions belonging to the space $\mathcal{H}_p$ as defined above. Assume that

$$\left. \begin{aligned} \lim_{y \to 0^+} u(x, y) &= f \\ \lim_{y \to 0^+} v(x, y) &= g \end{aligned} \right\} \text{ weakly in } \mathcal{D}'_{L^p} \tag{3.72}$$

then we have

$$\begin{aligned} Hf &= -g \\ Hg &= f \end{aligned} \tag{3.73}$$

Moreover

$$u(x, y) = \frac{1}{\pi} \left\langle f(t), \frac{y}{(t - x)^2 + y^2} \right\rangle \tag{3.74}$$

and

$$v(x, y) = \frac{1}{\pi} \left\langle f(t), \frac{t - x}{(t - x)^2 + y^2} \right\rangle \tag{3.75}$$

More clearly u and v are the only harmonic functions belonging to $\mathcal{H}_p$ satisfying (3.72) and (3.73). The convergence in (3.74) and (3.75) are interpreted in the weak topoplogy of $\mathcal{D}'_{L^p}$.

Proof. Since $u(x, y)$, $v(x, y)$ are conjugate harmonic functions in the upper half-plane, the function $F(z) = u + iv$ is an analytic funcion of $z = x + iy$, and by Theorem 7,

$$u(x, y) + iv(x, y) = \left\langle f(t) + ig(t), \frac{1}{2\pi i} \frac{1}{t - z} \right\rangle \tag{3.76}$$

$$= \left\langle f(t) + ig(t), \frac{1}{2\pi i} \frac{(t - x) + iy}{(t - x)^2 + y^2} \right\rangle \tag{3.77}$$

Letting $y \to 0^+$ in (3.77) and using Theorems 5 and 6, we get

$$f + ig = -\frac{1}{2i} H(f + ig) + \frac{1}{2}(f + ig)$$

Therefore

$$(f + ig)i + H(f + ig) = 0$$

or equivalently

$$(Hf - g) + i(Hg + f) = 0 \tag{3.78}$$

Since f and g are limits of harmonic functions, they assign real values to the real-valued elements of $\mathcal{D}_{L^{p'}}$, $1 < p < \infty$. Operating the left-hand side functionals on (3.78) on real-valued elements of $\mathcal{D}_{L^p}$, we get

$$\left.\begin{array}{l} Hf - g = 0 \\ f + Hg = 0 \end{array}\right\} \tag{3.79}$$

Since the functionals involved in (3.79) are linear, (3.79) also holds on complex-valued elements of $\mathcal{D}_{L^p}$.

Now equating the real and imaginary parts in (3.77), we get

$$u(x, y) = \left\langle f(t), \frac{1}{2\pi} \frac{y}{(t-x)^2 + y^2} \right\rangle + \left\langle g(t), \frac{1}{2\pi} \frac{t-x}{(t-x)^2 + y^2} \right\rangle$$

$$= \left\langle f(t), \frac{1}{2\pi} \frac{y}{(t-x)^2 + y^2} \right\rangle$$

$$+ \left\langle (Hf)(t), \frac{1}{2\pi} \frac{t-x}{(t-x)^2 + y^2} \right\rangle \qquad \text{[by (79)]}$$

$$= \frac{1}{2\pi} \left\langle f(t), \frac{y}{(t-x)^2 + y^2} + H\left\{ \frac{t-x}{(t-x)^2 + y^2} \right\} \right\rangle$$

$$= \frac{1}{2\pi} \left\langle f(t), \frac{y}{(t-x)^2 + y^2} + \frac{y}{(t-x)^2 + y^2} \right\rangle$$

$$= \frac{1}{\pi} \left\langle f(t), \frac{y}{(t-x)^2 + y^2} \right\rangle \qquad (3.80)$$

Similarly, by equating the imaginary parts in (3.77) and using (3.79), the Hilbert reciprocity relations, we get

$$v(x, y) = \frac{1}{\pi} \left\langle f(t), \frac{t-x}{(t-x)^2 + y^2} \right\rangle \qquad (3.81)$$

It is easy to verify that u and v as obtained in (3.80) and (3.81) are conjugate harmonic functions. An appeal to the structure formula for $f \in \mathcal{D}'_{L^p}$ shows that there exist constants C and C' such that

$$\sup_{x \in \mathbb{R}} |u(x, y)| \le C \frac{1}{y^{1/p}} \qquad \text{as } y \to \infty$$

$$\sup_{x \in \mathbb{R}} |v(x, y)| \le C' \frac{1}{y^{1/p}} \qquad \text{as } y \to \infty$$

$$\sup_{\substack{-\infty < x < \infty \\ y \ge \delta > 0}} |u(x, y)| = A_\delta < \infty$$

$$\sup_{\substack{-\infty < x < \infty \\ y \ge \delta > 0}} |v(x, y)| = B_\delta < \infty$$

Therefore $u(x, y)$ and $v(x, y)$ as defined by (3.67) and (3.68) belong to the space $\mathcal{H}_p$. The uniqueness of the conjugate harmonic functions u and v satisfying the desired result follows in view of the representation formulas (3.67) and (3.68), respectively. $\square$

Example 6. Solve the following Dirichlet boundary-value problems in the space $\mathcal{D}'_{L^p}$, (interpreting limits in $\mathcal{D}'_{L^p}$) $1 < p < \infty$.

$$\left(\frac{\partial^2}{\partial x^2} + \frac{\partial^2}{\partial y^2}\right) u(x, y) = 0, \qquad y > 0,\ u \in \mathcal{H}$$

such that

$$\lim_{y \to 0^+} u(x, y) = \frac{1}{\pi}\, \mathrm{p.v.}\left(\frac{1}{x}\right) \qquad \text{in } \mathcal{D}'_{L^p}$$

Solution. Rewriting the problem, we get

$$\nabla^2 u = 0, \qquad y > 0,\ u \in \mathcal{H}$$

$$\lim_{y \to 0^+} u(x, y) = H\delta$$

The obvious solution is

$$u(x, y) = \left\langle u(t),\, \frac{1}{\pi}\, \frac{t - x}{(t - x)^2 + y^2} \right\rangle$$

3.8. THE HILBERT PROBLEM FOR DISTRIBUTIONS IN $\mathcal{D}'_{L^p}$

3.8.1. Description of the Problem

Let us consider first a description of the classical Hilbert problem. Let $g(x)$ be a function defined on the real line. We wish to find a holomorphic function $G(z)$ analytic on the complex plane cut along the real line, such that

$$\lim_{y \to 0^+} G(x + iy) + \lim_{y \to 0^-} G(x + iy) = g(x) \qquad (3.82)$$

Denoting the two limit functions in (3.82) by γ_+ and γ_-, respectively, we have

$$\gamma_+ + \gamma_- = g(x) \qquad (3.83)$$

In many problems of mathematical physics slightly more difficult problems appear:

Given functions $g(x)$ and $k(x)$ defined on the real line, we wish to find a function $G(z)$ analytic on the complex plane cut along the real line such that

$$\gamma_+ + k(x)\gamma_- = \gamma \qquad (3.84)$$

where γ_+ and γ_- are the limits of $G(z)$ as described above. We explain formally how the problem in (3.84) can be reduced to a simpler form (3.83) by using the factorization technique. Essentially, the method consists in finding a function $k(z)$ analytic on the complex plane cut along the real axis such that

$$\log k(x) = \log K^+(x) + \log K^-(x) \qquad (3.85)$$

where

$$K^+(x) = \lim_{y \to 0^+} k(x + iy)$$

and

$$K^-(x) = \lim_{y\to 0^-} k(x+iy)$$

Note that the problem (3.85) can be solved by the same technique used in solving problem (3.83), and it also implies that

$$k(x) = K^+(x)K^-(x) \tag{3.86}$$

(3.86) is called the *factorization of $k(x)$*. Using the factorization (3.86), we can now reduce the problem (3.84) to the form

$$\gamma_+ + K^+(x)K^-(x)\gamma_- = \gamma$$

$$\frac{\gamma_+}{K^+(x)} + K^-(x)\gamma_- = \frac{\gamma}{K^+(x)} \tag{3.87}$$

Thus the problem (3.84) has formally been reduced to the form (3.83). Therefore our prime interest will be in solving problem (3.83), which we call the *Hilbert problem*. Lauwerier [58] considers the solution to the same problem, taking $g(x)$ as some regular, or even a singular, distribution with compact support. We will find a function $G(z)$ analytic in the complex plane cut along the real axis satisfying (3.83) where

$$\gamma_+ = \lim_{y\to 0^+} G(x+iy) \quad\quad \text{in the weak topology of } \mathcal{D}_{L^p}$$

and

$$\gamma_- = \lim_{y\to 0^-} G(x+iy) \quad\quad \text{in the weak topology of } \mathcal{D}_{L^p}$$

Mitrovic [61, 62, 63] considers the Hilbert problem in the space O'_α, $\alpha < 1$, but he interprets limits γ_+ and γ_- in the weak topology of $\mathcal{D}'$.

We now rephrase the distributional Hilbert problem.

3.8.2. The Hilbert Problem in $\mathcal{D}'_{L^p}$, $1<p<\infty$

Let $g(x) \in \mathcal{D}'_{L^p}$, $1<p<\infty$. We wish to find the analytic function $G(z)$ in the region $\operatorname{Re} z \neq 0$ such that

$$\gamma_+ + \gamma_- = \gamma \tag{3.88}$$

where

$$\gamma_+ = \lim_{y\to 0^+} G(x+iy) \quad\quad \text{in } \mathcal{D}'_{L^p} \text{ weakly} \tag{3.89}$$

$$\gamma_- = \lim_{y\to 0^-} G(x+iy) \quad\quad \text{in } \mathcal{D}'_{L^p} \text{ weakly} \tag{3.90}$$

Let us define

$$G(z) = \frac{1}{2\pi i} \left\langle g(t), \frac{1}{t-z} \right\rangle, \quad \text{Im } z > 0 \tag{3.91}$$

Therefore

$$G(z) = \frac{1}{2\pi i} \left\langle g(t), \frac{t-x}{(t-x)^2 + y^2} \right\rangle + \frac{1}{2\pi} \left\langle g(t), \frac{y}{(t-x)^2 + y^2} \right\rangle \tag{3.92}$$

Letting $y \to 0^+$ in (3.92), we get in the sense of weak limits in $\mathcal{D}'_{L^p}$,

$$\gamma_+ = -\frac{1}{2i} Hg + \frac{1}{2} g \tag{3.93}$$

Let us now define

$$G(z) = -\frac{1}{2\pi i} \left\langle g(t), \frac{1}{t-z} \right\rangle, \quad \text{Im } z < 0 \tag{3.94}$$

Therefore

$$G(z) = -\frac{1}{2\pi i} \left\langle g(t), \frac{t-x}{(t-x)^2 + y^2} \right\rangle - \frac{1}{2\pi} \left\langle g(t), \frac{y}{(t-x)^2 + y^2} \right\rangle \tag{3.95}$$

Letting $y \to 0^+$ in (3.95), we get in the sense of weak limits in $\mathcal{D}'_{L^p}$,

$$\gamma_- = +\frac{1}{2i} Hg + \frac{1}{2} g \tag{3.96}$$

Adding (3.93) and (3.96), we get

$$\gamma_+ + \gamma_- = g \tag{3.97}$$

If, however, we relax the conditions in (3.89) and (3.90) and interpret the corresponding limits in $\mathcal{D}'$, then there will be infinitely many solutions to the Hibert problem given by

$$G(z) = \frac{1}{2\pi i} \left\langle g(t), \frac{1}{t-z} \right\rangle + F(z), \quad \text{Im } z > 0 \tag{3.98}$$

$$= -\frac{1}{2\pi i} \left\langle g(t), \frac{1}{t-z} \right\rangle - F(z), \quad \text{Im } z < 0 \tag{3.99}$$

where $F(z)$ is an entire function.

If in addition we impose the condition that $F(z)$ is of finite degree at infinity, then

$$G(z) = \frac{1}{2\pi i} \left\langle g(t), \frac{1}{t-z} \right\rangle + P(z), \quad \text{Im } z > 0$$

$$= -\frac{1}{2\pi i} \left\langle g(t), \frac{1}{t-z} \right\rangle - P(z), \quad \text{Im } z < 0$$

where $P(z)$ is a polynomial.

Therefore the solution to (3.87) can be written in the form

$$\frac{g(z)}{k(z)} = \frac{1}{2\pi i}\left\langle\frac{g(t)}{k(t)},\frac{1}{t-z}\right\rangle, \qquad \operatorname{Im} z > 0$$

$$k(z)g(z) = \frac{1}{2\pi i}\left\langle\frac{g(t)}{k^+(t)},\frac{1}{t-z}\right\rangle, \qquad \operatorname{Im} z < 0$$

It is assumed that $\frac{g(t)}{k^+(t)}$ is a generalized function $\in (\mathcal{D}_{L^p})'$. The equation (3.84) can also be transformed to the form $G_+ - G_- = f$, as discussed in Chapter 2 and can be solved accordingly.

EXERCISES

1. Let $\mathcal{D}$ be the Schwartz testing function space of C^∞ functions on the real line with compact supports, and let H be the operator of classical Hilbert transformation. Show that

$$\mathcal{D} \cap H(\mathcal{D}) = \{0\}.$$

   Hint: Use $(\mathcal{F}H\varphi)(\sigma) = i\pi\operatorname{sgn}(\sigma)(\mathcal{F}\varphi)(\sigma)$;

$$\operatorname{sgn}(\sigma) = \begin{cases} \frac{\sigma}{|\sigma|} & \text{if } \sigma \neq 0 \\ 0 & \text{if } \sigma = 0 \end{cases}$$

2. Find the solution to the operator equation $Hy = y$ in $\mathcal{D}'_{L^p}$, $p > 1$ where H is the operator of the Hilbert transform defined on $\mathcal{D}'_{L^p}$. Use your result to show that the solution to the operator equation $y = Hy + f$ is unique in $\mathcal{D}'_{L^p}$.

3. Let H be the operator of the classical Hilbert transformation defined on L^p, $p > 1$. Show that $\|H\|_p = \nu(p)$, where

$$\nu(p) = \begin{cases} \tan\left(\frac{\pi}{2p}\right), & 1 < p \leq 2 \\ \cot\left(\frac{\pi}{2p}\right), & 2 \leq p < \infty \end{cases}$$

   See McLean and Elliott [49].

4. Prove that the only eigenvalues of the operator H are $\pm i$ and that the corresponding eigenfunctions are $f \mp iHf$ for any $f \in \mathcal{D}'_{L^p}$, $p > 1$. Are these the only eigenfunctions?

5. Laurent Schwartz has defined the Hilbert transform of $f \in \mathcal{D}'_{L^p}$, $p > 1$ by $-\frac{1}{\pi}f * p.v.\left(\frac{1}{x}\right)$, where $*$ represents the convolution operator. Prove that his definition of the Hilbert transform is equivalent to ours. This result has been proved by Carton-Lebrun [21].

4

THE HILBERT TRANSFORM OF SCHWARTZ DISTRIBUTIONS

4.1. INTRODUCTION

Chapter 3 presented the Hilbert transform of Schwartz distributions belonging to $\mathcal{D}'_{L^p}$, $1 < p < \infty$, using an analogue of Parseval's relation. However, this technique does not cover elements of $\mathcal{D}'$ that do not belong to $\mathcal{D}'_{L^p}$. The object of this chapter is to develop the theory of the Hilbert transform to the space $\mathcal{D}'$ by using a Parseval-type identity and demonstrate with examples the uses of this method. The technique used in this chapter is essentially the same as that followed by Ehrenpreis and Gel'fand and Shilov in extending the Fourier transform to distributions in $\mathcal{D}'$. Here is what we are going to do:

We denote by $H(\mathcal{D})$ the space of C^∞ functions defined on the real line whose every element is the Hilbert transform of an element of $\mathcal{D}$. We equip the space $H(\mathcal{D})$ with the topology transported from $\mathcal{D}$ into $H(\mathcal{D})$ by means of the operator H, thereby making the operator H defined by

$$(Hf)(x) = \lim_{\epsilon \to 0+} \frac{1}{\pi} \int_{|t-x|>\epsilon} \frac{f(t)}{x-t} dt \qquad (4.1)$$

as a homeomorphism from $\mathcal{D}$ onto $H(\mathcal{D})$. The Hilbert transform Hf of $f \in \mathcal{D}'$ is then defined to be an element of the space $H'(\mathcal{D})$ given by the relation

$$\langle Hf, \varphi \rangle = \langle f, -H\varphi \rangle \qquad \forall\, \varphi \in H(\mathcal{D})$$

With an appropriate interpretation of $H(Hf)$, $f \in H'(\mathcal{D})$, we show that

$$-H^2 f = f \qquad \forall\, f \in \mathcal{D}'$$

Applications of our results in solving some singular integro-differential equations are also given. In the next section we give an intrinsic definition of the space H(𝒟) and its topology. The method essentially consists in characterizing the space of holomorphic functions defined on the complex plane such that the Hilbert transform of its restriction to ℝ is in 𝒟, thereby giving also an intrinsic characterization of $H'(\mathcal{D})$.

4.2. THE TESTING FUNCTION SPACE H(𝒟) AND ITS TOPOLOGY

A C^∞ function φ defined on the real line belongs to the space H(𝒟) iff there exists a function $\psi \in \mathcal{D}$ such that

$$\varphi(x) = \frac{1}{\pi} \text{p.v.} \int_{-\infty}^{\infty} \frac{\psi(t)}{x-t} \, dt = (H\psi)(x) \tag{4.2}$$

The topology of the space H(𝒟) is the same as that transported from the space 𝒟 to H(𝒟) by means of the Hilbert transform H. Therefore a sequence $\{\varphi_\nu\}_{\nu=1}^\infty$ in H(𝒟) converges to zero in H(𝒟) if and only if the associated sequence $\{\psi_\nu\}_{\nu=1}^\infty$ converges to zero in 𝒟, where $H\psi_\nu = \varphi_\nu \ \forall \ \nu \in \mathbb{N}$.

We now prove some results that will be used in the sequel.

Theorem 1. Let H(𝒟) and $\mathcal{D}_{L^p}$, $1 < p < \infty$, be the spaces as defined before. Then

i. $H(\mathcal{D}) \subset \mathcal{D}_{L^p}$.

ii. Convergence of a sequence in H(𝒟) implies its convergence in $\mathcal{D}_{L^p}$.

Proof. (i) Let φ be an element of 𝒟 with support contained in the closed interval $[-a, a]$. Then by the results proved in Theorem 1 of Chapter 3 we have

$$(H\varphi)^{(k)} = \frac{1}{\pi} P \int_{-a}^{a} \frac{\varphi^{(k)}}{x-t} \, dt \tag{4.3}$$

Therefore, as $|x| \to \infty$,

$$(H\varphi)^{(k)}(x) = O\left(\left|\ln\left|\frac{x-a}{x+a}\right|\right|\right) = O\left(\frac{1}{x}\right)$$

Hence, if $\psi(x) = (H\varphi)(x)$, then $\psi(x) \in \mathcal{D}_{L^p}$. Since ψ is an arbitrary element of H(𝒟), it follows that $H(\mathcal{D}) \subset \mathcal{D}_{L^p}$. Again, since H is a homeomorphism from $\mathcal{D}_{L^p}$ onto itself, there exists a $\theta(x) \in \mathcal{D}_{L^p}$ satisfying

$$(H\theta)(x) = \psi(x)$$

Since $\mathcal{D}$ is dense in $\mathcal{D}_{L^p}$ [87, pp. 199–200], there exists a sequence $\{\theta_\nu\}_{\nu=1}^\infty$ in $\mathcal{D}$ tending to θ in $\mathcal{D}_{L^p}$ as $\nu \to \infty$. Now

$$\left\| \frac{d^k}{dx^k} \left(\psi(x) - (H\varphi_\nu)(x) \right) \right\|_p = \left\| H \left(\theta^{(k)} - \theta_\nu^{(k)} \right) \right\|_p$$

$$\leq C_p \left\| \theta^{(k)} - \theta_\nu^{(k)} \right\|_p \to 0 \qquad \text{as } \nu \to \infty$$

[99 Th. 101, pp. 132–133]. Therefore $H(\mathcal{D})$ is dense in $\mathcal{D}_{L^p}$. This completes the proof of part i.

(ii) If $\{\varphi_\nu\}_{\nu=1}^\infty$ is a sequence in $H(\mathcal{D})$ converging to zero in $H(\mathcal{D})$ as $\nu \to \infty$, then there exists a sequence $\{\psi_\nu\}$ in $\mathcal{D}$ tending to zero in $\mathcal{D}$ as $\nu \to \infty$ such that $H\psi_\nu = \varphi_\nu$. Using the properties of H, we have

$$\left\| \varphi_\nu^{(k)} \right\|_p = \left\| H\psi_\nu^{(k)} \right\|_p \leq C_p \left\| \psi_\nu^{(k)} \right\|_p \to 0 \qquad \text{as } \nu \to \infty \qquad (4.4)$$

this completes the proof of part ii. It can easily be seen that the space $H'(\mathcal{D})$ of ultradistributions consisting of continuous linear functionals on $H(\mathcal{D})$ contains the space $(\mathcal{D}_{L_p})'$; that is, $H'(\mathcal{D}) \supset (\mathcal{D}_{L_p})'$, $1 < p < \infty$. In the notation of Laurent Schwartz we can say that $H'(\mathcal{D}) \supset \mathcal{D}'_{L^q}$ where $\mathcal{D}'_{L^q}$ is defined to be the dual space of $\mathcal{D}_{L^p}$ and $\frac{1}{p} + \frac{1}{q} = 1$. $\square$

4.3. GENERALIZED HILBERT TRANSFORMATION

We now define the generalized Hilbert transformation Hf of $f \in \mathcal{D}'$ as an ultradistribution $Hf \in H'(\mathcal{D})$ such that

$$\langle Hf, \varphi \rangle = \langle f, -H\varphi \rangle \qquad \forall \, \varphi \in H(\mathcal{D}) \qquad (4.5)$$

where $H\varphi$ is the classical Hilbert transform defined by (4.1).

If $g \in H'(\mathcal{D})$, its Hilbert transform Hg is defined to be a Schwartz distribution by the relation

$$\langle Hg, \varphi \rangle = \langle g, -H\varphi \rangle \qquad \forall \, \varphi \in \mathcal{D} \qquad (4.6)$$

From (4.5) and (4.6) it follows that

$$\langle -H^2 g, \varphi \rangle = \langle g, \varphi \rangle \qquad \forall \, \varphi \in \mathcal{D} \qquad (4.7)$$

Therefore from (4.7) we have

$$-H^2 f = f \qquad \forall \, f \in \mathcal{D}'$$

If $f \in L^p$, $1 < p < \infty$, then it is easy to show that the Hilbert transform of the regular distribution generated by f is

$$\lim_{\epsilon \to 0+} \frac{1}{\pi} \int_{|t-x|>\epsilon} \frac{f(t)}{x - t} \, dt$$

Definition. The derivative of $g \in H'(\mathcal{D})$ is defined to be an ultradistribution g' belonging to $H'(\mathcal{D})$ by the relation

$$\langle g', \varphi \rangle = \langle g, -\varphi^{(1)} \rangle \qquad \forall \, \varphi \in H(\mathcal{D}) \tag{4.8}$$

Since $\varphi \in H(\mathcal{D})$ implies that $\varphi^{(1)} \in H(\mathcal{D})$, g' defined in (4.8) is a functional on $H(\mathcal{D})$. Linearity of g' is trivial, and the continuity of g' follows by virtue of the fact that $H(\mathcal{D})$ is closed with respect to differentiation and $g \in H'(\mathcal{D})$. So $g' \in H'(\mathcal{D})$.

Theorem 2. Let $f \in \mathcal{D}'$. Then

$$(Hf)^{(k)} = H(f^{(k)}) \qquad \forall \, k \in \mathbb{N}$$

where $f^{(k)}$ and $(Hf)^{(k)}$ are the kth derivatives of f and Hf, respectively.

The proof is easy and similar to that given in [70]. For this reason the details are omitted. As in the previous chapter we can easily show that $H\delta = +\frac{1}{\pi}\,\text{p.v.}(\frac{1}{x})$ and $H(\text{p.v.}(\frac{1}{x})) = -\pi\delta$. Let us now look at some examples that illustrate the applications of our results.

Example 1. Solve the following singular equation in $\mathcal{D}'$:

$$\frac{dy}{dx} + Hf' = \delta(x) \tag{4.9}$$

where $f \in H'(\mathcal{D})$.

Solution. Equation (4.9) can be written

$$\frac{d}{dx}(y + Hf) = \delta(x)$$
$$y + Hf = h(x) + C$$

or

$$y = -Hf + h(x) + C$$

This solution is meaningful, since $f \in H'(\mathcal{D})$ implies that $Hf \in \mathcal{D}'$.

Example 2. Find the distribution $y \in \mathcal{D}'$ satisfying the integral equation

$$y + Hy = f, \qquad f \in \mathcal{D}' \cap H'(\mathcal{D}) \tag{4.10}$$

Solution. Operating both sides of (4.10) by H, we get

$$Hy + H^2y = Hf$$

$$Hy - y = Hf$$

$$f - y - y = Hf$$

or

$$y = \frac{f - Hf}{2} \tag{4.11}$$

Note that (4.11) is meaningful in $\mathcal{D}'$, since the assumption $f \in \mathcal{D}' \cap H'(\mathcal{D})$ implies that $Hf \in \mathcal{D}'$.

4.4. AN INTRINSIC DEFINITION OF THE SPACE H($\mathcal{D}$) AND ITS TOPOLOGY

In the previous section we saw that the topology of the space H($\mathcal{D}$) was the one transported from the space $\mathcal{D}$ onto H($\mathcal{D}$) by means of the operator H, thereby making H a linear homeomorphism from $\mathcal{D}$ onto H($\mathcal{D}$). This simply means that if $\{\psi_\mu\}_{\mu=1}^\infty$ is a sequence in H($\mathcal{D}$) and $\{\varphi_\mu\}_{\mu=1}^\infty$ is the corresponding sequence in $\mathcal{D}$ such that $H\varphi_\mu = \psi_\mu$ for all $\mu \in \mathbb{N} \setminus \{0\}$ then $\lim_{\mu\to\infty} \psi_\mu = 0$ in H($\mathcal{D}$) iff $\lim_{\mu\to\infty} \varphi_\mu = 0$ in $\mathcal{D}$. This way of describing the topology of H($\mathcal{D}$) is not intrinsic. Even the description of the space H($\mathcal{D}$) that every element of H($\mathcal{D}$) is the classical Hilbert transform of some member of $\mathcal{D}$ is not quite intrinsic. In this section we will consider an intrinsic definition of the space H($\mathcal{D}$) and its topology.

Definition. The space Ψ. A holomorphic function $\Phi(z)$ defined on the complex plane is said to belong to Ψ iff

P1. $\Phi(z)$ is analytic on $C \setminus [a, b]$, the closed interval $[a, b]$ varies as Φ varies.
P2. $\Phi(z) = O\left(1/|z|\right)$ as $|z| \to \infty$.
P3. $\lim_{y\to 0^+} \Phi(x \pm iy) = \Phi^{\pm}(x)$ exists in $\mathcal{D}_{L^{p'}}$, $p' = p/(p-1)$, $1 < p < \infty$.

The fact that Ψ is nonempty can be seen by the fact that the function $k(z)$ defined by

$$k(z) = \int_{-\infty}^\infty \frac{\varphi(t)}{t - z}\, dt, \qquad \varphi \in \mathcal{D}$$

satisfies P1, P2, and P3, and so $k(z) \in \Psi$.

Theorem 3. Let $\Phi(z) \in \Psi$. Assume that $\Phi^+(x)$ and $\Phi^-(x)$ are limits of $\Phi(z)$ as $y \to 0^+$ and $y \to 0^-$, respectively, in $\mathcal{D}_{L^{p'}}$, $p' = p/(p-1)$, $1 < p < \infty$, and define the function $\psi(x)$ and $\varphi(x)$ by the relation

$$\psi(x) = \Phi^+(x) + \Phi^-(x) \tag{4.12}$$

$$\varphi(x) = \Phi^+(x) - \Phi^-(x) \tag{4.13}$$

Then $\psi(x) \in H(\mathcal{D})$, $\varphi(x) \in \mathcal{D}$, and

$$\Phi(z) = \frac{1}{2\pi i} \int_{-\infty}^{\infty} \frac{\varphi(t)}{t - z}\, dt \tag{4.14}$$

Moreover

$$\psi(x) = H\left(\frac{\varphi}{i}\right)(x) \tag{4.15}$$

Proof. Since $\Phi(z) = O\left(\frac{1}{|z|}\right)$, $|z| \to \infty$, the integral defined by

$$I(\epsilon, z) = \frac{1}{2\pi i} \int_{-\infty}^{\infty} \frac{\Phi(t + i\epsilon)}{t - z}\, dt, \qquad \operatorname{Im} z = y \neq 0$$

exists. Now using Cauchy integral theorem for the upper half-plane, we can show that

$$\frac{1}{2\pi i} \int_{-\infty}^{\infty} \frac{\varphi(t + i\epsilon)}{t - z}\, dt = \begin{cases} \Phi(z + i\epsilon), & \operatorname{Im} z > 0 \\ 0, & \operatorname{Im} z < 0 \end{cases} \tag{4.16}$$

Letting $\epsilon \to 0^+$ in (4.13) and using P3, we get

$$\frac{1}{2\pi i} \int_{-\infty}^{\infty} \frac{\varphi^+(t)}{t - z}\, dt = \begin{cases} \Phi(z), & \operatorname{Im} z > 0 \\ 0, & \operatorname{Im} z < 0 \end{cases} \tag{4.17}$$

Similarly by performing integration in the lower half-plane and by letting $\epsilon \to 0^-$, we can prove that

$$\frac{1}{2\pi i} \int_{-\infty}^{\infty} \frac{\varphi^-(t)}{t - z}\, dt = \begin{cases} 0, & \operatorname{Im} z > 0 \\ -\Phi(z), & \operatorname{Im} z < 0 \end{cases} \tag{4.18}$$

Combining (4.17) and (4.18) we get,

$$\Phi(z) = \frac{1}{2\pi i} \int_{-\infty}^{\infty} \frac{\varphi^+(t) - \varphi^-(t)}{t - z}\, dt, \qquad \operatorname{Im} z \neq 0 \tag{4.19}$$

Since $\Phi(z)$ is analytic outside the closed interval $[a, b]$, it follows that

$$\Phi^+(x) = \Phi^-(x) \qquad \forall\, x \notin [a, b]$$

In other words, $\varphi(x)$ defined by (4.13) must be zero outside $[a, b]$. But $\varphi \in \mathcal{D}_{L^{p'}}$, therefore $\varphi \in \mathcal{D}$. Using (4.19), we have

$$\Phi(z) = \frac{1}{2\pi i} \int_{-\infty}^{\infty} \frac{\varphi(t)}{t - z}\, dt$$

Thus

$$\Phi(z) = \frac{1}{2\pi i} \int_{-\infty}^{\infty} \frac{\varphi(t)(t-x)}{(t-x)^2 + y^2} \, dt + \frac{1}{2\pi} \int_{-\infty}^{\infty} \frac{\varphi(t)y}{(t-x)^2 + y^2} \, dt \qquad (4.20)$$

By letting $y \to 0\pm$, respectively, we get

$$\Phi^+(x) = -\frac{1}{2i}(H\varphi)(x) + \frac{1}{2}\varphi(x) \qquad (4.21)$$

and

$$\Phi^-(x) = -\frac{1}{2i}(H\varphi)(x) - \frac{1}{2}\varphi(x) \qquad (4.22)$$

Adding (4.21) and (4.22), we get

$$\Phi^+(x) + \Phi^-(x) = -\frac{1}{i}(H\varphi)(x)$$

or in view of (4.12) we get

$$\psi(x) = -\frac{1}{i}(H\varphi)(x) \in H(\mathcal{D})$$

This completes the proof of Theorem 3. □

Corollary. There is a one to one correspondence between the space Ψ and the testing function space $H(\mathcal{D})$.

Proof. We first show that there is one-to-one correspondence between Ψ and $\mathcal{D}$. In view of the result (4.14) of Theorem 3, it follows that for given $\Phi(z) \in \Psi$, there exists a $\varphi \in \mathcal{D}$ satisfying

$$\Phi(z) = \frac{1}{2\pi i} \int_{-\infty}^{\infty} \frac{\varphi(t)}{t-z} \, dt \qquad (4.23)$$

From (4.23), $\varphi(t) \equiv 0$ implies that $\Phi(z) \equiv 0$. Therefore $\Phi(z) \not\equiv 0$ implies that $\varphi(t) \not\equiv 0$, so the correspondence given by (4.23) is one to one.

Again, if $\varphi \in \mathcal{D}$, then the function

$$\frac{1}{2\pi i} \int_{-\infty}^{\infty} \frac{\varphi(t)}{t-z} \, dt \in \Psi$$

That is, for given $\varphi \in \mathcal{D}$ we can find a function $F(z)$ in Ψ. Clearly the correspondence between Ψ and $\mathcal{D}$ is one to one. But the correspondence between $\mathcal{D}$ and $H(\mathcal{D})$ is also one to one; therefore the correspondence between Ψ and $H(\mathcal{D})$ is one to one. □

4.5. THE INTRINSIC DEFINITION OF THE SPACE H($\mathcal{D}$)

In view of Theorem 3 and its corollary, we note that the testing function space H($\mathcal{D}$) consists of all C^∞ functions $\Psi(x)$ of the form $-i(\varphi^+(x) + \varphi^-(x))$, where $\Phi(x + iy)$ belongs to the space Ψ and the limits $\varphi^\pm(x)$ are taken in the $\mathcal{D}_{L^{p'}}$ sense as $y \to \pm 0$ (respectively), $p' = \left(\frac{p}{p-1}\right)$.

4.5.1. The Intrinsic Definition of the Topology of H($\mathcal{D}$)

We can now define the convergence in H($\mathcal{D}$) as follows: A sequence $\{\psi_\mu\}_{\mu=1}^\infty$ in H($\mathcal{D}$) converges to zero in H($\mathcal{D}$) iff all complex-valued functions $\varphi_\mu(z)$ in Ψ that generate $\psi_\mu(x)$ in accordance with Theorem 3 are analytic outside a fixed interval $[a, b]$. Note that this condition enforces that the corresponding function $\varphi_\mu(x)$, satisfying

$$\frac{-1}{\pi i} H \varphi_\mu = \psi_\mu$$

have their supports contained in the closed bounded interval $[a, b]$. $\gamma_k(\psi_\mu) \to 0$ as $\mu \to \infty$ for all $k \in \mathbb{N}$, where $\gamma_k(\varphi) = \left\| \varphi^{(k)} \right\|_{p'}$, $p' = p/(p - 1)$, $1 < p < \infty$.

Discussion. In view of (3.24) and the boundedness of the operator H, we have

$$\gamma_k(\varphi_\mu) = \gamma_k(H\psi_\mu) \leq \left(\frac{C_p}{\pi}\right) \gamma_k(\psi_\mu) \to 0$$

as $\mu \to \infty$ if $\psi_\mu \to 0$ in H($\mathcal{D}$). Therefore the corresponding sequence $\{\varphi_\mu\}_{\mu=1}^\infty$ in $\mathcal{D}$ converges to zero uniformly on every compact subset of $\mathbb{R}$ [87, p. 200] along with all its derivatives. Likewise $\{\varphi_\mu\}_{\mu=1}^\infty$ converges in $\mathcal{D}$. That is, if $H\varphi_\mu = \psi_\nu$, then $\psi_\nu \to 0$ in H($\mathcal{D}$) implies that $\varphi_\mu \to 0$ in $\mathcal{D}$. The fact that $\{\varphi_\mu\}_{\mu=1}^\infty$ converges to zero in $\mathcal{D}$ implies that $\psi_\nu \to 0$ in H($\mathcal{D}$) is trivial to prove. So the mode of convergence of the sequence $\{\psi_\mu\}_{\mu=1}^\infty$ in H($\mathcal{D}$) to zero can be intrinsically defined by Theorem 1.

4.6. A GEL'FAND-SHILOV TECHNIQUE FOR THE HILBERT TRANSFORM

Gel'fand and Shilov in [44, pp. 151–154] have extended the Hilbert transform to generalized functions of the space Φ' and have proved the inversion formula $-H^2 f = f$ for all $f \in \Phi'$ where elements of the testing function space Φ belong to the Fourier transform space of a testing function space Ψ (i.e., $\Phi = \mathcal{F}(\Psi)$) where Ψ is equipped with the topology generated by a countable set of norms and the topology of Φ is that transported from the space Ψ by the operator of Fourier transformation $\mathcal{F}$. They claimed that a regular distribution generated by a locally integrable function $f(x)$ satisfying the asymptotic order of $f(x) = O(|x|^{1-\epsilon})$, $x \to \pm\infty$, $0 < \epsilon < 1$, is Hilbert transformable according to their scheme. As it stands, their claim is incorrect. Nevertheless, their technique is ingeneous and has helped us to extend the Hilbert transform to a space of ultradistributions containing the space S' of tempered distributions. We will consider this extension of the Gel'fand-Shilov technique in the next section. The technique of Gel'fand and Shilov given below is a simplified version of their technique discussed in [44, pp. 151–154].

4.6.1. Gel'fand-Shilov Testing Function Space Ψ

The testing function space Ψ consists of functions $\psi(s)$ $(-\infty < s < \infty)$ possessing the following properties:

1. The functions $s^k \psi(s)$ is absolutely integrable on the line $-\infty < s < \infty$ for any $k = 0, 1, 2, 3, \ldots$, and $\psi'(s)$ is bounded on the real line.
2. $\psi(s)$ is continuous and has a continuous derivative $\psi'(s)$ on each of the half-lines $-\infty < s \leq 0,\ 0 \leq s < \infty$; the functions $\psi(s)$ and $\psi'(s)$ may have a discontinuity of the first kind at the point $s = 0$.
3. $s^k \psi'(s)$ is absolutely integrable on the line $-\infty < s < \infty$ for any $k = 0, 1, 2, 3, \ldots$.

It is now a simple exercise to show that

$$\max_{-\infty < s < \infty} |\psi(s)| < \infty$$

Gel'fand and Shilov introduce a topology on the space Ψ by introducing the following countable set of norms:

$$\|\psi(s)\|_k = \int_{-\infty}^{\infty} |s^k \psi(s)|\, ds + \int_{-\infty}^{\infty} |s^k \psi'(s)|\, ds + \max_{-\infty < s < \infty} |\psi(s)|$$

$$+ \max_{-\infty < s < \infty} |\psi'(s)| \qquad \text{[110, pp. 7–21]}$$

With this topology the space Ψ is a complete countable multinormed space. They have used an ingeneous technique to show that the Fourier transform of each $\psi \in \Psi$ is infinitely differentiable and satisfies the asymptotic order $\mathcal{F}(\Psi)(x) = O\left(\frac{1}{|x|}\right)$, $x \to \pm\infty$. Another elementary proof of this fact can be given as follows:

$$(\mathcal{F}\psi)(x) = \int_{-\infty}^{\infty} \psi(s) e^{isx}\, dx$$

$$\int_{-\infty}^{\infty} \frac{\partial}{\partial x}\left(\psi(s) e^{isx}\right) ds = \int_{-\infty}^{\infty} is\psi(s) e^{isx}\, ds$$

Therefore

$$\left|\int_{-\infty}^{\infty} \frac{\partial}{\partial x}\left(\psi(s) e^{isx}\right) ds\right| \leq \int_{-\infty}^{\infty} |s\psi(s)|\, ds$$

and

$$\int_{-\infty}^{\infty} \frac{\partial}{\partial x}\left(\psi(s)e^{isx}\right) ds$$

is uniformly convergent. Hence using the standard result on the switch in the order of integration and differention, one can show that

$$\frac{d}{dx}(\mathcal{F}\psi)(x) = \int_{-\infty}^{\infty} i\psi(s)se^{isx}ds$$

Using similar argument and the method of induction, one can show that

$$\left(\frac{d}{dx}\right)^k (\mathcal{F}\psi)(x) = \int_{-\infty}^{\infty} (is)^k \psi(s)e^{isx}ds$$

Clearly $\varphi(x) = (\mathcal{F}\psi)(x)$ is infinitely differentiable on the real line.

We now prove that $(\mathcal{F}\psi(x)) = O\left(\frac{1}{|x|}\right)$, $x \to \pm\infty$. We first show that $\lim_{s \to \pm\infty} \psi(s) = 0$.

For $a > 0$ we have

$$\psi(N) = \psi(a) + \int_a^N \psi'(s) ds$$

for $\psi(s)$ is absolutely integrable in the real time $(-\infty, \infty)$. Letting $N \to \infty$, we have

$$\lim_{N \to \infty} \psi(N) = \psi(a) + \int_a^{\infty} \psi'(s) ds$$

Therefore $\lim_{s \to \infty} \psi(s)$ exists. Again, since $\int_a^{\infty} |\psi(s)| ds$ exists and $\lim_{s \to \infty} \psi(s)$ exists, it follows that $\lim_{s \to \infty} \psi(s) = 0$. Similarly it can be shown that $\lim_{s \to -\infty} \psi(s) = 0$. Using the technique of integration by parts for improper integrals, we get

$$(\mathcal{F}\psi)(x) = \frac{i}{x} \int_{-\infty}^{\infty} \psi'(s)e^{isx}ds + \frac{(\psi(0-) - \psi(0+))}{ix}$$

$$|(\mathcal{F}\psi)(x)| \le \frac{1}{|x|} \int_{-\infty}^{\infty} |\psi'(s)| ds + \frac{1}{|x|} |\psi(0-) - \psi(0+)|$$

Hence

$$(\mathcal{F}\psi)(x) = O\left(\frac{1}{|x|}\right), \qquad x \to \pm\infty.$$

If $\psi(s) \in \Psi$, then $\psi(-s)$ also belong to Ψ:

$$\left(\mathcal{F}(\psi(-t))\right)(x) = \int_{-\infty}^{\infty} \psi(-t)e^{+itx}dt = \int_{-\infty}^{\infty} \psi(t)e^{-itx}dt = \varphi(-x)$$

Therefore, if $\varphi(t) \in \Phi$, then $\varphi(-t) \in \Phi$, and we have

$$\mathcal{F}(\Psi) = \Phi \quad \text{and} \quad \mathcal{F}(\Phi) = \Psi$$

In other words, Φ and Ψ are Fourier transforms of each other.

We now wish to show that Φ is closed with respect to the operation of the Hilbert transformation H. Let $\varphi \in \Phi$, with $\mathcal{F}(H\varphi) = i \operatorname{sgn}(x)(\mathcal{F}\varphi)(x)$. Note that Ψ is closed with respect to multiplication by $\operatorname{sgn}(x)$. Therefore

$$H\Phi \in \Phi$$

since Φ and Ψ are duals of each other. This completes the proof of the fact that Φ is closed with respect to the operation of the classical Hilbert transformation H.

4.6.2. The Topology of the Space Φ

The topology of the space Φ is the one transported from the space Ψ by means of the operation of the Fourier transformation. A sequence $\{\psi_\nu\}_{\nu=1}^\infty$ (where $\varphi_\nu = \mathcal{F}\psi_\nu$) converges to the identically zero function as $\nu \to \infty$ in the topology of the space Ψ iff $\|\psi_\nu\|_k \to 0$ as $\nu \to \infty$ for each $k = 0, 1, 2, 3, \ldots$.

It will be an interesting problem to define the space Φ and its topology in an intrinsic way. Gel'fand and Shilov have given an easy proof of the inversion formula

$$H^2\varphi = -\varphi \qquad \forall\, \varphi \in \Phi$$

as follows: Given

$$\mathcal{F}(H\varphi) = i \operatorname{sgn}(x)\mathcal{F}\varphi \qquad \text{a.e. } \forall\, \varphi \in \Phi$$

Now $H\varphi \in \Phi$ for all $\varphi \in \Phi$. Therefore

$$\mathcal{F}(H^2\varphi) = \mathcal{F}(H(H\varphi)) = i \operatorname{sgn}(x)\mathcal{F}(H\varphi)$$
$$= i \operatorname{sgn}(x)\, i \operatorname{sgn}(x)(\mathcal{F}\varphi)$$
$$= -\mathcal{F}\varphi \qquad \text{a.e.}$$

Then

$$\mathcal{F}(H^2\varphi + \varphi) = 0$$

and $H^2\varphi = -\varphi$ everywhere $\forall\, \varphi \in \Phi$, since $\varphi, H\varphi, H^2\varphi$ are all continuous functions. If $f \in \varphi'$, its Hilbert transform H^*f was defined by Gel'fand and Shilov [44] as an element of Φ' satisfying the following relation:

$$\langle H^*f, \varphi \rangle = \langle f, H\varphi \rangle \qquad \forall\, \varphi \in \Phi$$

where $H\varphi$ is the classical Hilbert transformation of the function φ.

It is easy to show that

$$(H^*)^2 f = -f \qquad \forall\, f \in \Phi'$$

since

$$\langle H^*(H^* f), \varphi \rangle = \langle H^* f, H\varphi \rangle$$
$$= \langle f, H^2 \varphi \rangle$$
$$= \langle f, -\varphi \rangle \qquad \forall\, \varphi \in \Phi$$

Therefore

$$H^{*2} f = -f \qquad \forall\, f \in \Phi'$$

They claimed that a function f that is locally integrable and satisfies the asymptotic order $f(x) = O[x^{1-\epsilon}]$, $x \rightarrow \pm\infty$ for $0 < \epsilon < 1$, is Hilbert transformable according to their definition [44, pp. 153–154]. As it stands, this claim is wrong, for the integral $\int_{-\infty}^{\infty} f(x)(H\varphi)(x)\,dx$ is divergent for all nonzero $\varphi \in \Phi$. We will now extend the Hilbert transform to a space of generalized functions that is closed with respect to multiplication by polynomial functions and that also contains the regular generalized function generated by locally integrable functions $f(x)$ satisfying the asymptotic order $f(x) = O(x^{1-\epsilon})$, $x \rightarrow \infty$, $0 < \epsilon < 1$.

4.7. AN EXTENSION OF THE GEL'FAND-SHILOV TECHNIQUE FOR THE HILBERT TRANSFORM

In this section we have extended the technique of Gel'fand and Shilov for the Hilbert transform to a space of the ultradistributions, where multiplication by a polynomial is permissible. Thus in this section we will be able to solve linear singular differential equations with polynomial coefficients by using the Hilbert transform technique developed. Our extension technique will be based upon the classical relation

$$((\mathcal{F}H)\varphi)(x) = i\,\mathrm{sgn}(x)(\mathcal{F}\varphi)(x) \qquad \forall\, \varphi \in S$$

Using Parseval's relation for Fourier transform, we have

$$\int_{-\infty}^{\infty} (\mathcal{F}H\varphi)(x)\psi(x)\,dx = \int_{-\infty}^{\infty} (H\varphi)(x)(\mathcal{F}\psi)(x)\,dx \qquad \forall\, \varphi, \psi \in S$$
$$\int_{-\infty}^{\infty} i\,\mathrm{sgn}(x)(\mathcal{F}\varphi)(x)\psi(x)\,dx = \int_{-\infty}^{\infty} (H\varphi)(x)(\mathcal{F}\psi)(x)\,dx$$

Dividing through by i and using Parseval's relation for Fourier transform, again we have

$$\int_{-\infty}^{\infty} \varphi(x) \mathcal{F}\big(\text{sgn}(x)\psi(x)\big)\, dx = -i \int_{-\infty}^{\infty} (H\varphi)(x)(\mathcal{F}\psi)(x)\, dx$$

In the duality notations this result can be rewritten as

$$\langle \varphi, \mathcal{F}\big(\text{sgn}\, x\psi(x)\big) \rangle = \langle H\varphi, -i\mathcal{F}\big(\psi(x)\big) \rangle \qquad \forall\, \varphi, \psi \in S \qquad (4.24)$$

where the operators $\mathcal{F}$ and H are defined as follows:

$$\mathcal{F}(\varphi)(x) = \int_{-\infty}^{\infty} \varphi(t) e^{+itx}\, dt \qquad \forall\, \varphi \in S \qquad (4.25)$$

$$(Hf)(x) = +\frac{1}{\pi} \lim_{\epsilon \to 0+} \int_{|x-t|>\epsilon} \frac{f(t)}{x-t}\, dt \qquad (4.26)$$

The relation (4.24) will be important to us. By using the analogue of the relation (4.24), we will extend the Hilbert transform H and its inverse H^{-1} to a certain space of ultradistributions.

4.7.1. The Testing Function Space S_1

The set S_1 is defined to be the set of C^∞ functions defined on $\mathbb{R}$ satisfying $S_1 \subset S$, where S is the testing function space of Schwartz tempered distributions such that each $\varphi \in S_1$ is zero in an open interval containing the origin. The topology of S_1 is that induced on it by S. Clearly S_1 is a locally convex, Hausdorff topological vector space but is not sequentially complete. The space S_1 is not dense in S either.

4.7.2. The Testing Function Space Z_1

An infinitely differentiable complex-valued function φ defined on the real line is said to belong to the space Z_1 iff it is the Fourier transform of an element $\varphi \in S_1$. Since the operator $\mathcal{F}$ of the Fourier transformation is a homeomorphism from S onto itself, it follows that Z_1 is a subspace of S. We equip Z_1 with the topology induced upon it by the topology of S. The space Z_1 is also a locally convex Hausdorff topological vector space but is not sequentially complete.

We now will show in brief that Z_1 is not dense in S. Assume that it is so. Then

$$\bar{Z}_1 = S$$
$$\implies \overline{\mathcal{F}(S_1)} = S$$
$$\implies \mathcal{F}(\bar{S}_1) = S$$
$$\implies \bar{S}_1 = \mathcal{F}^{-1}(S) = S \qquad (S_1 \text{ is dense in } S)$$

which is a contradiction. In that case Theorem 1 as stated and proved in [76] is wrong.*
However, the convergence of a sequence to zero in Z_1 also implies the convergence of
the same sequence to zero in the space S. Therefore the restriction of $f \in S'$ to Z_1 is
in Z_1', but there may not be one-to-one correspondence between Z_1' and S'. However,
since any tempered distribution is a continuous linear functional on Z_1, we can say
that $Z_1' \supset S'$. Thus any distribution of compact support is an ultradistribution in Z_1'.

4.7.3. The Hilbert Transform of Ultradistributions in Z_1'

In analogy to the relation (4.24), we now define the Hilbert transform Hf of $f \in Z_1'$
by

$$\langle Hf, -i\mathcal{F}(\varphi) \rangle = \langle f, \mathcal{F}(\text{sgn}(x)\varphi(x)) \rangle \qquad \forall\, \varphi \in S_1 \qquad (4.27)$$

The relation (4.27) defines Hf as a continuous linear functional over the testing
function space Z_1. In fact Hf, the Hilbert transform of $f \in Z_1'$, is the element of Z_1'
that assigns the same number to $-i\mathcal{F}(\varphi)$ as f assigns to $\mathcal{F}(\text{sgn}(x)\varphi(x))$ for all

$$\varphi \in S_1 \qquad (4.28)$$

The inversion formula $-H^2 f = f$ can be proved as follows:

$$
\begin{aligned}
\langle H^2 f, -i\mathcal{F}(\varphi) \rangle &= \langle H(Hf), -i\mathcal{F}(\varphi) \rangle \\
&= \langle Hf, \mathcal{F}(\text{sgn}(x)\varphi) \rangle \\
&= \left\langle Hf, -i\mathcal{F}\left(\frac{\text{sgn}(x)\varphi}{-i}\right) \right\rangle \\
&= \left\langle f, \mathcal{F}\left(\text{sgn}(x)\left(\frac{\text{sgn}(x)\varphi}{-i}\right)\right) \right\rangle \\
&= \langle f, i\mathcal{F}(\varphi) \rangle \qquad \forall\, \varphi \in S \\
&= \langle -f, -i\mathcal{F}(\varphi) \rangle
\end{aligned}
$$

Therefore

$$H^2 f = -f \qquad \forall\, f \in Z_1' \qquad (4.29)$$

Note that our technique has the advantage that (4.29) is derived without using
the corresponding classical inversion formula. In Example 5 we will show that
$H(t^n f)(x) = x^n (Hf)(x)$ for all $n \in \mathbb{N}$ and $f \in Z_1'$. By this fact it can be shown
that the Hilbert transform of the ultradistribution generated by a polynomial $p(t)$
is a zero ultradistribution, a fact that can also be verified by manipulations. Since
more general results will be proved in the next chapter, the proofs of these facts are

---

*This error was pointed out by C. Carton-Lebrun.

being omitted here. It now follows that the inversion formula (4.29) establishes the uniqueness theorem for the operator H in Z_1' modulo the space of ultradistributions whose Hilbert transform is zero. The class of such distributions will be called the *zero ultradistribution*. Unlike the other spaces of the Hilbert transformable generalized functions, it is harder to prove some of the basic formulas in the space Z_1'. I give some examples below.

Example 3. Show that $(H\delta)(x) = (1/\pi)\,\mathrm{p.v.}(1/x)$ when $\delta \in Z_1'$.

Solution. By definition (4.27), we have

$$\langle H\delta, -i\mathcal{F}(\varphi)\rangle = \langle \delta, \mathcal{F}(\mathrm{sgn}(x)\varphi)\rangle \tag{4.30}$$

$$= \left\langle \delta(t), \int_{-\infty}^{\infty} \mathrm{sgn}(x)\varphi(x)e^{+ixt}\,dx \right\rangle$$

$$= \int_{-\infty}^{\infty} \mathrm{sgn}(x)\varphi(x)\,dx \tag{4.31}$$

Now

$$\left\langle \frac{i}{\pi}\,\mathrm{p.v.}\frac{1}{x}, \mathcal{F}(\varphi)\right\rangle = \left\langle \mathcal{F}\left(\frac{i}{\pi}\,\mathrm{p.v.}\frac{1}{x}\right), \varphi\right\rangle$$

$$= \langle -\mathrm{sgn}(x), \varphi(x)\rangle$$

$$= \int_{-\infty}^{\infty} -\mathrm{sgn}(x)\varphi(x)\,dx \tag{4.32}$$

From (4.31) and (4.32), we have

$$\left\langle \frac{1}{\pi}\,\mathrm{p.v.}\frac{1}{x}, -i\mathcal{F}(\varphi)\right\rangle = \langle H\delta, i\mathcal{F}(\varphi)\rangle \qquad \forall\,\varphi \in S_1$$

namely $H\delta = (1/\pi)\,\mathrm{p.v.}(1/x)$ in Z_1'.

Further generalization of the spaces Z_1, Z_1' will be discussed in the next chapter. The spaces Z_1 and S_1 are closed with respect to differentiation. Thus the distributional differentiation on Z_1' can be defined as

$$\langle Df, \mathcal{F}(\varphi)\rangle = \langle f, -D\mathcal{F}(\varphi)\rangle \qquad \forall\,\varphi \in S_1 \tag{4.33}$$

So, if $f \in Z_1'$, then $Df \in Z_1'$.

Example 4. Show that $(Hf)' = Hf'$ for all $f \in Z_1'$.

Solution.

$$\langle Hf', -i\mathcal{F}(\varphi(x))\rangle = \langle f', \mathcal{F}(\text{sgn}(x)\varphi(x))\rangle$$

$$= \left\langle f, -\frac{d}{dt}\mathcal{F}(\text{sgn}(x)\varphi(x))\right\rangle$$

$$= \langle f, -\mathcal{F}(+ix\,\text{sgn}(x)\varphi(x))\rangle$$

$$= \langle f, -\mathcal{F}(ix\,\text{sgn}(x)\varphi(x))\rangle$$

$$= \langle Hf, i\mathcal{F}(ix\varphi(x))\rangle$$

$$= \langle Hf, \mathcal{F}(-x\varphi(x))\rangle$$

$$= \left\langle Hf, i\frac{d}{dt}\mathcal{F}(\varphi(x))\right\rangle$$

$$= \left\langle \frac{d}{dt}Hf, -i\mathcal{F}(\varphi(x))\right\rangle$$

That is,

$$\langle Hf', -i\mathcal{F}(\varphi(x))\rangle = \langle (Hf)', -i\mathcal{F}(\varphi(x))\rangle \qquad \forall\,\varphi \in S_1$$

Therefore $Hf' = (Hf)'$. Using induction, we can show that $Hf^{(k)} = (Hf)^{(k)}$.

Example 5. For $f \in Z_1'$ show that

$$H(t^m f) = x^m(Hf)(x) \qquad \forall\,m \in \mathbb{N} \tag{4.34}$$

Solution. We will prove the result (4.34) for $m = 1$, since the general result can be proved by induction:

$$\langle H(tf), -i\mathcal{F}(\varphi(x))\rangle = \langle tf, \mathcal{F}(\text{sgn}(x)\varphi(x))\rangle$$

$$= \langle f, t\mathcal{F}(\text{sgn}(x)\varphi(x))\rangle$$

$$= \langle f, \mathcal{F}(+i\,\text{sgn}(x)\varphi'(x))\rangle$$

$$= \langle Hf, -i\mathcal{F}(+i\varphi'(x))\rangle$$

$$= \langle Hf, -ix\mathcal{F}(\varphi)\rangle$$

$$= \langle xHf, -i\mathcal{F}(\varphi)\rangle$$

Thus

$$\langle H(tf), -i\mathcal{F}(\varphi)\rangle = \langle xHf, -i\mathcal{F}(\varphi)\rangle \qquad \forall\,\varphi \in S_1$$

Therefore

$$H(tf)(x) = xHf \qquad \forall f \in Z_1'$$

Remark. It may be noted that the spaces S_1 and Z_1 lack the completeness property with respect to the topology induced on them by S. However, by modifying their definitions slightly, we can define new testing function spaces S_η and Z_η as follows: For $\eta > 0$, we define the testing function space S_η as the space of C^∞ functions defined on $\mathbb{R}$ such that $S_\eta \subset S$ each $\varphi \in S_\eta$ vanishes in the open interval $(-\eta, \eta)$. The space Z_η is defined by

$$Z_\eta = \mathcal{F}(S_\eta)$$

The spaces S_η and Z_η both are subspaces of S, and both are sequentially complete. As such, S_η and Z_η both are Frechet spaces. Duals of S_η and Z_η will be denoted by S_η' and Z_η', respectively, and all the preceeding results and examples are valid for S_η' and Z_η' too.

While dealing with some specific problems such as boundary-value problems, it is a useful idea to have a testing function space which is a Frechet space.

To this end we can also construct the testing function space S_0 as the strict inductive limit of the sequence $\{S_{\eta_i}\}_{i=1}^\infty$ where $\eta_1 > \eta_2 > \cdots$ and $\lim_{i \to \infty} \eta_i = 0$. That is, we define $S_0 = \bigcup_{i=1}^\infty S_{\eta_i}$. Since $S_{\eta_1} \subset S_{\eta_2} \subset S_{\eta_3} \cdots$ and the topology of S_{η_k} is stronger than that induced on S_{η_k} by $S_{\eta_{k+1}}$ for each $k = 1, 2, \ldots$, we can say that a sequence $\{\varphi_\nu\}_{\nu=1}^\infty$ in S_0 converges to zero in S_0 iff

1. $\varphi_\nu \in S_{\eta_k}$ for some k and all $\nu \in \mathbb{N}$.
2. $\varphi_\nu \to 0$ as $\nu \to \infty$ in S_{η_k}.

Clearly the convergence of the sequence to zero in S_{η_k} implies convergence of the same sequence to zero in $S_{\eta_{k+1}}$ and the convergence to zero in $S_{\eta_{k+2}}$, and so on.

Example 6. Solve in Z_η', $\eta > 0$,

$$xy + Hy = f' \qquad \text{where } f \in Z_\eta' \tag{4.35}$$

Solution. Operating on both sides of (4.35) with H, we get

$$H(xy) + H^2 y = Hf'$$

$$xHy - y = (Hf)'$$

$$x[f' - xy] - y = (Hf)'$$

$$(x^2 + 1)y = xf' - (Hf)'$$

or

$$y = \frac{x}{x^2 + 1} f' - \frac{1}{x^2 + 1} (Hf)'$$

4.8. DISTRIBUTIONAL HILBERT TRANSFORMS IN n-DIMENSIONS

Our objective in this section is to extend the Hilbert transform and its inversion formula along with many distributional results to n-dimensions. The theory of the n-dimensional Hilbert transform, in general, turns out to be very complicated. Since our definition of the n-dimensional Hilbert transform will be expressed in terms of the properties of the corresponding n-dimensional Fourier transform, our results turn out to be quite easy to obtain. A more efficient and the organized work on the n-dimensional Hilbert transform will be presented in the next chapter. I now give some definitions and preliminaries.

4.8.1. The Testing Function Space $S_1(\mathbb{R}^n)$

The testing function space $S_1(\mathbb{R}^n)$ is defined to be the space of complex-valued infinitely differentiable functions defined on $\mathbb{R}^n$ such that $S_1(\mathbb{R}^n) \subset S(\mathbb{R}^n)$, and each $\varphi \in S_1(\mathbb{R}^n)$ vanishes in an open ball centred at the origin (the ball may depend on φ). The topology of $S_1(\mathbb{R}^n)$ is the same as that induced on $S_1(\mathbb{R}^n)$ by $S(\mathbb{R}^n)$. Clearly $S_1(\mathbb{R}^n)$ is a locally convex Hausdorff topological vector space but is not sequentially complete. The dual of $S_1(\mathbb{R}^n)$ is denoted by $S_1'(\mathbb{R}^n)$.

4.8.2. The Testing Function Space $Z_1(\mathbb{R}^n)$

The testing function space $Z_1(\mathbb{R}^n)$ is defined to be the space of infinitely differentiable complex-valued functions on $\mathbb{R}^n$ such that

$$Z_1(\mathbb{R}^n) = \mathcal{F}\big(S_1(\mathbb{R}^n)\big)$$

That is,

$$Z_1(\mathbb{R}^n) = \big\{ \varphi \in S(\mathbb{R}^n) : \varphi(x) = \mathcal{F}(\psi)(x) \; \forall \; \psi \in S_1(\mathbb{R}^n) \big\}$$

where $\mathcal{F}$ is the Fourier transform operation in $\mathbb{R}^n$. Since $\mathcal{F}$ is a homeomorphism from $S(\mathbb{R}^n)$ onto itself it follows that

$$Z_1(\mathbb{R}^n) \subset S(\mathbb{R}^n)$$

The space $Z_1(\mathbb{R}^n)$ is also a locally convex Hausdorff topological vector space but it is not sequentially complete. The restriction of $f \in S'(\mathbb{R}^n)$ to $Z_1(\mathbb{R}^n)$ is in $Z_1'(\mathbb{R}^n)$. It is in this sense that we write

$$Z_1'(\mathbb{R}^n) \supset S'(\mathbb{R}^n)$$

but there is not one to one correspondence between $Z_1'(\mathbb{R}^n)$ and $S'(\mathbb{R}^n)$. In other words, for an element in $Z_1'(\mathbb{R}^n)$ there may be more than one element of $S'(\mathbb{R}^n)$ that may correspond, since the space $Z_1(\mathbb{R}^n)$ is not dense in $S(\mathbb{R}^n)$. This fact can be easily proved by reasoning similar to that given in Chapter 2.

The completeness of the testing function space is considered to be quite an important property especially when we have to deal with partial differential equation problems. We therefore construct the following testing function space.

4.8.3. The Testing Function Space $S_\eta(\mathbb{R}^n)$

The testing function space $S_\eta(\mathbb{R}^n)$ consists of infinitely differentiable complex-valued functions defined on $\mathbb{R}^n$ such that every element of $S_\eta(\mathbb{R}^n)$ belongs to $S(\mathbb{R}^n)$ and vanishes in the region $|x_1| < \eta, |x_2| < \eta, \ldots, |x_n| < \eta$. The topology of $S_\eta(\mathbb{R}^n)$ is the same as that induced by $S(\mathbb{R}^n)$. It can easily be seen that $S_\eta(\mathbb{R}^n)$ is a sequentially complete locally convex Hausdorff topological vector space. Clearly $S_\eta(\mathbb{R}^n) \subset S(\mathbb{R}^n)$ and the restriction of an element of $f \in S'(\mathbb{R}^n)$ to $S_\eta(\mathbb{R}^n)$ is in $S'_\eta(\mathbb{R}^n)$.

4.8.4. The Testing Function Space $Z_\eta(\mathbb{R}^n)$

The testing function space $Z_\eta(\mathbb{R}^n)$ is defined to be the space of infinitely differentiable complex-valued functions on $\mathbb{R}^n$ such that for every $\varphi \in Z_\eta(\mathbb{R}^n)$ there exists a $\psi \in S_\eta(\mathbb{R}^n)$ satisfying

$$\varphi(x) = (\mathcal{F}\psi)(x)$$

That is,

$$Z_\eta(\mathbb{R}^n) = \mathcal{F}\big(S_\eta(\mathbb{R}^n)\big)$$

where $\mathcal{F}$ is the Fourier transform operator on $\mathbb{R}^n$. The space $Z_\eta(\mathbb{R}^n) \subset S(\mathbb{R}^n)$ and the topology of $Z_\eta(\mathbb{R}^n)$ is the same as that induced on $Z_\eta(\mathbb{R}^n)$ by $S(\mathbb{R}^n)$. $Z_\eta(\mathbb{R}^n)$ is clearly a Frechet space. The restrictions of $f \in S'(\mathbb{R}^n)$ to $Z'_\eta(\mathbb{R}^n)$ is in $Z'_\eta(\mathbb{R}^n)$.

4.8.5. The Strict Inductive Limit Topology of $Z_\eta(\mathbb{R}^n)$

Let $\{\eta_i\}_{i=1}^\infty$ be a sequence of diminishing positive numbers tending to zero. Then clearly

$$Z_{\eta_1} \subset Z_{\eta_2} \subset Z_{\eta_3} \subset \cdots$$

The topology of Z_{η_i} is stronger than the topology induced on Z_{η_i} by $Z_{\eta_{i+1}}$. We must now define the countable union space Z_0 by

$$Z_0 = \bigcup_{i=1}^\infty Z_{\eta_i} \tag{4.36}$$

A sequence $\{\varphi_\nu\}_{\nu=1}^\infty$ belonging to $Z_0(\mathbb{R}^n)$ is said to converge to zero in $Z_0(\mathbb{R}^n)$ iff it belongs to some Z_{η_i} and converges to zero as $\nu \to \infty$ in the topology of Z_{η_i}. The topology defined in this way on Z_0 is said to be the strict inductive topology of the sequence $Z_{\eta_i}(\mathbb{R}^n)$. The elements of $Z'_0(\mathbb{R}^n)$ will be called ultradistributions. We say that a sequence $\{\varphi_\nu\}_{\nu=1}^\infty$ is a Cauchy sequence in $Z_0(\mathbb{R}^n)$ iff for all $\nu \in \mathbb{N} \setminus \{0\}$, $\varphi_\nu \in Z_{\eta_i}(\mathbb{R}^n)$ for some $i \in \mathbb{N}$, and $\{\varphi_m(x) - \varphi_n(x)\}$ tends to zero in Z_{η_i} as m and $n \to \infty$ independently of each other. It is easy to see that $Z_0'(\mathbb{R}^n)$ is a sequentially complete locally convex Hausdorff topological vector space.

Analogously we define the space

$$S_0 = \bigcup_{i=1}^\infty S_{\eta_i}$$

The topology, the completeness concept, and the convergence of a sequence in S_0 can be defined in the same way as in Z_0. We now define the Hilbert transform Hf of an ultradistribution $f \in Z_0'(\mathbb{R}^n)$ as an ultradistribution on $Z_0(\mathbb{R}^n)$ by the relation

$$\langle Hf, -i\mathcal{F}(\varphi) \rangle = \langle f, \mathcal{F}(\operatorname{sgn}(x)\varphi) \rangle \qquad \forall \, \varphi \in S_0(\mathbb{R}^n) \tag{4.37}$$

It is quite evident from (4.37) that Hf defined by (4.37) is a linear functional on $Z_0(\mathbb{R}^n)$. If $\{\varphi_\nu\}_{\nu=1}^\infty$ is a sequence $\to 0$ in $S_0(\mathbb{R}^n)$, then $\mathcal{F}(\operatorname{sgn}(x)\varphi_\nu) \to 0$ in $Z_0(\mathbb{R}^n)$ as $\nu \to \infty$. Therefore $Hf \in Z_0'(\mathbb{R}^n)$.

Theorem 4. Let $Z_0'(\mathbb{R}^n)$ be the space of ultradistribution defined in Section 3.5, let $P_k(t) = \sum_{0 \le |i| \le k} a_i t^i$, where

$$t^i = t_1^{i_1} t_2^{i_2} \cdots t_n^{i_n}, \qquad i_j \ge 0$$

and let

$$|i| = \sum_{j=1}^n i_j$$

be a k-degree (finite) polynomial. Then there exists an ultradistribution belonging to $Z_0'(\mathbb{R}^n)$ that corresponds to this polynomial and the Hilbert transform of the ultradistribution generated by $P_k(t)$ is zero.

Proof.   1. We generate an ultradistribution from the polynomial $P_k(t)$ using the relation

$$\langle P_k(t), \varphi \rangle = \int_{\mathbb{R}^n} P_k(t)\varphi(t)\, dt \qquad \forall \, \varphi \in Z_0(\mathbb{R}^n) \tag{4.38}$$

The integral in the right-hand side of (4.38) clearly exists as $\varphi \in Z_0(\mathbb{R}^n)$. The linearity of the functional $P_k(t)$ defined by (4.38) is trivial. Continuity of the functional $P_k(t)$ can as well be proved by using the properties of $S(\mathbb{R}^n)$.

2. For $P_k(t) = t^k$, where k is an arbitrary n-tuple with each component in $\mathbb{N}$, the general result can be proved by applying the result

$$\langle Ht^m, i\mathcal{F}(\varphi)\rangle$$

$$= \langle t^m, \mathcal{F}(\operatorname{sgn}(x)\varphi)\rangle \qquad \forall \, \varphi \in S_0(\mathbb{R}^n) \qquad (4.39)$$

$$= \int_{\mathbb{R}^n} t^m \mathcal{F}(\operatorname{sgn}(x)\varphi)(t)\, dt$$

$$= \int_{\mathbb{R}^n} dt \, t^m \int_{\mathbb{R}^n} \operatorname{sgn}(x)\varphi(x)e^{+ixt}\, dx$$

$$= \lim_{N \to \infty} \int_{-N}^{N}\int_{-N}^{N} \cdots \int_{-N}^{N} dt \, (i)^m \int_{\mathbb{R}^n} \operatorname{sgn}(x)\varphi^{(m)}(x)e^{+ixt}\, dx$$

$$= (2i)^{(m-n)} \lim_{N \to \infty} \int_{\mathbb{R}^n} \operatorname{sgn}(x)\varphi^{(m)}(x)\frac{\sin(Nx_1)}{x_1}\frac{\sin(Nx_2)}{x_2}$$

$$\cdots \frac{\sin(Nx_m)}{x_m}\, dx \qquad \begin{array}{l}\text{(by switching the order}\\ \text{of integration)}\end{array} \qquad (4.40)$$

It can be easily proved that the right-hand-side expression in (4.40) tends to zero as $N \to \infty$. $\square$

We now state some results without proof that can be proved quite easily.

For $\delta \in Z_0'(\mathbb{R}^n)$

$$(H\delta)(t) = \frac{(1)}{\pi^n}\, \text{p.v.}\, \left(\frac{1}{x}\right) = \frac{(1)}{\pi^n}\, \text{p.v.}\, \left[\frac{1}{x_1 x_2 \cdots x_n}\right]$$

These symbols will be explained in greater detail in the next chapter.

$$Hf^{(m)} = (Hf)^{(m)} \qquad \forall \, m = 0, 1, 2, 3, \ldots$$

where

$$\langle f^{(m)}, \varphi\rangle = (-1)^{(m)}\langle f, \varphi^{(m)}\rangle \qquad \forall \, \varphi \in Z_0(\mathbb{R}^n)$$

$$H(t^m f)(x) = x^m(Hf)(x)$$

where

$$t^m = t_1^{m_1} t_2^{m_2} \cdots t_n^{m_n}$$

Note that

$$\frac{\delta(x)}{1 + |x|^2} \notin Z_0'(\mathbb{R}^n)$$

since $Z_0(\mathbb{R}^n)$ is not closed with respect to multiplication by $(1/1 + |x|^2)$. But $(1 + x^2)f \in Z_0'(\mathbb{R}^n)$ if $f \in Z_0'(\mathbb{R}^n)$ because $Z_0(\mathbb{R}^n)$ is closed with respect to multiplication by polynomial functions.

Example 7. Let us find the Hilbert transform of

$$(1 + x^2)\delta(x) \in Z_0'(\mathbb{R}^n)$$

Solution. Customarily we have

$$(1 + x^2)\delta(x) = (1 + 0^2)\delta(x) = \delta(x)$$

Therefore

$$H\left[(1 + x^2)\delta(x)\right] = H\delta = \frac{1}{\pi^n} \text{ p.v. } \left[\frac{1}{x}\right]$$

But, if we make use of the formulas derived in this section, we get

$$H\left[(1 + x^2)\delta(x)\right] = (1 + t^2)H\delta = (1 + t^2)\frac{1}{\pi^n} \text{ p.v. } \frac{1}{t}$$

We got an additional term like

$$t^2 \frac{1}{\pi^n} \text{ p.v. } \frac{1}{t}$$

By a method analogous to that used in Theorem 4, we can show that

$$t^2 \frac{1}{\pi^n} \text{ p.v. } \frac{1}{t} \equiv \frac{t}{\pi^n}, \qquad \text{(ultradistribution)}$$

By using a method similar to one used in proving Theorem 6, one can show that the ultradistribution generated by a polynomial is zero ultradistribution. Therefore t/π^n represents a zero ultradistribution. Hence

$$H\left[(1 + x^2)\delta(x)\right] = \frac{1}{\pi^n} \text{ p.v. } \left[\frac{1}{x}\right]$$

The inversion formula (4.29) can be extended in $Z'(\mathbb{R}^n)$ as

$$H^2 f = (-1)^n f \qquad \forall f \in Z'(\mathbb{R}^n)$$

EXERCISES

1. **(a)** Prove that $f(x) = x$ defined on the real line is absolutely continuous on the real line.

   (b) Prove that an absolutely continuous function $f(x)$ on the real line is uniformly continuous there but is not bounded on the real line.

   Suggestion: Try the counterexample $f(x) = x$.

2. Prove that a function $f(x)$ defined as

$$f(x) = \begin{cases} \dfrac{\sin x}{x}, & x \neq 0 \\ 1, & x = 0 \end{cases}$$

is a C^∞ function on the real line and that the function $f(x)$ does not belong to the testing function space Φ, even though $f(x)$ and each of its derivatives is $O\left(\frac{1}{|x|}\right)$, $|x| \to \infty$.

Hint: Prove that the function $g(x)$ defined by

$$g(x) = \frac{1}{2\pi} \int_{-\infty}^{\infty} \frac{\sin t}{t} e^{-itx} dt$$

is discontinuous at the points $x = \pm 1$. It is easy to see that

$$g(x) = \frac{1}{4\pi} \int_{-\infty}^{\infty} \frac{2 \sin t}{t} \cos tx\, dt$$

$$= \frac{1}{\pi} \int_{-\infty}^{\infty} \frac{\left[\sin\{(1+x)t\} + \sin\{(1-x)t\}\right]}{t} dt, \text{ etc.}$$

3. Prove that the function $f(x) = e^{-|x|}$, $-\infty < x < \infty$ belongs to the testing function space Ψ. Prove that the function $g(x)$ defined by

$$g(x) = \begin{cases} e^{-(1/x^2)-x^2}, & x \neq 0 \\ 0, & x = 0 \end{cases}$$

also belongs to the testing function space Ψ.

4. Let D be the Schwartz testing function space consisting of C^∞ functions with compact supports. Let Q be the space of functions obtained by multiplying each element of D by $\mathrm{sgn}(x)$ where

$$\mathrm{sgn}(x) = \begin{cases} 1 & \text{if } x > 0 \\ 0 & \text{if } x = 0 \\ -1 & \text{if } x < 0 \end{cases}$$

That is, $Q = \{f(x);\ f(x) = \mathrm{sgn}(x)\varphi(x),\ \varphi \in D\}$. Prove that $Q \subset \Psi$.

Prove that

$$\{f(x);\ f(x) = \text{sgn}(x)\varphi(x),\ \varphi(x) \in S\} \subset \Psi$$

where S is Schwartz testing function space of rapid descent.

5. Give an intrinsic definition of the space Φ. *Hint:* Let $\varphi(x)$ be a C^∞ function defined on the real line such that

$$\varphi^{(k)}(x) = O\left(\frac{1}{|x|}\right), \qquad x \to \pm\infty,\ k = 0, 1, 2, 3, \ldots$$

Here $\varphi^{(0)}(x) = \varphi(x)$. Add some more conditions to φ to prove that $\varphi(x) \in \Phi$:

$$\left(\mathcal{F}^{-1}(\varphi)\right)(x) \in \Psi$$

6. Give an intrinsic definition of the topology of the space Φ (open problem).
7. Let $\varphi \in S_0(R)$. Prove that

$$\lim_{n \to \infty} \int_{-\infty}^{\infty} \frac{\sin nx}{x} \varphi(x) \, \text{sgn}(x) \, dx = 0$$

Note that $\lim_{n \to \infty} \frac{\sin nx}{\pi x} = \delta(x)$ in D'. [See Zemanian [109].]

8. Gel'fand and Shilov defined the space Ψ of all functions $\psi(s)$ $(-\infty < s < \infty)$, possessing the following properties:

(i) The function $s^k \psi(s)$ is absolutely continuous on the line $-\infty < s < \infty$ for all $k = 0, 1, 2, 3, \ldots$.

(ii) $\psi(s)$ is continuous and has a continuous derivative $\psi'(s)$ on each of the half-lines $-\infty < s \le 0, 0 \le s < \infty$; the function $\psi(s)$ and $\psi'(s)$ may have a discontinuity of the first kind at the point $s = 0$.

(iii) $s^k \psi'(s)$ is absolutely integrable on the line $-\infty < s < \infty$ for each $k = 0, 1, 2, 3, \ldots$.

Prove or disprove that

(a) $\displaystyle \sup_{-\infty < s < \infty} |\psi(s)| < \infty$

(b) $\displaystyle \sup_{-\infty < s < \infty} |\psi'(s)| < \infty$

(c) Prove that this space, Ψ defined by Gel'fand and Shilov, are different from that defined in this chapter. Illustrate this fact by giving a specific example.

5

n-DIMENSIONAL HILBERT TRANSFORM

5.1. GENERALIZED *n*-DIMENSIONAL HILBERT TRANSFORM AND APPLICATIONS

This chapter extends the distributional Hilbert transform H and the corresponding inversion formula $-\mathrm{H}^2 = I$ to *n*-dimensions. It derives many related results and demonstrates their applications in finding solutions to some singular integral equations in the space $\mathcal{D}'_{L^p}$.

5.1.1. Notation and Preliminaries

As usual we denote by $\mathbb{R}^n$ and $\mathbb{C}^n$ the real and complex *n*-dimensional Euclidean spaces, respectively. An element of $\mathbb{R}^n$ and $\mathbb{C}^n$ is an *n*-tuple $x = (x_1, x_2, \ldots, x_n)$ and $z = (z_1, z_2, \ldots, z_n)$, respectively, $z_j = x_j + iy_j$ $(j = 1, 2, \ldots, n)$. We also write

$$\|x\| = \left(\sum_{j=1}^n x_j^2 \right)^{1/2} \quad \|z\| = \left(\sum_{j=1}^n |z_j|^2 \right)^{1/2}$$

$$D^\alpha = \frac{\partial^{|\alpha|}}{\partial x_1^{\alpha_1} \ldots \partial x_n^{\alpha_n}} \quad |\alpha| = \alpha_1 + \alpha_2 + \cdots + \alpha_n$$

$$x^\alpha = x_1^{\alpha_1} x_2^{\alpha_2} \ldots x_n^{\alpha_n} \quad \alpha! = \alpha_1! \alpha_2! \ldots \alpha_n!$$

5.1.2. The Testing Function Space $\mathcal{D}_{L^p}(\mathbb{R}^n)$

An infinitely differentiable function φ defined over $\mathbb{R}^n$ is said to belong to the space $\mathcal{D}_{L^p}(\mathbb{R}^n)$ iff $D^\alpha \varphi(x)$ belongs to $L^p(\mathbb{R}^n)$ for each $|\alpha| = 0, 1, 2, 3, \ldots$; we introduce a topology on $\mathcal{D}_{L^p}(\mathbb{R}^n)$ with the help of the sequence of seminorms γ_m on $\mathcal{D}_{L^p}(\mathbb{R}^n)$ as follows: For each $\varphi \in \mathcal{D}_{L^p}(\mathbb{R}^n)$ we define

$$\gamma_{|\alpha|}(\varphi) = \left(\int_{\mathbb{R}^n} |D^\alpha \varphi(x)|^p \, dx \right)^{1/p} \tag{5.1}$$

where $|\alpha| = 0, 1, 2, 3, \ldots$. Since γ_0 is a norm the collection of seminorms γ_m is separating [110, p. 8]. Therefore the space $\mathcal{D}_{L^p}(\mathbb{R}^n)$ is a countably multinormed space. We denote by $\mathcal{D}'_{L^p}(\mathbb{R}^n)$ the dual of the space $\mathcal{D}_{L^{p'}}(\mathbb{R}^n)$ where $p' = p/(p-1)$, $p > 1$. We say that a sequence $\{\varphi_\mu\}_{\mu=1}^\infty$ converges to φ in $\mathcal{D}_{L^p}(\mathbb{R}^n)$ if, for each $|m| = 0, 1, 2, \ldots$,

$$\gamma_{|m|}(\varphi_\mu - \varphi) \to 0 \qquad \text{as } \mu \to \infty$$

The topology of $\mathcal{D}_{L^p}(\mathbb{R}^n)$ is generated by the seminorms $\gamma_{|m|}$ in the usual manner [110, pp. 8–14]. The Schwartz testing function space $\mathcal{D}(\mathbb{R}^n)$ is dense in $\mathcal{D}_{L^p}(\mathbb{R}^n)$ [101]. We state here the structure formula for $f \in \mathcal{D}'_{L^p}(\mathbb{R}^n)$, which can be proved in view of similar results proved for spaces $K(M_p)$ [44, pp. 109–110]. See also [87, p. 201].

Theorem 1. Structure Formula. If $f \in \mathcal{D}'_{L^p}(\mathbb{R}^n)$, $1 < p < \infty$, then f is equal to the finite linear combination of the derivatives of function in $L^p(\mathbb{R}^n)$. That is, for each $f \in \mathcal{D}'_{L^p}(\mathbb{R}^n)$ there exist f_α's in $L^p(\mathbb{R}^n)$, satisfying

$$\langle f, \varphi \rangle = \sum_{|\alpha|=0}^{k} (-1)^{|\alpha|} \int_{\mathbb{R}^n} D^\alpha \, \varphi(x) f_\alpha(x) \, dx \quad \forall \varphi \in \mathcal{D}_{L^{p'}}(\mathbb{R}^n) \tag{5.2}$$

5.1.3. The Test Space $X(\mathbb{R}^n)$

The space $X(\mathbb{R}^n)$ consists of all test functions $\varphi(x_1, x_2, \ldots, x_n)$ on $\mathbb{R}^n$ of the form

$$\varphi(x) = \sum_{\mu=1}^{k} \varphi_{\mu_1}(x_1) \varphi_{\mu_2}(x_2) \ldots \varphi_{\mu_n}(x_n) \tag{5.3}$$

where $\varphi_{\mu_j}(x_j) \in \mathcal{D}(\mathbb{R}) \; \forall j = 1, 2, \ldots, n$ and $\forall \mu = 1, 2, 3, \ldots, k$, where k is some finite number and μ_j are positive integers $\leq k$ such that $\mu_i \neq \mu_j$ if $i \neq j$. We consider the relative topology on $X(\mathbb{R}^n)$ induced by the Schwartz testing function space $\mathcal{D}(\mathbb{R}^n)$.

Lemma 1. The test space $X(\mathbb{R}^n)$ is dense in the space $\mathcal{D}(\mathbb{R}^n)$.

Proof. Let $\varphi(x) \in \mathcal{D}(\mathbb{R}^n)$ with support contained in $B(a) = \{x \in \mathbb{R}^n : |x_i| \leq a, 1 \leq i \leq n\}$, $a > 0$. By Weierstrass' approximation theorem [106, pp. 68–71], for an integer $m > 0$ there exists a polynomial $P_m(x_1, x_2, \ldots, x_n)$ such that in $B(2a) = \{x \in \mathbb{R}^n : |x_i| \leq 2a, 1 \leq i \leq n\}$, it differs from $\varphi(x_1, x_2, \ldots, x_n)$ by less than $\frac{1}{m}$ together with all its derivatives up to order m.

Thus

$$|\varphi^{(k)}(x) - P_m^{(k)}(x_1, x_2, x_3, \ldots, x_n)| < \frac{1}{m} \tag{5.4}$$

$$\forall |k| = 0, 1, 2, \ldots, m \text{ and } |x_i| \leq 2a$$

Since $\varphi(x) = 0$ outside $|x_i| \geq a, i = 1, 2, 3, \ldots, n$, it follows that

$$|P_m^{(k)}(x_1, x_2, \ldots, x_m)| < \frac{1}{m} \tag{5.5}$$

for

$$a \leq |x_i| \leq 2a, \qquad i = 1, 2, 3, \ldots, n$$

and $|k| = 0, 1, 2, \ldots, m$.

Now, let $e(x) \in \mathcal{D}(\mathbb{R})$, such that $e(x) = 1$ for $|x| \leq a$ and zero for $|x| \geq 2a$ such that $0 \leq e(x) \leq 1$ for all $x \in \mathbb{R}$. Then

$$\{P_m(x)E(x)\}_{m=1}^{\infty} \tag{5.6}$$

where $E(x) = e(x_1)e(x_2) \cdots e(x_n)$ converges to $\varphi(x)$ in $\mathcal{D}$ and it can be easily noted that

$$P_m(x)E(x) \in X(\mathbb{R}^n) \qquad \text{for each } m \in \mathbb{N}$$

Since each element of this sequence and the functions $\varphi(x)$ have their supports contained in $|x_i| \leq 2a, i = 1, 2, \ldots, n$, we need only show that the sequence converges uniformly to φ along with all its derivatives to the corresponding derivative of φ in the region $|x_i| \leq 2a, i = 1, 2, \ldots, n$. Now

$$|\varphi(x) - P_m(x)E(x)| = |\varphi(x) - P_m(x) + P_m(x)(1 - E(x))|$$
$$\leq |\varphi(x) - P_m(x)| + |P_m(x)| |1 - E(x)| \tag{5.7}$$
$$\leq \frac{1}{m} + \frac{1}{m}$$

This is because $1 - E(x)$ vanishes in the region $|x_i| \leq a, i = 1, 2, 3, \ldots, n$ and $1 - E(x) \leq 1$ in the region $a \leq |x_i| \leq 2a, i = 1, 2, 3, \ldots, n$.

Therefore, letting $m \to \infty$ in (5.7), we can prove that the sequence $\{P_m(x)E(x)\}_{m=1}^{\infty}$ tends to $\varphi(x)$ uniformly on $\mathbb{R}^n$.

Next consider the sequence of the first derivative:

$$|\varphi^{(1)}(x) - D\{P_m(x)E(x)\}|$$
$$= |\varphi^{(1)}(x) - P_m^{(1)}(x)| + |P_m^{(1)}(x)(1 - E(x))| + |P_m^{(1)}(x)DE(x)|$$
$$\leq \frac{1}{m} + \frac{1}{m} + \frac{A}{m} \to 0 \qquad \text{as } m \to \infty$$

for every x satisfying $|x_i| \leq 2a$, $i = 1, 2, \ldots, n$, where

$$A = \max_{\substack{i=1,\ldots,n \\ x \in \mathbb{R}^n}} \left| \frac{\partial}{\partial x_i} E(x) \right|$$

Therefore the first derivative of the sequence (5.6) converges uniformly to the corresponding first derivative of φ as $m \to \infty$. Carrying similar argument and using the techniques of induction we can show that the sequence (5.6) converges to φ in $\mathcal{D}$ as $m \to \infty$. □

Corollary 1. The space $X(\mathbb{R}^n)$ is dense in the space $\mathcal{D}_{L^p}(\mathbb{R}^n)$.

Proof. Let $\varphi \in \mathcal{D}_{L^p}(\mathbb{R}^n)$. Since $\mathcal{D}(\mathbb{R}^n)$ is dense in $\mathcal{D}_{L^p}(\mathbb{R}^n)$ [87], there exists a sequence $\{\psi_\mu\}_{\mu=1}^\infty$ in $\mathcal{D}(\mathbb{R}^n)$ such that

$$\psi_\mu \xrightarrow{\mathcal{D}_{L^p}(\mathbb{R}^n)} \varphi \qquad \text{as } \mu \to \infty \tag{5.8}$$

Since $X(\mathbb{R}^n)$ is dense in $\mathcal{D}(\mathbb{R}^n)$ by Lemma 1, for each fixed μ there exists $\varphi_{\mu,\nu} \in X(\mathbb{R}^n)$ such that

$$\varphi_{\mu,\nu} \xrightarrow{\mathcal{D}(\mathbb{R}^n)} \psi_\mu \qquad \text{as } \nu \to \infty \tag{5.9}$$

Since the convergence of a sequence in $\mathcal{D}(\mathbb{R}^n)$ implies its convergence in $\mathcal{D}_{L^p}(\mathbb{R}^n)$ [87], it follows that

$$\varphi_{\mu,\nu} \xrightarrow{\mathcal{D}_{L^p}(\mathbb{R}^n)} \psi_\mu \qquad \text{as } \nu \to \infty.$$

We now have the following situation:

$$\varphi_{1,\nu} \to \psi_1$$
$$\varphi_{2,\nu} \to \psi_2$$
$$\vdots \qquad \vdots$$
$$\varphi_{\mu,\nu} \to \psi_\mu$$
$$\vdots \qquad \vdots$$

and $\psi_\mu \xrightarrow{\mathcal{D}(\mathbb{R}^n)} \varphi$. Since the space $\mathcal{D}_{L^p}(\mathbb{R}^n)$ is metrizable with metric given by

$$\rho(\varphi, \psi) = \sum_{m=0}^\infty \frac{1}{2^m} \frac{\gamma_m(\varphi - \psi)}{1 + \gamma_m(\varphi - \psi)}$$

Therefore, by using the result [88, p. 8] there exists a subsequence of $\{\varphi_{\mu,\nu}\}_{\mu,\nu=1}^\infty$ that converges to φ. Therefore $X(\mathbb{R}^n)$ is dense in $\mathcal{D}_{L^p}(\mathbb{R}^n)$. □

5.2. THE HILBERT TRANSFORM OF A TEST FUNCTION IN $X(\mathbb{R}^n)$

Let

$$\varphi(x) = \sum_{\mu=1}^{k} c_\mu \prod_{i=1}^{n} \varphi_\mu(x_i)$$

where each of $\varphi_\mu(x_i)$ belongs to $\mathcal{D}(\mathbb{R})$. Then the function φ belongs to $X(\mathbb{R}^n)$, and conversely, only a function belonging to the space $X(\mathbb{R}^n)$ has this representation.

We define the n-dimensional Hilbert transform of φ as

$$(H\varphi)(x) = \lim_{\max \epsilon_i \to 0} \frac{1}{\pi^n} \int_{\substack{|x_i - t_i| > \epsilon_i \\ i=1,2,\dots,n}} \prod_{i=1}^{n} \frac{\varphi(t)}{(x_i - t_i)} \, dt \tag{5.10}$$

$$(H\varphi)(x) = \lim_{\max \epsilon_i \to 0^+} \left(\frac{1}{\pi^n} \right) \int_{\substack{|t_i - x_i| > \epsilon_i > 0 \\ i=1,2,\dots,n}} \frac{\sum_{\mu=1}^{k} c_\mu \prod_{i=1}^{n} \varphi_\mu(t_i)}{\prod_{i=1}^{n} [(x_i - t_i)]} \, dt \tag{5.11}$$

$$= \sum_{\mu=1}^{k} c_\mu \prod_{i=1}^{n} \left[\lim_{\epsilon_i \to 0^+} (1/\pi) \int_{|t_i - x_i| > \epsilon_i > 0} \frac{\varphi_\mu(t_i)}{(x_i - t_i)} \right] \tag{5.12}$$

$$= \sum_{\mu=1}^{k} c_\mu \prod_{i=1}^{n} \big(H_i \varphi_\mu(t_i) \big)(x_i) \tag{5.13}$$

Each of the terms $(H_i \varphi_\mu)(x_i) \in \mathcal{D}_{L^p}$, $p > 1$. It follows that $(H\varphi)(x)$ given by (5.13) belongs to $\mathcal{D}_{L^p}(\mathbb{R}^n)$, $p > 1$, whenever $\varphi \in X(\mathbb{R}^n)$.

Applying the operator H on both sides of (5.13), we get

$$(H^2 \varphi)(t) = \sum_{\mu=1}^{k} c_\mu \prod_{i=1}^{n} (H_i{}^2) \varphi_\mu(t_i)$$

$$= (-1)^n \sum_{\mu=1}^{k} c_\mu \prod_{i=1}^{n} \varphi_\mu(t_i) = (-1)^n \varphi(t)$$

It follows that

$$H^2 = (-1)^n I \qquad \text{in } X(\mathbb{R}^n) \tag{5.14}$$

We now prove that the operator

$$H : X(\mathbb{R}^n) \to \mathcal{D}_{L^p}(\mathbb{R}^n), \qquad p > 1$$

is a bounded linear mapping. Linearity of H is trivial. The boundedness property of H is a special case of a more general result [99].

Theorem 2. Let H be the mapping defined by (5.12). Then H is a linear continuous injection of $X(\mathbb{R}^n)$ into $\mathcal{D}_{L^p}(\mathbb{R}^n)$.

Since $X(\mathbb{R}^n)$ is dense in $\mathcal{D}_{L^p}(\mathbb{R}^n)$, the continuous injection $H : X(\mathbb{R}^n) \to \mathcal{D}_{L^p}(\mathbb{R}^n)$ extends uniquely to the whole space $\mathcal{D}_{L^p}(\mathbb{R}^n)$ by the limiting process as follows: For $\varphi \in \mathcal{D}_{L^p}(\mathbb{R}^n)$, there exists a sequence $\{\varphi_\mu\}_{\mu=1}^{\infty}$ in $X(\mathbb{R}^n)$ such that

$$\varphi_\mu \xrightarrow{\mathcal{D}_{L^p}(\mathbb{R}^n)} \varphi \qquad \text{as } \mu \to \infty.$$

We define $H\varphi$ on $(\mathbb{R}^n)$ by the relation

$$(H\varphi)(x) = \lim_{\mu \to \infty} [H\varphi_\mu](x) \tag{5.15}$$

The existence of the limit (5.15) follows from the continuity of the operator H on $\mathcal{D}_{L^p}(\mathbb{R}^n)$. To prove the uniqueness of the limit, assume that there exist two sequences: $\{\varphi_\mu\}_{\mu=1}^{\infty}$ and $\{\psi_\mu\}$ in $X(\mathbb{R}^n)$ that converge to φ in $\mathcal{D}_{L^p}(\mathbb{R}^n)$. Then, by definition, we have, for all $|k| = 0, 1, 2, \ldots,$

$$\gamma_{|k|}(\varphi - \varphi_\mu) \to 0 \qquad \text{as } \mu \to \infty$$

$$\gamma_{|k|}(\varphi - \psi_\nu) \to 0 \qquad \text{as } \nu \to \infty$$

Now

$$\begin{aligned}
\gamma_{|k|}(H\varphi_\mu - H\psi_\nu) &= \gamma_{|k|}\big(H(\varphi_\mu - \psi_\nu)\big) \\
&\leq C_{p1}C_{p2}\cdots C_{pn}\, \gamma_{|k|}(\varphi_\mu - \psi_\nu) \\
&\leq C_p\big[\gamma_{|k|}(\varphi_\mu - \varphi) + \gamma_{|k|}(\varphi - \psi_\mu)\big] \to 0
\end{aligned}$$

as $\mu, \nu \to \infty\ \forall\, |k| \in \mathbb{N}$.

Therefore the limit (5.15) exists and is unique. In Section 5.5 it will be shown that limit in the right-hand side of (5.15) is

$$\lim_{\max \epsilon_i \to 0} \frac{1}{\pi^n} \int_{\substack{|x_i - t_i| > \epsilon_i > 0 \\ i = 1,2,3,\ldots}} \frac{\varphi(t)dt}{\displaystyle\prod_{i=1}^{n}(x_i - t_i)}$$

The linearity of H is obvious. As a matter of fact we can now prove

Theorem 3. Let H be a mapping defined by (5.15). Then for $1 < p < \infty$, H is a linear homeomorphism from $\mathcal{D}_{L^p}$ onto itself and

$$H^2 = (-1)^n I \qquad \text{on } \mathcal{D}_{L^p}(\mathbb{R}^n),\ p > 1 \tag{5.16}$$

or

$$H^{-1} = (-1)^n H \tag{5.17}$$

Proof. Since (5.16) holds on $X(\mathbb{R}^n)$, it must hold on $\mathcal{D}_{L^p}(\mathbb{R}^n)$, $p > 1$, by the corollory to Lemma 1 and Theorem 1.

In view of (5.16),

$$H\varphi = 0 \Longrightarrow \varphi = 0$$

So H defined by (5.15) is one to one, and that this is also onto follows from (5.16). Hence from (5.16) it also follows that

$$H^{-1} = (-1)^n H$$

Since H is linear and continuous, so is H^{-1}. This completes the proof of the theorem. $\square$

Theorem 4. Let $\varphi \in \mathcal{D}_{L^p}(\mathbb{R}^n)$. Then

$$D^k[H\varphi] = H[D^k\varphi] \qquad \forall |k| \in \mathbb{N} \tag{5.18}$$

Proof. We will prove the result for the test functions in $X(\mathbb{R}^n)$, and the result for the test functions in $\mathcal{D}_{L^p}(\mathbb{R}^n)$ can be proved by limiting process and by the fact that the operator

$$D^k : \mathcal{D}_{L^p}(\mathbb{R}^n) \longrightarrow \mathcal{D}_{L^p}(\mathbb{R}^n), \qquad p > 1 \tag{5.19}$$

is continuous $\forall |k| \in \mathbb{N}$. Let

$$\varphi = \sum_{\mu=1}^{r} c_\mu \left(\prod_{i=1}^{n} \varphi_{\mu,i} \right) \in X(\mathbb{R}^n)$$

Then

$$D^k(H\varphi)(x) = \sum_{\mu=1}^{r} \frac{c_\mu \partial^{|k|}}{\partial x_1^{k_1} \partial x_2^{k_2} \cdots \partial x_n^{k_n}} \prod_{i=1}^{n} [H_i \varphi_{\mu,i}] (x)$$

$$= \sum_{\mu=1}^{r} c_\mu \prod_{i=1}^{n} \left(\frac{\partial^{k_i}}{\partial x_i^{k_i}} H_i \varphi_{\mu,i} \right) (x)$$

Now using Theorem 1 of Chapter 3, we have

$$\frac{\partial^{k_i}}{\partial x_i^{k_i}} [H_i(\varphi_{\mu,i})] (x_i) = H_i \left[\varphi_{\mu,i}^{(k_i)} \right] (x_i) \qquad \forall i \in \mathbb{N}$$

Therefore

$$D^k[H\varphi](x) = \sum_{\mu=1}^{r} c_\mu \prod_{i=1}^{n} \left[H_i \varphi_{\mu,i}^{(k_i)} \right] (x)$$

$$= H \left(\sum_{\mu=1}^{r} c_\mu \left[\prod_{i=1}^{n} \varphi_{\mu,i}^{(k_i)} \right] \right) (x)$$

$$= HD^k \varphi \qquad \forall \varphi \in X(\mathbb{R}^n)$$

This completes the proof of the theorem. $\square$

Theorem 5. 5 Let $f \subset L^p(\mathbb{R}^n), p > 1, n > 1$. Define the operator H^* on $L^p(\mathbb{R}^n)$ by

$$(H^*f)(x) = \frac{1}{\pi^n} \lim_{\substack{\max \epsilon_i \to 0}} \int_{\substack{|t_i - x_i| > \epsilon_i > 0 \\ i=1,2,...,n}} \frac{f(t)\,dt}{\prod_{i=1}^{n}(x_i - t_i)} \qquad (5.20)$$

Then the limit in (5.20) exists in $L^p(\mathbb{R}^n)$ sense and almost everywhere sense, and

$$\left\| H^*f \right\|_p \le C_p \|f\|_p \qquad (5.21)$$

where C_p is a constant independent of f.

Proof. Theorem 5 is a very special case of more general result proved in [57]. ☐

Corollary 2. $H^{*2} = (-1)^n I$ on $L^p(\mathbb{R}^n), p > 1$.

Proof. Proof is trivial, since $H^* = H$ and

$$H^2 = (-1)^n I \qquad \text{on } X(\mathbb{R}^n) \qquad (5.22)$$

Now $X(\mathbb{R}^n)$ is dense on $L^p(\mathbb{R}^n)$. Therefore in view of the continuity of H, (5.22) is true on $L^p(\mathbb{R}^n)$. Therefore $H^{*2} = (-1)^n I$ on $L^p(\mathbb{R}^n)$. ☐

Corollary 3. Let φ be an arbitrary element of $\mathcal{D}_{L^p}(\mathbb{R}^n), p > 1$. Then

$$D^k(H^*\varphi) = H^*(D^k\varphi) \qquad \forall |k| \in \mathbb{N} \qquad (5.23)$$

Proof. This is true for H by Theorem 4 on $\mathcal{D}_{L^p}(\mathbb{R}^n), p > 1$. But $H = H^*$ on $\mathcal{D}_{L^p}(\mathbb{R}^n)$. Therefore (5.23) must be true. ☐

5.2.1. The Hilbert Transform of Schwartz Distributions in $\mathcal{D}_{L^p}'(\mathbb{R}^n), p > 1$

If $f \in \mathcal{D}_{L^p}'(\mathbb{R}^n)$, then the generalized Hilbert transform Hf of f is defined by the relation

$$\langle Hf, \varphi \rangle = \langle f, (-1)^n H\varphi \rangle \qquad \forall \varphi \in \mathcal{D}_{L^{p'}}(\mathbb{R}^n), \ p' = \frac{p}{p-1}, \ 1 < p < \infty \qquad (5.24)$$

when $H\varphi$ is defned by (3.17). That is, the Hilbert transform Hf of $f \in \mathcal{D}_{L^p}'(\mathbb{R}^n)$ is a linear functional that assigns the same number to $\varphi \in \mathcal{D}_{L^p}(\mathbb{R}^n)$ as f assigns to $(-1)^n H\varphi$. It is easy to show from (5.24) that Hf defined by (5.24) belongs to $\mathcal{D}_{L_p}'(\mathbb{R}^n)$.

Corollary 4. The operator $H : \mathcal{D}_{L^p}'(\mathbb{R}^n) \to \mathcal{D}_{L^p}'(\mathbb{R}^n)$ as defined by (5.24) is a linear isomorphism from $\mathcal{D}_{L^p}'(\mathbb{R}^n)$ onto itself and

$$H^2 f = (-1)^n f \qquad \forall f \in \mathcal{D}_{L^p}'(\mathbb{R}^n)$$

The result follows easily in view of (4.22) and the corollory to Theorem 1.

Corollary 5. The operator $H : \mathcal{D}'_{L^p}(\mathbb{R}^n) \to \mathcal{D}'_{L^p}(\mathbb{R}^n)$ as defined by (5.26) is a linear homeomorphism with respect to the strong as well as the weak topology introduced on $\mathcal{D}'_{L^p}(\mathbb{R}^n)$.

Proof. The proof is similar to that given in Chapter 2 for the analogous case $n = 1$. □

Definition. The distribution $\delta \in \mathcal{D}'_{L^p}(\mathbb{R}^n)$, $p > 1$ is defined by

$$\langle \delta, \varphi \rangle = \varphi(0) \qquad \forall \varphi \in \mathcal{D}_{L^p}(\mathbb{R}^n)$$

$$\overset{\Delta}{=} \varphi(0, 0, 0, \ldots, 0)$$

Linearity and continuity of δ is evident.

Definition. We now define the Schwartz distribution $\text{p.v.}\frac{1}{x}$ belonging to $\mathcal{D}'_{L^p}(\mathbb{R}^n)$, $n > 1$, as follows:

$$\left\langle \text{p.v.}\frac{1}{x}, \varphi(x) \right\rangle = \lim_{\max \epsilon_i \to 0+} \int_{\substack{|x_i| > \epsilon_i \\ i=1,2,\ldots,n}} \frac{\varphi(x)}{x_1 x_2 \cdots x_n} \, dx \qquad \forall \varphi \in \mathcal{D}_{L^{p'}}(\mathbb{R}^n), \ 1 < p < \infty$$

(5.25)

If $\varphi_\nu \to 0$ in $\mathcal{D}_{L^{p'}}(\mathbb{R}^n)$, then

$$\left| \left\langle \text{p.v.}\frac{1}{x}, \varphi_\nu(x) \right\rangle \right| = |(H\varphi_\nu)(0)| \to 0 \qquad \text{as } \nu \to \infty$$

In view of the fact that the order of integration in the right-hand side of (5.25) can be switched, we can write

$$\text{p.v.}\frac{1}{x} = \text{p.v.}\left(\frac{1}{\prod_{i=1}^n x_i} \right)$$

or

$$\text{p.v.}\frac{1}{x} = \text{p.v.}\frac{1}{(x_1 x_2 \cdots x_n)}$$

Definition. Let $f \in \mathcal{D}'_{L^p}(\mathbb{R}^n)$. Then we define the generalized function $D^k f \in \mathcal{D}'_{L^p}(\mathbb{R}^n)$ by the relation

$$\langle D^k f, \varphi \rangle = (-1)^{|k|} \langle f, D^k \varphi \rangle \qquad \forall \varphi \in \mathcal{D}_{L^{p'}}, \ p' = \frac{p}{p-1}, \ |k| \in \mathbb{N}$$

(5.26)

Theorem 6. If $f \in \mathcal{D}'_{L^p}(\mathbb{R}^n)$, $p > 1$, then $D^k(Hf) = H(D^k f)$ for all $|k| = 0, 1, 2, \ldots$.

Proof. For $\varphi \in \mathcal{D}_{L^{p'}}(\mathbb{R}^n)$ we have

$$\begin{aligned}
\langle HD^k f, \varphi \rangle &= \langle D^k f, (-1)^n H\varphi \rangle \\
&= (-1)^{|k|} \langle f, D^k(-1)^n H\varphi \rangle \\
&= (-1)^{|k|} \langle f, (-1)^n HD^k \varphi \rangle \\
&= (-1)^{|k|} \langle Hf, D^k \varphi \rangle \\
&= \langle D^k Hf, \varphi \rangle \qquad \square
\end{aligned}$$

5.3. SOME EXAMPLES

Example 1. Let us solve the following equation in the space $\mathcal{D}'_{L^p}(\mathbb{R}^n)$.

$$Hy = Hf + \delta(x_1, x_2, \ldots, x_n) \tag{5.27}$$

Solution. Operating both sides of (5.24) by H, we get

$$H^2 y = H^2 f + H\delta$$

$$(-1)^n y = (-1)^n f + H\delta \tag{5.28}$$

Now

$$\begin{aligned}
\langle H\delta, \varphi \rangle &= \langle \delta, (-1)^n H\varphi \rangle \qquad \forall \varphi \in \mathcal{D}_{L^{p'}}(\mathbb{R}^n) \\
&= \left\langle \delta(x), \lim_{\max \epsilon_i \to 0+} (-1)^n \frac{1}{\pi^n} \int_{\substack{|x_i - t_i| > \epsilon_i \\ i=1,2,\ldots,n}} \frac{\varphi(t)\, dt}{\prod_{i=1}^{n}(x_i - t_i)} \right\rangle \\
&= \left\langle \text{p.v.} \frac{1}{t\pi^n}, \varphi \right\rangle
\end{aligned}$$

Therefore

$$H\delta = \frac{1}{\pi^n} \text{p.v.} \left(\frac{1}{t} \right)$$

Therefore (5.28) reduces to

$$(-1)^n y = (-1)^n f + \frac{1}{\pi^n} \text{p.v.} \left(\frac{1}{t} \right)$$

or

$$y = f + \left(-\frac{1}{\pi} \right)^n \text{p.v.} \frac{1}{t}$$

Example 2. Let us now consider the solution to the integral equation (operator equation)

$$y = Hy + f(x), \qquad f \in \mathcal{D}'_{L^p}(\mathbb{R}^n) \qquad (5.29)$$

Solution. Operating both sides of (5.29) by H, we get

$$Hy = H^2y + Hf$$

or

$$Hy = (-1)^n y + Hf$$

or

$$y - f = (-1)^n y + Hf$$

If n is odd, then

$$y = \frac{f + Hf}{2}$$

If n is even, then

$$Hf = -f$$

If f given in (5.29) is such that $Hf = -f$ is satisfied, then (5.29) is satisfied for all $y \in \mathcal{D}'_{L^p}\mathbb{R}^n$. If, however, f given in (5.29) does not satisfy $Hf = -f$, then there does not exist a solution to (5.29) in $\mathcal{D}'_{L^p}(\mathbb{R}^n)$.

If f given in (5.26) is of the form

$$f(x) = \prod_{i=1}^{n}(H_i g_i)(x_i) - \prod_{i=1}^{n} g_i(x_i)$$

where $g_1, g_2, \ldots, g_n \in L^p$ and $H_1, H_2, \ldots, H_n$ are all one-dimensional Hilbert transforms,

$$(H_i g_i)(x_i) = \frac{1}{\pi} \lim_{\epsilon \to 0+} \int_{|t_i - x_i| > \epsilon} \frac{g_i(t_i)}{x_i - t_i} dt_i$$

then clearly $Hf = -f$ is satisfied.

Again, if f is of the form

$$f(x) = \prod_{i=1}^{n}(H_i g_i)(x_i) + \prod_{i=1}^{n} g_i(x_i)$$

then $Hf = f$. Clearly a nonzero function of this form cannot, in general, satisfy the equation $Hf = -f$. Hence the existence of a nonzero f not satisfying $Hf = -f$ is also demonstrated. For further details of such functions, see [80].

If $f \in L^p(\mathbb{R}^n)$, $p > 1$, then the generalized Hilbert transform Hf of f is given by

$$(Hf)(x) = \lim_{\max \epsilon_i \to 0+} (1/\pi^n) \int_{\substack{|t_i - x_i| > \epsilon_i \\ i = 1, 2, \ldots, n}} \frac{f(t)\, dt}{(x_1 - t_1)(x_2 - t_2) \cdots (x_n - t_n)}$$

5.4. GENERALIZED (n + 1)-DIMENSIONAL DIRICHLET BOUNDARY-VALUE PROBLEMS

Let $f(x)$ be a piecewise continuous and bounded function on the real line $\mathbb{R}$. It is a classical result that the solution to the Dirichlet boundary-value problem

$$\frac{\partial^2 u}{\partial x^2} + \frac{\partial^2 u}{\partial y^2} = 0 \tag{5.30}$$

$$\lim_{y \to 0^+} u(x, y) = f(x)$$

in the upper half-plane $\Omega = (z = x + iy, y > 0)$ is

$$u(x, y) = \frac{y}{\pi} \int_{-\infty}^{\infty} \frac{f(t)}{|t - z|^2} \, dt \qquad [6, \text{p. } 45] \tag{5.31}$$

We extend the notion of Harmonic function to the space $\mathbb{R}^{(n+1),+} = \{(x, y) : x \in \mathbb{R}^n, y > 0\}$. We call $u(x, y)$ a harmonic function in an open region of $\mathbb{R}^{(n+1),+}$, if it is infinitely differentiable at each point of the region and satisfies

$$u_{yy}(x, y) + \sum_{i=1}^{n} u_{x_i x_i}(x, y) = 0$$

In this chapter we exploit the above definition of harmonic functions to solve the Dirichlet boundary-value problem in the space $\mathbb{R}^{(n+1),+}$ with a distributional boundary condition. As it turns out, our solution is quite constructive, and its n-dimensional case is an extension of the corresponding classical Dirichlet boundary-value problem.

Let us consider the Poisson kernel for $\mathbb{R}^{(n+1),+}$, given by

$$P_y(x) = \frac{y}{C_n \left[y^2 + \sum_{i=1}^{n} x_i^2 \right]^{(n+1)/2}}, \qquad y > 0 \tag{5.32}$$

where

$$C_n = \frac{\pi^{(n+1)/2}}{\Gamma((n+1)/2)}$$

The following properties of $P_y(x)$ are well known:

$$\int_{\mathbb{R}^n} P_y(x) \, dx = 1 \qquad \text{for each } y > 0 \tag{5.33}$$

Let $\varphi \in L^q(\mathbb{R}^n)$, $q > 1$, then $U_\varphi(x, y)$ defined by

$$U_\varphi(x, y) = \int_{\mathbb{R}^n} \varphi(x - t) P_y(t) \, dt$$

is harmonic in the space $\mathbb{R}^{n+1,+}$ and

$$\left\| U_\varphi(x, y) \right\|_{L^q} \leq \left\| \varphi(x) \right\|_{L^q}, \qquad y > 0 \tag{5.34}$$

Also $U_\varphi(x, y)$ converges to $\varphi(x)$ in the norm of $L^q(\mathbb{R}^n)$ as $y \to 0^+$ [105, pp. 25–62].

Theorem 7. Let $f \in \mathcal{D}'_{L^p}(\mathbb{R}^n)$, $1 < p < \infty$. Then for all $\varphi \in \mathcal{D}_{L^{p'}}(\mathbb{R}^n)$, $p' = p/(p-1)$,

$$\langle\langle f(t), P_y(x-t)\rangle, \varphi(x)\rangle \to \langle f, \varphi\rangle \qquad \text{as } y \to 0^+$$

In other words, $\langle f(t), P_y(x-t)\rangle \to f(t)$ in $\mathcal{D}'_{L^p}(\mathbb{R}^n)$ as $y \to 0^+$.

Proof. Using the structure formula for f, we have

$$I(y) = \langle\langle f(t), P_y(x-t)\rangle, \varphi(x)\rangle$$

$$= \sum_{|\alpha|=0}^{r} (-1)^{|\alpha|}\langle\langle f_{|\alpha|}(t), D_t^\alpha P_y(x-t)\rangle, \varphi(x)\rangle \qquad (5.35)$$

where $f_\alpha \in L^p(\mathbb{R}^n)$. Now, using Fubini's theorem and integration by parts, we get

$$I(y) = \sum_{|\alpha|=0}^{r} \left\langle f_\alpha(t), \int_{\mathbb{R}^n} P_y(x-t)\varphi^{(\alpha)}(x)\, dx \right\rangle \qquad (5.36)$$

Let

$$U_\varphi(t, y) = \int_{\mathbb{R}^n} P_y(x-t)\varphi^{(\alpha)}(x)\, dx$$

Then it follows from (5.36) that for each fixed $y > 0$, $u_\varphi(t, y)$ belongs to $C^\infty(\mathbb{R}^{(n+1),+})$ and that

$$\left\|U_\varphi(t, y)\right\|_{L_p} \le \left\|\varphi^{(\alpha)}(t)\right\|_{L^p} < \infty, \qquad (5.37)$$

$$\left\|U_\varphi(t, y) - \varphi^{(\alpha)}(t)\right\|_{L^p} \to 0 \qquad \text{as } y \to 0^+ \qquad (5.38)$$

However,

$$D_t^\beta U_\varphi(t, y) = \int_{\mathbb{R}^n} P_y(x-t)\varphi^{(\alpha+\beta)}(x)\, dx \qquad \forall\, |\beta| = 0, 1, 2, 3, \ldots$$

Therefore

$$\gamma_{|\beta|}\left(U_\varphi(t, y) - \varphi^{(\alpha)}(t)\right) \to 0 \qquad \text{as } y \to 0^+ \;\forall\, |\beta| = 0, 1, 2, \ldots$$

$$\int_{\mathbb{R}^n} P_y(x-t)\varphi^{(\alpha)}(x)\, dx \to \varphi^{(\alpha)}(t) \qquad \text{as } y \to 0^+ \qquad (5.39)$$

From (5.36) and (5.39),

$$\lim_{y \to 0^+} I(y) = \sum_{|\alpha|=0}^{r} \langle f_\alpha(t), \varphi^{(\alpha)}(t)\rangle = \langle f, \varphi\rangle \qquad \forall \varphi \in \mathcal{D}_{L^{p'}}$$

Therefore

$$\langle f(t), P_y(x-t)\rangle \to f(t) \qquad \text{in } \mathcal{D}'_{L^p}(\mathbb{R}^n) \text{ as } y \to 0^+ \qquad \square$$

Example 3. Consider the Dirichlet boundary-value problem

$$U_{yy}(x, y) + \sum_{i=1}^{n} U_{x_i x_i}(x, y) = 0, \qquad y > 0$$

$$\lim_{y \to 0^+} U(x, y) = f(x) \qquad \text{in } \mathcal{D}'_{L^p}(\mathbb{R}^n), \ p > 1 \qquad (5.40)$$

Solution. Let us define a function $u(x, y)$ by

$$U(x, y) = \langle f(t), P_y(x - t) \rangle, \qquad y > 0$$

Using the structure formula for f, we get

$$U(x, y) = \sum_{|\alpha|=0}^{r} (-1)^{|\alpha|} \langle f_\alpha(t), D_t^\alpha P_y(x - t) \rangle$$

$$= \sum_{|\alpha|=0}^{r} (-1)^{|\alpha|} U_\alpha(x, y)$$

where

$$U_\alpha(x, y) = \langle f_{|\alpha|}(t), D_t^\alpha P_y(x - t) \rangle$$

By (5.33), $u_\alpha(x, y)$ is harmonic in $\mathbb{R}^{(n+1)}$; therefore $u(x, y)$ is harmonic. In view of Theorem 7, $u(x, y)$ satisfies the distributional boundary condition (5.40).

5.5. THE HILBERT TRANSFORM OF DISTRIBUTIONS IN $\mathcal{D}'_{L^p}(\mathbb{R}^n)$, $p > 1$, ITS INVERSION AND APPLICATIONS

Pandey and Chaudhary [77] in 1983 developed the theory of Hilbert transform of Schwartz distribution space $(\mathcal{D}_{L^p})'$, $p > 1$, in one dimension using Parseval's types of relations for one-dimensional Hilbert transform. They noted that their theory coincides with the corresponding theory for the Hilbert transform developed by Schwartz [87] through the technique of convolution in one dimension.

The corresponding theory for the Hilbert transform in n-dimension is considerably harder and will be successfully accomplished in this book. As I show here, we can also develop the n-dimensional theory of Hilbert transform to $\mathcal{D}'(\mathbb{R}^n)$ by a method analogous to that used by Ehrenpreis [36] to extend the theory of Fourier transform to $\mathcal{D}'$. We can also exploit the result proved in Theorem 8 to give an intrinsic definition of the space $H(\mathcal{D}(\mathbb{R}^n))$ and its topology. Some applications of these results to solve singular integral equations will be discussed. A related boundary-value problem and its solutions will also be discussed.

5.5.1. The *n*-Dimensional Hilbert Transform

If $f \in L^p(\mathbb{R}^n)$, $p > 1$, then it is well known that its Hilbert transform $(Hf)(x)$ defined by

$$(Hf)(x) = \frac{1}{\pi^n} \lim_{i} \max \epsilon_i \to 0^+ \int_{\substack{|t_i - x_i| > \epsilon_i \\ i = 1,2,3,\dots,n}} \frac{f(t)\, dt}{\prod_{i=1}^{n}(x_i - t_i)} \tag{5.41}$$

exists a.e. and $(Hf)(x) \in L^p(\mathbb{R}^n)$. It is also known that there exists a constant $C_p > 0$ independent of f satisfying

$$\|(Hf)(x)\|_p \le C_p \|f\|. \tag{5.42}$$

The existence of the integral in (5.41) and its boundedness property as stated in (5.42) was proved by Riesz and Tichmarsh [99] for $n = 1$. The results for $n > 1$ were proved by several authors such as Kokilashvili [57] among others. Riesz and Tichmarsh also obtained the following formula:

$$(H^2 f)(x) = -f(x) \qquad \text{a.e.} \tag{5.43}$$

for the one-dimensional Hilbert transform.

In this section I generalize the above inversion formula for $n > 1$ to the space $L^p(\mathbb{R}^n)$, $p > 1$ and then to Schwartz distribution spaces $\mathcal{D}'_{L^p}(\mathbb{R}^n)$ and $\mathcal{D}'(\mathbb{R}^n)$.

5.5.2. Schwartz Testing Functions Space $\mathcal{D}(\mathbb{R}^n)$

The space $\mathcal{D}(\mathbb{R}^n)$, $n \ge 1$ is the Schwartz testing function space of C^∞ functions defined on $\mathbb{R}^n$ having compact support, and the C^∞ functions defined on $\mathbb{R}$ with compact support will be denoted by $\mathcal{D}$ or $\mathcal{D}(\mathbb{R})$. The topology of $\mathcal{D}(\mathbb{R}^n)$ is the same as defined by Schwartz [87]. Accordingly a sequence $\{\varphi_m\}_{m=1}^{\infty}$ converges to zero in $\mathcal{D}(\mathbb{R}^n)$ if and only if

1. $\varphi_1, \varphi_2, \varphi_3, \dots$ have their support contained in a compact set K.
2. $\varphi_m^{(k)}(x) \to 0$ as $m \to \infty$ uniformly for each $|k| = 0, 1, 2, \dots$ on arbitrary compact subset of $\mathbb{R}^n$.

The space $X(\mathbb{R}^n)$ is defined to be the collection of $\varphi \in \mathcal{D}(\mathbb{R}^n)$, which are finite sums of the form

$$\varphi(x) = \sum \varphi_{m_1}(x_1)\varphi_{m_2}(x_2) \cdots \varphi_{m_n}(x_n) \tag{5.44}$$

where $\varphi_{m_i} \in \mathcal{D}$, for all $i = 1, 2, \dots, n$. We have the following well-known result:

Lemma 1. The space $X(\mathbb{R}^n)$ is dense in the space $L^p(\mathbb{R}^n)$, $p > 1$ with respect to the norm topology of $L^p(\mathbb{R}^n)$ [106, p. 71].

5.5.3. The Inversion Formula

If $\varphi \in X(\mathbb{R}^n)$ and φ has the representation (5.44), then

$$(H\varphi)(x) = \sum \prod_{i=1}^{n} (H_i \varphi_{m_i})(x_i) \qquad (5.45)$$

where $H_i(\varphi_{m_i}) \triangleq \hat{\varphi}_{m_i}$, the classical one-dimensional Hilbert transform of φ_{m_i} defined by

$$(H_i \varphi_{m_i})(x_i) = \frac{1}{\pi} P \int_{\mathbb{R}} \frac{\varphi_{m_i}(t_i)\, dt_i}{(x_i - t_i)} = \hat{\varphi}_{m_i}(x_i).$$

We are now ready to prove our inversion theorem.

Theorem 8. Let H be the operator of the classical Hilbert transform as defined by (5.41) in n-dimensions. Then for all $f \in L^p(\mathbb{R}^n)$,

$$(H^2 f)(x) = (-1)^n f(x) \qquad \text{a.e.} \qquad (5.46)$$

Proof. Equations (5.41) and (5.43) imply that the inversion formula (5.46) is valid for the subspace $X(\mathbb{R}^n)$ of $L^p(\mathbb{R}^n)$. To prove it on $L^p(\mathbb{R}^n)$, let us assume that $f \in L^p(\mathbb{R}^n)$ and that $\{\varphi_j\}_{j=1}^{\infty}$ is a sequence in $X(\mathbb{R}^n)$ tending to f in $L^p(\mathbb{R}^n)$ as $j \to \infty$. Such a sequence exists by Lemma 1. Then

$$\|H^2 f - (-1)^n f\|_p = \|H^2 f - (-1)^n f - (H^2 \varphi_j - (-1)^n \varphi_j)\|_p$$

$$= \|H^2(f - \varphi_j) - (-1)^n(f - \varphi_j)\|_p \qquad (5.47)$$

Now $H : L^p(\mathbb{R}^n) \to L^p(\mathbb{R}^n)$ is a bounded linear operator [57]. Therefore H^2 is also a bounded linear operator from $L^p(\mathbb{R}^n)$ into itself. By (5.7),

$$\|H^2 f - (-1)^n f\|_p \leq K_p \|f - \varphi_j\|_p \to 0 \qquad \text{as } j \to \infty$$

Hence

$$H^2 f = (-1)^n f \qquad (5.48)$$

in the $L^p(\mathbb{R}^n)$ sense, and so a.e. as well.

The Testing Function Space $\mathcal{D}_{L^p}(\mathbb{R}^n)$. A complex-valued function defined on $\mathbb{R}^n$ belongs to the space $\mathcal{D}_{L^p}(\mathbb{R}^n)$, $p > 1$ if and only if

$$\varphi \in C^{\infty}(\mathbb{R}^n),$$

$$\varphi^{(k)} \in L^p(\mathbb{R}^n), \qquad \forall\, |k| \in \mathbf{N},$$

where

$$\varphi^{(k)}(t) = D^k \varphi(t)$$

$$= D_{t_1}^{k_1} d_{t_2}^{k_2} \cdots D_{t_n}^{k_n} \varphi(t)$$

$$D_{t_i} \varphi = \frac{\partial \varphi}{\partial t_i}, \qquad i = 1, 2, 3, \ldots, n$$

$$k = (k_1, k_2, \ldots, k_n)$$

and

$$|k| = \sum_{i=1}^{n} k_i, \qquad k_i \in \mathbf{N}, \ i = 1, 2, \ldots, n \qquad \square$$

5.5.4. The Topology on the Space $\mathcal{D}_{L^p}(\mathbb{R}^n)$

The topology over $\mathcal{D}_{L^p}(\mathbb{R}^n)$ is generated by the separating collection of seminorms $\{\gamma_{(k)}\}$, $|k| \in \mathbf{N}$, where

$$\gamma_{(k)}(\varphi) = \left(\int_{\mathbb{R}^n} |\varphi^{(k)}(t)|^p \, dt \right)^{1/p} \tag{5.49}$$

See [110]. Therefore a sequence φ_i converges to φ in $\mathcal{D}_{L^p}(\mathbb{R}^n)$ as $j \to \infty$ if and only if

$$\gamma_{(k)}(\varphi_j - \varphi) \to 0 \qquad \text{as } j \to \infty \ \forall \ |k| \in \mathbb{N}.$$

A sequence φ_j is said to be a Cauchy sequence in $\mathcal{D}_{L^p}(\mathbb{R}^N)$ if and only if for all $|k| \in \mathbb{N}$,

$$\gamma_{(k)}(\varphi_m - \varphi_n) \to 0 \qquad \text{as } m, n \to \infty$$

independently of each other.

The space $\mathcal{D}_{L^p}(\mathbb{R}^n)$ $(1 < p < \infty)$ is a sequentially complete, locally convex Hausdorff topological vector space [55].

Note If $\varphi \in \mathcal{D}_{L^p}(\mathbb{R}^n)$, then $\varphi^{(k)}(x) \to 0$ as $|x| \to \infty$ for each $|k| \in \mathbb{N}$ [87]. If φ_j is a sequence tending to zero in $\mathcal{D}_{L^p}(\mathbb{R}^n)$ as $j \to \infty$, then for each $|k| \in \mathbb{N}$,

$$\varphi_j^{(k)}(x) \to 0 \qquad \text{uniformly on } \mathbf{R}^n \text{ as } j \to \infty.$$

This result is well known [87, p. 200].

Theorem 9. The operator H of n-dimensional Hilbert transform as defined by (5.1) is a homeomorphism from $\mathcal{D}_{L^p}(\mathbb{R}^n)$ onto itself.

Proof. The result is well known for $n = 1$; see [99]. We use this fact to prove the result for $n > 1$. For $\varphi(t)$ in $\mathcal{D}_{L^p}(\mathbb{R}^n)$, $p > 1$, let us define

$$(H_i\varphi)(t_1, t_2, \ldots, t_{i-1}, x_i, t_{i+1}, \ldots, t_n) \tag{5.50}$$

$$= \frac{1}{\pi} P \int_{\mathbb{R}^n} \frac{\varphi(t_1, t_2, \ldots, t_{i-1}, y_i, t_{i+1}, \ldots, t_n)}{x_i - y_i} \, dy_i$$

$$= \overline{\varphi}(t_1, t_2, \ldots, t_{i-1}, x_i, t_{i+1}, \ldots, t_n)$$

It is easy to see that if $f \in \mathcal{D}_{L^p}(\mathbb{R}^n)$, then

$$
\begin{aligned}
(Hf)(x) &= (H_1 H_2 \cdots H_{i-1} H_i H_{i+1} \cdots H_n f)(x) \\
&= (H_i (H_1 H_2 \cdots H_{i-1} H_{i+1} \cdots H_n) f)(x)
\end{aligned}
$$

(operators $H_1, H_2, H_3, \ldots$ are commutative).

Therefore, for $\varphi \in \mathcal{D}_{L^p}(\mathbb{R}^n)$, $p > 1$, we have

$$
(H\varphi)(x) = H_i\big(\overline{\varphi}(x_1, x_2, \ldots, x_{i-1}, t_i, x_{i+1}, \ldots, x_n)\big)
$$

where

$$
\overline{\varphi}(x_1, x_2, \ldots, x_{i-1}, t_i, x_{i+1}, \ldots, x_n)
$$

$$
= \frac{1}{(\pi)^{n-1}} \left[P \int_{\mathbb{R}^{n-1}} \frac{\varphi(y_1, y_2, \ldots, y_{i-1}, t_i, y_{i+1}, \ldots, y_n)}{\prod_{\substack{j=1 \\ j \neq i}}^{n} (x_j - y_j)} dy_1 \cdots dy_{i-1} dy_{i+1} \cdots dy_n \right]
$$

By successive application of Theorem 2 for $n = 1$, it follows that

$$
\overline{\varphi}(x_1, x_2, \ldots, x_{i-1}, t_i, x_{i+1}, \ldots, x_n) \in \mathcal{D}_{L^p}(\mathbb{R}^n)
$$

When $x_1, x_2, \ldots, x_{i-1}, x_{i+1}, \ldots, x_n$, are kept fixed, then it follows that

$$
\frac{\partial}{\partial x_i}(H\varphi)(x) = H_i \frac{\partial}{\partial t_i}(\overline{\varphi}(x_1, x_2, \ldots, x_{i-1}, t_i, x_{i+1}, \ldots, x_n) \tag{5.51}
$$

$$
= H_i H_1 H_2, \ldots, H_{i-1} H_{i+1} \ldots H_n \frac{\partial}{\partial t_i} \varphi(t_1, \ldots, t_n)
$$

$$
= H\left(\frac{\partial \varphi}{\partial t_i}\right), \qquad i = 1, 2, \ldots, n
$$

By successive application of this result, it can be shown that

$$
D^k(H\varphi)(x) = H(D^k \varphi)(x) \tag{5.52}
$$

Therefore, using (5.42), we have

$$
\|D^k(H\varphi)(x)\|_p = \|H(D^k \varphi)(x)\|_p \le C_p \|D^k \varphi\|_p
$$

Hence

$$
\varphi \in \mathcal{D}_{L^p}(\mathbb{R}^n) \Rightarrow H\varphi \in \mathcal{D}_{L^p}(\mathbb{R}^n) \tag{5.53}
$$

In view of the inversion formula (5.6), we have

$$
H\varphi = 0 \Rightarrow \varphi = 0 \tag{5.54}
$$

In other words, H is one to one.

The fact that H is onto follows by the same inversion formula; for if $\varphi \in \mathcal{D}_{L^p}(\mathbb{R}^n)$, we have

$$
H[(H\varphi)(-1)^n] = \varphi \tag{5.55}
$$

Note that $(-1)^n H\varphi \in \mathcal{D}_{L^p}(\mathbb{R}^n)$. Therefore H^{-1} exists, and using (5.46), we have

$$H^{-1} = (-1)^n H \qquad (5.56)$$

Since H is linear and continuous, in view of (5.56) H^{-1} is also linear and continuous. $\square$

5.5.5. The *n*-Dimensional Distributional Hilbert Transform

For $p > 1$, assume that $f \in \mathcal{D}_{L^p}(\mathbb{R}^n)$ and $g \in \mathcal{D}_{L^q}(\mathbb{R}^n)$, where $\frac{1}{p} + \frac{1}{q} = 1$. Then it is easy to show that

$$\int_{\mathbb{R}^n} (Hf)(x)g(x)\,dx = \int_{\mathbb{R}^n} f(x)(-1)^n (Hg)(x)\,dx \qquad (5.57)$$

In the adjoint notation (5.57) can written

$$\langle Hf, g \rangle = \langle f, (-1)^n Hg \rangle \qquad (5.58)$$

We are motivated by the equation (5.57) to define the Hilbert transform of distribution in *n*-dimension.

In conformity with the notation used by Laurent Schwartz, we will denote $\mathcal{D}'_{L^p}(\mathbb{R}^n)$, $p > 1$, or some time abbreviated as $\mathcal{D}'_{L^p}$ as the dual space of $\mathcal{D}_{L^q}(\mathbb{R}^n)$ where

$$\frac{1}{p} + \frac{1}{q} = 1$$

Definition. For $f \in \mathcal{D}'_{L^p}(\mathbb{R}^n)$, we define the *n*-dimensional Hilbert transform Hf of f as an element of $\mathcal{D}'_{L^p}(\mathbb{R}^n)$ satisfying

$$\langle Hf, \varphi \rangle = \langle f, (-1)^n H\varphi \rangle \qquad \forall \varphi \in \mathcal{D}_{L^q}(\mathbb{R}^n) \qquad (5.59)$$

$H\varphi$ in (5.59) stands for the classical *n*-dimensional Hilbert transform of φ.

It can be easily shown that the functional Hf defined by (5.59) is linear and continuous on $\mathcal{D}_{L^q}(\mathbb{R}^n)$.

Example 4. Find $H\delta$ where $\delta \in \mathcal{D}'_{L^p}(\mathbb{R}^n)$.

From the definition (5.59) we have

$$\langle H\delta, \varphi \rangle = \langle \delta, (-1)^n H\varphi \rangle$$

$$= \left\langle \delta, (-1)^n \frac{1}{\pi^n} P \int_{\mathbb{R}^n} \frac{\varphi(t)\,dt}{(x_1 - t_1)\cdots(x_n - t_n)} \right\rangle$$

$$= \frac{1}{\pi^n} \frac{(-1)^n}{(-1)^n} P \int_{\mathbb{R}^n} \frac{\varphi(t)\,dt}{t_1 t_2 \cdots t_n}$$

$$= \left\langle \frac{1}{\pi^n} p.v. \left(\frac{1}{t_1 t_2 \cdots t_n} \right), \varphi \right\rangle \qquad \forall \varphi \in \mathcal{D}_{L^p}(\mathbb{R}^n)$$

Therefore

$$H\delta = \frac{1}{\pi^n} p.v. \left(\frac{1}{t_1 t_2 \cdots t_n} \right) \triangleq \frac{1}{\pi^n} p.v. \left[\frac{1}{t} \right]. \tag{5.60}$$

Example 5. Find

$$H \left(p.v. \left[\frac{1}{t} \right] \right)$$

Operating both sides of (5.20) by H, we get

$$H^2\delta = \frac{1}{\pi^n} H \left(p.v. \left[\frac{1}{t} \right] \right)$$

Hence

$$H \left(p.v. \left[\frac{1}{t} \right] \right) = (-\pi)^n \delta$$

Since the operators $H_1, H_2, \ldots, H_n$ as defined in Section 5.5 are commutative, we see that

$$p.v. \left(\frac{1}{t_1 t_2 \cdots t_n} \right) = p.v. \left(\frac{1}{t_{i_1} t_{i_2} \cdots t_{i_n}} \right)$$

where $i_1, i_2, \ldots, i_n$ is a permutation of $1, 2, \ldots, n$.

5.5.6. Calculus on $\mathcal{D}'_{L^p}(\mathbb{R}^n)$

Let $f \in \mathcal{D}'_{L^p}(\mathbb{R}^n)$. Then the distributional differentiation of $\mathcal{D}'_{L^p}(\mathbb{R}^n)$ is defined as

$$\langle D^k f, \varphi \rangle = \langle f, (-1)^{|k|} D^k \varphi \rangle \quad \forall \varphi \in \mathcal{D}_{L^q}(\mathbb{R}^n), \ q = \frac{p}{p-1}, \ p > 1 \tag{5.61}$$

Now we prove the following:

Theorem 10. Let $f \in \mathcal{D}'_{L^p}(\mathbb{R}^n)$. Then

$$D^k H f = H D^k f$$

Proof.

$$\begin{aligned}
\langle D^k H f, \varphi \rangle &= \langle H f, (-1)^{|k|} D^k \varphi \rangle, \forall \varphi \in \mathcal{D}_{L^q}(\mathbb{R}^n) \\
&= \langle f, (-1)^n H (-1)^{|k|} D^k \varphi \rangle \\
&= \langle D^k f, (-1)^n H \varphi \rangle \\
&= \langle H D^k f, \varphi \rangle
\end{aligned}$$

Hence Theorem 10 is established. $\square$

Example 6. Solve in $\mathcal{D}'_{L^p}(\mathbb{R}^n)$ the operator equation

$$y = Hy + f \tag{5.62}$$

where $f \in \mathcal{D}'_{L^p}(\mathbb{R}^n)$, $n > 1$.

Solution. Operating both sides of (5.62) by H, and applying the Inversion Theorem 1, and using (5.22), we get

$$y[1 - (-1)^n] = f + Hf \tag{5.63}$$

For the case where n is odd,

$$(5.63) \Rightarrow y = \frac{Hf + f}{2} \tag{5.64}$$

For the case where n is even,

$$(5.63) \Rightarrow Hf = -f \tag{5.65}$$

Therefore solution to (5.62) does not exist if

$$Hf \neq -f \tag{5.66}$$

If $Hf = -f$ is satisfied, then there exists infinitely many solutions, and in this case

$$y = \frac{f}{2} \quad \text{is a solution to (5.62)}$$

If g_i are such that they satisfy

$$Hg_i = g_i \tag{5.67}$$

then

$$y = \frac{f}{2} + \sum_{i=1}^{m} c_i g_i \tag{5.68}$$

where c_i's are arbitrary constant, satisfy (5.65).

The fact that there exists nonzero solutions to $Hy = y$ (n even) follows easily, since

$$y = \varphi_1(y_1)\varphi_2(y_2) \cdots \varphi_n(y_n)$$
$$+ (H_1\varphi_1)(y_1)(H_2\varphi_2)(y_2) \cdots (H_n\varphi_n)(y_n),$$

where $\varphi_i \in \mathcal{D}$, satisfies $Hy = y$ when n is even, and

$$y = (H_1\varphi_1)(y_1) \cdots (H_n\varphi_n)(y_n) - \varphi_1(y_1) \cdots \varphi_n(y_n)$$

satisfies

$$Hy = -y$$

There do exists nonzero y's not satisfying

$$Hy = -y$$

when n is even.

As an example, if we choose

$$y = \prod_{i=1}^{n}(H_i\varphi_i)(y_i) + \prod_{i=1}^{n}\varphi_i(y_i)$$

where $\varphi_i \in \mathcal{D}$ such that $y \neq 0$, then it does not satisfy $Hy = -y$ when n is even. It is still an open problem to determine the whole class of solutions to

$$y = Hy + f$$

when $Hf = -f$ is satisfied for n even.

5.5.7. The Testing Function Space $H(\mathcal{D}(\mathbb{R}^n))$

A complex-valued C^∞ function φ defined on R^n belongs to the space $H(\mathcal{D}(\mathbb{R}^n))$ if and only if $\varphi(x)$ is the n-dimensional Hilbert transform of some $\psi(t)$ in $\mathcal{D}(\mathbb{R}^n)$. Hence $\varphi \in H(\mathcal{D}(\mathbb{R}^n)) \Leftrightarrow$ there exists $\psi(t)$ in $\mathcal{D}(\mathbb{R}^n)$ such that

$$\varphi(x) = \frac{1}{\pi^n} P \int_{\mathbb{R}^n} \frac{\psi(t)}{x - t} dt = H\psi \tag{5.69}$$

where the integral is being taken in the Cauchy principal value sense and $(x - t)$ in (5.69) is interpreted as

$$\prod_{i=1}^{n}(x_i - t_i)$$

The topology of $H(\mathcal{D}(\mathbb{R}^n))$ is the same as that transported from the space $\mathcal{D}(\mathbb{R}^n)$ to $H(\mathcal{D}(\mathbb{R}^n))$ by means of the Hilbert transform H. Therefore a sequence φ_n in $H(\mathcal{D}(\mathbb{R}^n))$ converges to zero in $H(\mathcal{D}(\mathbb{R}^n))$ if and only if its associated sequence ψ_n converges to zero in $\mathcal{D}(\mathbb{R}^n)$, where $H\psi_n = \varphi_n$ for all $n \in \mathbb{N}$.

Theorem 11. Let $H(\mathcal{D}(\mathbb{R}^n))$ and $\mathcal{D}_{L^p}(\mathbb{R}^n)$ be the spaces defined as before. Then

i. $H(\mathcal{D}(\mathbb{R}^n)) \subset \mathcal{D}_{L^p}(\mathbb{R}^n)$ and $H(\mathcal{D}(\mathbb{R}^n))$ is dense in $\mathcal{D}_{L^p}(\mathbb{R}^n)$.
ii. Convergence of a sequence in $H(\mathcal{D}(\mathbb{R}^n))$ implies its convergence in $\mathcal{D}_{L^p}(\mathbb{R}^n)$. Hence the restriction of any $f \in \mathcal{D}'_{L^p}(\mathbb{R}^n)$ to $H(\mathcal{D}(\mathbb{R}^n))$ is in $H'(\mathcal{D}(\mathbb{R}^n))$. Therefore

$$H'(\mathcal{D}(\mathbb{R}^n)) \supset \mathcal{D}'_{L^p}(\mathbb{R}^n)$$

Proof. (i) Since $\mathcal{D}(\mathbb{R}^n)$ is dense in $\mathcal{D}_{L^p}(\mathbb{R}^n)$ and

$$H : \mathcal{D}_{L^p}(\mathbb{R}^n) \xrightarrow{\text{onto}} \mathcal{D}_{L^p}(\mathbb{R}^n)$$

is a homeomorphism, we conclude that $H(\mathcal{D}(\mathbb{R}^n))$ is dense in $\mathcal{D}_{L^p}(\mathbb{R}^n)$. See also [57].

(ii) Let $\psi_j \to 0$ in $H(\mathcal{D}(\mathbb{R}^n))$. Then there exists a sequence $\psi_j \to 0$ in $\mathcal{D}(\mathbb{R}^n)$ as $j \to \infty$ such that $H\varphi_j = \psi_j$. Now using (5.42) and (5.44), we have

$$\|\psi_j^{(k)}\|_p \le C_p \|\psi_j^{(k)}\|_p \to 0 \qquad \text{as } j \to \infty \qquad \square$$

Remark In view of the Inversion Theorem 1,

$$H : H(\mathcal{D}(\mathbb{R}^n)) \to \mathcal{D}(\mathbb{R}^n)$$

is linear and continuous.

5.5.8. The *n*-Dimensional Generalized Hilbert Transform

The generalized Hilbert transform Hf of $f \in \mathcal{D}'(\mathbb{R}^n)$ is defined to be an ultradistribution $Hf \in H'(\mathcal{D}(\mathbb{R}^n))$ such that

$$\langle Hf, \varphi \rangle = \langle f, (-1)^n H\varphi \rangle \qquad \forall \varphi \in H(\mathcal{D}(\mathbb{R}^n)) \tag{5.70}$$

where $H\varphi$ is the classical Hilbert transform defined by (5.1). If $g \in H'(\mathcal{D}(\mathbb{R}^n))$, its Hilbert transform Hg is defined to be a Schwartz distribution by the relation

$$\langle Hg, \varphi \rangle = \langle g, (-1)^n H\varphi \rangle, \qquad \varphi \in \mathcal{D}(\mathbb{R}^n)$$

Let $g = Hf$ for some $f \in \mathcal{D}'(\mathbb{R}^n)$. Then

$$\langle H^2 f, \varphi \rangle = \langle Hf, (-1)^n H\varphi \rangle, \qquad \forall \varphi \in \mathcal{D}(\mathbb{R}^n) \tag{5.71}$$
$$= \langle f, H^2 \varphi \rangle$$
$$= \langle f, (-1)^n \varphi \rangle$$
$$\Rightarrow H^2 = (-1)^n I \text{ on } \mathcal{D}'(\mathbb{R}^n)$$

where I is the identity operator.

Definition. The derivation $D^k g$ of an ultradistribution $g \in H'(\mathcal{D}(\mathbb{R}^n))$ is defined as

$$\langle D^k g, \varphi \rangle = \langle g, (-1)^{|k|} D^k \varphi \rangle \tag{5.72}$$

for every $\varphi \in H(\mathcal{D}(\mathbb{R}^n))$.

Theorem 12. Let $f \in \mathcal{D}'(\mathbb{R}^n)$. Then

$$(Hf)^{(k)} = H(f^{(k)}) \tag{5.73}$$

Proof.

$$\langle D^k Hf, \varphi \rangle = \langle Hf, (-1)^{|k|} D^k \varphi \rangle \qquad \forall \varphi \in H(\mathcal{D}(\mathbb{R}^n))$$
$$= \langle f, (-1)^{|k|+n} HD^k \varphi \rangle$$
$$= \langle f, (-1)^{|k|+n} D^k H\varphi \rangle \qquad \text{(from (5.52))}$$
$$= \langle D^k f, (-1)^n H\varphi \rangle$$
$$= \langle HD^k f, \varphi \rangle \qquad \square$$

Example 7. Solve in $\mathcal{D}'(\mathbb{R}^n)$,

$$\frac{\partial y}{\partial x_1} + H \frac{\partial f}{\partial x_1} = \delta(x)$$

We rewrite the equation in the form

$$\frac{\partial}{\partial x_1}[y + Hf] = \delta(x) = \delta(x_1) * \delta(x_2) * \cdots * \delta(x_n)$$

Then

$$y + Hf = h(x_1) * \delta(x_2) * \cdots * \delta(x_n) + C(x_2, x_3, \ldots, x_n)$$

5.5.9. An Intrinsic Definition of the Space $H(\mathcal{D}(\mathbb{R}^n))$ and Its Topology

This section gives an intrinsic definition of the space $H(\mathcal{D}(\mathbb{R}^n))$ and its topology. We now have some lemmas to be used in the sequel.

Lemma 1. Let $\{\varphi_\nu\}_{\nu=1}^\infty$ be a sequence of functions tending to zero in $\mathcal{D}_{L^p}(\mathbb{R}^n)$ as $\nu \to \infty$:

$$\gamma_{(k)}(\varphi_\nu) \to 0 \qquad \text{as } \nu \to \infty \; \forall \, |k| \in \mathbb{N}$$

Then for each $|k| = 0, 1, 2, \ldots,$

$$\varphi_\nu^{(k)} \to 0 \qquad \text{as } \nu \to \infty \text{ uniformly } \forall x \in \mathbb{R}^n$$

Proof. The lemma is well known [87], but a very simple proof can be given as follows:

$$\varphi^{(k)}(x) = \langle \delta(t), \varphi^{(k)}(x - t) \rangle \qquad \forall \varphi \in \mathcal{D}_{L^p}(\mathbb{R}^n)$$

In view of the boundness property of generalized functions, there exists a constant $C > 0$, an $r = (r_1, r_2, \ldots, r_n)$, and $|r| = r_1 + r_2 + \cdots + r_n$ such that

$$|\varphi^{(k)}(x)| \leq C\gamma'_{|r|}(\varphi^{(k)}(x - t)) \qquad \text{[110, p. 8–19]}$$
$$\leq C\gamma'_{|r|}(\varphi^{(k)}(t))$$

where

$$\gamma'_{|0|} = \gamma_{|0|} \quad \text{and} \quad \gamma'_{|r|} = \max_{|j| \le |r|} \gamma_{(j)}$$

Therefore

$$|\varphi_\nu^{(k)}(x)| \le C\gamma'_{|r|}(\varphi_\nu^{(k)}(t)) \to 0 \qquad \text{as } \nu \to \infty$$

independently of x. $\square$

This proof makes use of the fact that $\delta(x) \in \mathcal{D}_{L^p}(\mathbb{R}^n)$, which amounts to proving Lemma 1. An independent proof can be given by Schwartz technique as developed in Chapter 7.

Lemma 2. Let $\varphi(t) \in \mathcal{D}(\mathbb{R}^n)$. Then as $y \to 0^+$ (i.e., $y_i \to 0^+$ for all $i = 1, 2, 3, \ldots, n$),

i. In $\mathcal{D}_{L^p}(\mathbb{R}^n)$, $p > 1$,

$$\frac{1}{\pi^n} \int_{\mathbb{R}^n} \varphi(t) \frac{y_1}{(t_1 - x_1)^2 + y_1^2} \frac{y_2}{(t_2 - x_2)^2 + y_2^2} \tag{5.74}$$
$$\cdots \frac{y_n}{(t_n - x_n)^2 + y_n^2} dt \to \varphi(x)$$

ii. In $\mathcal{D}_{L^p}(\mathbb{R}^n)$, $p > 1$,

$$\int_{\mathbb{R}^n} \varphi(t) \prod_{i=1}^n \left[\frac{(t_i - x_i)}{(t_i - x_i)^2 + y_i^2} \right] dt \to P \int_{\mathbb{R}^n} \frac{\varphi(t)}{\prod_{i=1}^n (t_i - x_i)} dt \tag{5.75}$$

iii. In $\mathcal{D}_{L^p}(\mathbb{R}^n)$, $m = 1, 2, \ldots, n$, $(p > 1)$,

$$\frac{1}{\pi^{n-m}} \int_{\mathbb{R}^n} \varphi(t) \prod_{i=1}^m \left[\frac{(t_i - x_i)}{(t_i - x_i)^2 + y_i^2} \right] \times \prod_{i=m+1}^n \left[\frac{y_i}{(t_i - x_i)^2 + y_i^2} \right] dt$$
$$\to (H_m \cdots H_3 H_2 H_1 \varphi)(x_1, x_2, \ldots, x_m, x_{m+1}, \ldots, x_n) \tag{5.76}$$

iv. In $\mathcal{D}_{L^p}(\mathbb{R}^n)$, $m = 1, 2, \ldots, n$,

$$\frac{1}{\pi^{n-m}} \int_{\mathbb{R}^n} \varphi(t) \prod_{i=1}^m \left[\frac{(t_{l_i} - x_{l_i})}{(t_{l_i} - x_{l_i})^2 + y_{l_i}^2} \right] \times \prod_{i=m+1}^n \left[\frac{y_{l_i}}{(t_{l_i} - x_{l_i})^2 + y_{l_i}^2} \right] dt$$
$$\to (H_{l_n} H_{l_2} \cdots H_{l_1} \varphi)(\cdots x_{l_1} \cdots x_{l_2} \cdots x_{l_n} \cdots) \tag{5.77}$$

Proof. (i) For the proof see [48, p. 400].

(ii) Denoting the left-hand-side expression in (5.77) by $\pi^n F(x)$, we see that

$$F^{(k)}(x) = \frac{1}{\pi^n} \int_{\mathbb{R}^n} \varphi^{(k)}(t) \prod_{i=1}^{n} \frac{(t_i - x_i)}{(t_i - x_i)^2 + y_i^2} dt$$

By successive application of Fubini's theorem and [99, thm. 101, p. 132], it follows that

$$\|F^{(k)}(x)\| \leq c_p^n \|\varphi^{(k)}(x)\|_p$$

where c_p is a constant independent of φ and $y_1, y_2, \ldots, y_n$.

Since the space $X(\mathbb{R}^n)$ is dense in $\mathcal{D}_{L^p}(\mathbb{R}^n)$, $p > 1$, it is easy to show that

$$\|F^{(k)}(x) - H\varphi^{(k)}(x)\|_p \to 0 \qquad \text{as } y_1, y_2, \ldots, y_n \to 0$$

A much more general result is proved in [48].

(iii) follows as a result of (i), and (iv) is only an elementary variation of (iii) and can be proved similarly. $\square$

Lemma 3. Let $z_i \in \mathbb{C}$ for $j = 1, 2, 3, \ldots, n$, where $z_j = x_j + iy_j$ and $x_i, y_i \in \mathbb{R}$. For $\varphi(t) \in \mathcal{D}(\mathbb{R}^n)$, define a function F as a mapping from $\mathbb{C}^n$ to $\mathbb{C}$ by

$$F(z) = \int_{\mathbb{R}^n} \frac{\varphi(t)}{\prod_{i=1}^{n}(t_i - z_i)} dt$$

if $y_i \neq 0$ for all $i = 1, 2, \ldots, n$, and

$$F(z_1, z_2, \ldots, z_{i-1}, x_i, z_{i+1}, \ldots, z_n) = \frac{1}{2}[F(z_1, z_2, \ldots, z_{i-1}, x_i^+, z_{i+1}, \ldots, z_n) \qquad (5.78)$$
$$+ F(z_1, z_2, \ldots, z_{i-1}, x_i^-, z_{i+1}, \ldots, z_n)]$$

if $y_i = 0$, for i, $1 \leq i \leq n$. Then $\lim_{y \to 0^+} F(z)$ converges uniformly to

$$\sum (i\pi)^{n-1} H_{j_1} H_{j_2} \cdots H_{j_l} \varphi \qquad \forall x \in \mathbb{R}^n$$

Proof. Since $z_j = x_j + iy_j$, $\forall j = 1, 2, \ldots, n$,

$$F(z) = \int_{\mathbb{R}^n} \varphi(t) \left[\prod_{i=1}^{n} \frac{(t_j - x_j) + iy_j}{(t_j - x_j)^2 + y_j^2} \right] dt$$

as $y \to 0^+$, in view of Lemma 2 (ii),

$$F(z) = \sum \int_{\mathbb{R}^n} \varphi(t) \left[\prod_{m=1}^{l} \frac{(t_{j_m} - x_{j_m})^+}{(t_{j_m} - x_{j_m})^2 + y_{j_m}^2} \right]$$
$$\times \left[\prod_{m=1}^{n-l} \frac{i^{n-1} y_{j'_m}}{(t_{j'_m} - x_{j'_m})^2 + y_{j'_m}^2} \right] dt$$

and the result follows in view of Lemma 2 (iv). Now, for our central problem of defining the space $H(\mathcal{D}(\mathbb{R}^n))$ intrinsically, we need the following:

Definition. A holomorphic function $\psi(z)$ defined on the complex n-space $\mathbb{C}^n$ belongs to the space Ψ if and only if the following properties hold:

P1. $\psi(z)$ is holomorphic outside the intervals $a_i \leq x_i \leq b_i$, $i = 1, 2, 3, \ldots, n$ (the interval depending upon $\psi(z)$).

P2. $\psi^{(k)}(z) = 0\left(\dfrac{1}{|z_1||z_2|\cdots|z_n|}\right)$, as $|z_i| \to \infty$, $i = 1, 2, \ldots, n$, for each fixed k satisfying $|k| = 0, 1, 2, 3, \ldots$.

P3. (a) For each fixed $|k| = 0, 1, 2, 3 \ldots$, $\psi^{(k)}(z)$ converges uniformly $\forall x \in \mathbb{R}^n$ as $y \to 0^+$.

(b) For each fixed $|k| = 0, 1, 2, 3, \ldots$, $\psi^{(k)}(z)$ converges uniformly $\forall x \in \mathbb{R}^n$ as $y \to 0^-$.

P4.

$$\psi(z_1, z_2, \ldots, z_{i-1}, x_i, z_{i+1}, \ldots, z_n)$$

$$= \frac{1}{2}[\psi(z_1, z_2, \ldots, z_{i-1}, x_i^+, z_{i+1}, \ldots, z_n)$$

$$+ \psi(z_1, z_2, \ldots, z_{i-1}, x_i^-, z_{i+1}, \ldots, z_n)], \qquad i = 1, 2, 3, \ldots, n$$

where

$$\psi(z_1, z_2, \cdots, z_{i-1}, x_i^{\pm}, z_{i+1}, \cdots, z_n) = \lim_{y_i \to 0^{\pm}} \psi(z_1, z_2, \cdots, z_i, \cdots, z_n)$$

Theorem 13. A necessary and sufficient condition that a function $\psi(z)$ defined on the complex n-space $\mathbf{C}^n$ belongs to the space Ψ is that there exists a $\varphi(t) \in \mathcal{D}(\mathbb{R}^n)$ satisfying

$$\psi(z) = \int_{\mathbb{R}^n} \frac{\varphi(t)}{\prod\limits_{i=1}^{n}(t_i - z_i)} dt \qquad \text{Im } z_i \neq 0, \ \forall i = 1, 2, 3, \ldots, n \qquad (5.79)$$

$$= p.v. \int_{\mathbb{R}^n} \frac{\varphi(t)}{(t_1 - z_1) \cdots (t_i - x_i) \cdots (t_n - z_n)} dt \qquad (5.80)$$

when $\text{Im } z_i = 0$ for some i, $1 \leq i \leq n$.

Proof. *Necessity:* If $\psi(z) \in \Psi$ in view of the properties **P1**, $\psi(z)$ as a function of $x \in \mathbb{R}^n$ is a member of $\mathcal{D}_{L^p}(\mathbb{R}^n)$ for a fixed $y \neq 0$ (i.e., for each component of $y \in \mathbb{R}^n$ nonzero). Now from **P1** and **P2** it follows that if $\{y_m\}_{m=1}^{\infty}$ is an arbitrary sequence in $\mathbb{R}^n$ such that $\|y_m\| \to 0$ as $m \to \infty$, then

$$\|\psi^{(k)}(x + iy_m) - \psi^{(k)}(x + iy_l)\|_p \to 0$$

as $l, m \to \infty$ independently of each other. Therefore $\{\psi(x + iy_m)\}_{m=1}^{\infty}$ is a Cauchy sequence in $\mathcal{D}_{L^p}(\mathbb{R}^n)$, $p > 1$. Since $\mathcal{D}_{L^p}(\mathbb{R}^n)$ is sequentially complete, there exists a function $\psi_+(x)$ in $\mathcal{D}_{L^p}(\mathbb{R}^n)$ such that

$$\lim_{m \to \infty} \psi(x + iy_m) = \psi_+(x) \qquad \text{in } \mathcal{D}_{L^p}(\mathbb{R}^n), \ p > 1$$

Since $\{y_m\}$ is an arbitrary sequence in $\mathbb{R}^n$ tending absolutely to zero, it follows that

$$\lim_{y \to 0^+} \psi(x + iy) = \psi_+(x) \qquad \text{in } \mathcal{D}_{L^p}(\mathbb{R}^n) \tag{5.81}$$

Similar arguments show the existence of a function $\psi_-(x)$ in $\mathcal{D}_{L^p}(\mathbb{R}^n)$ satisfying

$$\lim_{y \to 0^-} \psi(x + iy) = \psi_-(x) \qquad \text{in } \mathcal{D}_{L^p}(\mathbb{R}^n), \ p > 1 \tag{5.82}$$

and hence is the uniform limit [86] with respect to every $x \in \mathbb{R}^n$.

In quite a similar way it can be shown that

$$\psi(z_1, z_2, \ldots, z_{i-1}, x_i^{\pm}, z_{i+1}, \ldots, z_n) \in \mathcal{D}_{L^p}(\mathbb{R}^n)$$

for each fixed $z_j \in \mathbb{C}$, $1 \leq j \leq n$ and $j \neq i$. Therefore

$$\psi(z_1, z_2, \ldots, z_{i-1}, x_i, z_{i+1}, \ldots, z_n) \tag{5.83}$$

$$= \frac{1}{2}[\psi(z_1, z_2, \ldots, x_i^+, \ldots, z_n) + \psi(z_1, z_2, \ldots, x_i^-, \ldots, z_n)]$$

belongs to $\mathcal{D}_{L^p}(\mathbb{R}^n)$, $p > 1$ for fixed $y_1, y_2, \ldots, y_{i-1}, y_{i+1}, \ldots, y_n \neq 0$, where $y_j = \text{Im } z_j$, $1 \leq j \leq n$, $j \neq i$. Since $\psi(z)$ is analytic outside the interval $[a_i, b_i]$ on the X_i-axis,

$$\psi(z_1, z_2, \cdots, x_i^+, \cdots, z_n) - \psi(z_1, z_2, \cdots, x_i^-, \cdots, z_n)] = 0$$

outside $[a_i, b_i]$ on the X_i real line, $\forall i = 1, 2, \ldots, n$. Using Cauchy's integral theorem, it can be shown that

$$\frac{1}{2\pi i} \int_{-\infty}^{\infty} \frac{1}{t_t - z_j} \psi(z_1, z_2, \ldots, (t_j + i\epsilon_j), z_{j+1}, \ldots, z_n) \, dt_j$$

$$= \psi(z_1, z_2, \ldots, z_j + i\epsilon_j, \ldots, z_n), \qquad \text{Im } z_j > 0$$

$$= 0, \qquad \text{Im } z_j < 0 \tag{5.84}$$

Letting $\epsilon_j \to 0^+$ in (5.44), we have

$$\frac{1}{2\pi i} \int_{-\infty}^{\infty} \frac{1}{t_j - z_j} \psi(z_1, \ldots, z_{j-1}, t_j^+, \ldots, z_n) \, dt_j \tag{5.85}$$

$$= \psi(z_1, z_2, \ldots, z_j, \ldots, z_n), \qquad \text{Im } z_j > 0$$

$$= 0, \qquad \text{Im } z_j < 0$$

Similarly we show that

$$\frac{1}{2\pi i} \int_{-\infty}^{\infty} \frac{1}{t_j - z_j} \psi(z_1, z_2, \ldots, z_{j-1}, t_j^-, z_{j+1} \ldots, z_n) \, dt_j \tag{5.86}$$

$$= \psi(z_1, z_2, \ldots, z_j, \ldots, z_n), \qquad \operatorname{Im} z_j < 0$$

$$= 0, \qquad \operatorname{Im} z_j > 0$$

Therefore, combining (5.85) and (5.86), we get

$$\frac{1}{2\pi i} \int_{-\infty}^{\infty} \frac{1}{t_j - z_j} [\psi(z_1, z_2, \ldots, z_{j-1}, t_j^+, z_{j+1} \ldots, z_n) \tag{5.87}$$

$$- \psi(z_1, z_2, \ldots, z_{j-1}, t_j^-, z_{j+1} \ldots, z_n)] \, dt_j$$

$$= \psi(z_1, z_2, \ldots, z_n), \qquad \operatorname{Im} z_j \neq 0, \ 1 \le j \le n$$

In view of Lemmas 2 and 3 and **P4**, it follows that

$$\psi(z_1, z_2, \ldots, z_{j-1}, x_j, z_{j+1}, \ldots, z_n) \tag{5.88}$$

$$= \frac{1}{2\pi i} P \int_{-\infty}^{\infty} \frac{1}{t_j - x_j} [\psi(z_1, z_2, \ldots, z_{j-1}, t_j^+, z_{j+1}, \ldots, z_n)$$

$$- \psi(z_1, z_2, \ldots, z_{j_1}, t^-, z_{j+1}, \ldots, z_n)] \, dt_j$$

$$= p.v. \int_{-\infty}^{\infty} \frac{\theta(z_1, z_2, \ldots, z_{j-1}, t_j, z_{j+1}, \ldots, z_n)}{x_j - t_j} \, dt_j \tag{5.89}$$

where

$$-2\pi i \theta(z_1, z_2, \ldots, t_j, \ldots, z_n) \tag{5.90}$$

$$= \psi(z_1, z_2, \ldots, t_j^+, \ldots, z_n) - \psi(z_1, z_2, \ldots, t_j^-, \ldots, z_n)$$

Clearly $\theta(z_1, z_2, \ldots, t_j, \ldots, z_n) = 0$ when $t_j \notin [a_j, b_j]$. Exploiting Lemmas 2, 3, and **P4** once again, we can prove that

$$\psi(z_1, z_2, \ldots, z_{j-1}, x_j, z_{j+1}, \ldots, z_{l-1}, x_l, z_{l+1}, \ldots, z_n)$$

$$= p.v. \int_{-\infty}^{\infty} \int_{-\infty}^{\infty} \frac{\eta(z_\ell) \, dt_j \, dt_\ell}{(x_j - t_j)(x_\ell - t_\ell)z}$$

where $z_{j\ell} = (z_1, z_2, \ldots, z_n)$ with z_j and z_ℓ replaced by x_j and x_ℓ, respectively, for a suitable $\eta(z_1, z_2, \ldots, t_j, \ldots, t_l, \ldots, z_n)$ vanishing whenever $t_j \notin [a_j, b_j]$ and $t_l \notin [a_l, b_l]$. Using similar arguments, one can show that there exists $\varphi(x) \in \mathcal{D}(\mathbb{R}^n)$ with support contained in $a_i \le t_i \le b_i \ \forall i = 1, 2, \ldots, n$, such that

$$\psi(x_1, x_2, \ldots, x_n) = p.v. \int_{\mathbb{R}^n} \frac{\varphi(t)}{\prod_{i=1}^n (x_i - t_i)} \, dt \tag{5.91}$$

Now using (5.87) and repeating the technique of contour integration, and so on, as used in deducing (5.87), it can shown that there exists $\varphi(t) \in \mathcal{D}(\mathbb{R}^n)$ satisfying

$$\psi(z) = \int_{\mathbb{R}^n} \frac{\varphi(t)}{\prod_{i=1}^{n}(z_j - t_j)}\, dt \qquad (5.92)$$

when $\operatorname{Im} z_j \neq 0 \; \forall j$, $1 \leq j \leq n$. It can be seen during the course of derivation that φ's used in (5.91) and (5.92) are the same. This completes the proof of necessity.

Sufficiency: Assume that $\varphi(t) \in \mathcal{D}(\mathbb{R}^n)$, and define a function $\psi(z)$ and a mapping from $\mathbb{C}^n$ to $\mathbb{C}$ by relation

$$\psi(z) = \int_{\mathbb{R}^n} \frac{\varphi(t)}{\prod_{i=1}^{n}(t_j - z_j)}\, dt \qquad (5.93)$$

when $\operatorname{Im} z_j \neq 0 \; \forall j$, $1 \leq j \leq n$;

$$= p.v. \int_{\mathbb{R}^n} \frac{\psi(t)}{(t_1 - z_1) \cdots (t_{j-1} - z_{j-1}(t_j - x_j) \cdots (t_n - z_n)}\, dt \qquad (5.94)$$

when $\operatorname{Im} z_j = 0$ for some j, $1 \leq j \leq n$.

The support of $\varphi(t)$ is contained in $a_i \leq t_i \leq b_i$, $i = 1, 2, \ldots, n$. From (5.93) and (5.94), it follows quite easily that **P1, P2, P3,** and **P4** hold. $\square$

To demonstrate one-to-one correspondence between the space Ψ and $H(\mathcal{D}(\mathbb{R}^n))$, we can define the space $H(\mathcal{D}(\mathbb{R}^n))$ in genuinely intrinsic way as follows:

Theorem 14. A C^∞ function $\psi(x)$ defined on $\mathbb{R}^n$ is said to belong to the space $H(\mathcal{D}(\mathbb{R}^n))$ if and only if there exists a holomorphic function $\psi(z)$ defined on $\mathbb{C}^n$ satisfying **P1, P2, P3,** and **P4**. In other words, $\psi(x) \in H(\mathcal{D}(\mathbb{R}^n))$ if and only if $\psi(x)$ can be extended uniquely as a holomorphic function satisfying **P1, P2, P3,** and **P4**.

The convergence of a sequence $\{\psi_m(x)\}_{m=1}^{\infty}$ to zero in the space $H(\mathcal{D}(\mathbb{R}^n))$ can be defined in an intrinsic way as follows:

A sequence $\{\psi_m\}_{m=1}^{\infty}$ in $H(\mathcal{D}(\mathbb{R}^n))$ converges to zero in $H(\mathcal{D}(\mathbb{R}^n))$ if and only if

i. The associated functions $\psi_m(z)$ in accordance with Theorem 1 are analytic outside a closed n-box $\prod_{j=1}^{n}[a_j, b_j]$ of $\mathbb{R}^n$, or else $\psi_m(x)$ is analytic outside a fixed closed n-box $\prod_{j=1}^{n}[a_j, b_j]$.

ii. $\psi_m(x) \to 0$ in $\mathcal{D}_{L^p}(\mathbb{R}^n)$ as $m \to \infty$.

Proof. Clearly, if $\{\psi_m(x)\}_{m=1}^{\infty}$ is a sequence in $\mathcal{D}(\mathbb{R}^n)$ tending to zero in $\mathcal{D}(\mathbb{R}^n)$ as $m \to \infty$ and

$$\psi_m(x) = p.v. \int_{\mathbb{R}^n} \frac{\varphi_m(t)}{\prod_{j=1}^{n}(t_j - x_j)}\, dt \qquad (5.95)$$

$$\psi_m(z) = p.v. \int_{\mathbb{R}^n} \frac{\varphi_m(t)}{\prod_{j=1}^{n}(t_j - z_j)}\, dt, \qquad \operatorname{Im} z_i \neq 0 \; \forall i = 1, 2, \ldots, n$$

then $\psi_m(z)$ is analytic outside the closed intervals $a_j \leq x_j \leq b_j$, $j = 1, 2, \ldots, n$, and

$$D^k \psi_m(x) = p.v. \int_{\mathbb{R}^n} \frac{D_t^k \varphi_m(t)}{\prod_{j=1}^n (t_j - x_j)} \, dt$$

Therefore

$$\|D^k \psi_m(x)\|_p \leq c_p \|\psi_m^{(k)}\|_p \to 0 \qquad \text{as } m \to \infty.$$

Hence conditions i and ii are satisfied.

If however, conditions i and ii are assumed, then there exist closed intervals $a_j \leq t_j \leq b_j$ containing the supports of all

$$\varphi_m(x) = \left(-\frac{1}{\pi^2} \right)^n \int_{\mathbb{R}^n} \frac{\psi_m(x)}{\prod_{j=1}^n (t_j - x_j)} \, dt$$

Therefore

$$\|\varphi_m^{(k)}(x)\|_p \leq \frac{1}{\pi^{2n}} c_p \|\psi_m^{(k)}\|_p \to 0 \qquad \text{as } m \to \infty$$

That is, $\varphi_m(x) \to 0$ in $\mathcal{D}_{L^p}(\mathbb{R}^n)$ as $m \to \infty$. By Lemma 1, $\varphi_m(x) \to 0$ uniformly $\forall x \in \mathbb{R}^n$ as $m \to \infty$. By property i, all $\varphi_m(x)$ have support in a fixed n-box $\prod_{j=1}^n [a_j, b_j]$. Therefore, if $\psi_m(x) \to 0$ in $H(\mathcal{D}(\mathbb{R}^n))$ as $m \to \infty$, then the associated sequence $\{\varphi_m\}_{m=1}^\infty$ tends to zero in $\mathcal{D}(\mathbb{R}^n)$ as $m \to \infty$.

We have proved that $\varphi_m \to 0$ in $\mathcal{D}(\mathbb{R}^n)$ as $m \to \infty \Leftrightarrow \psi_m \to 0$ in $H(\mathcal{D}(\mathbb{R}^n))$ as $m \to \infty$. Thus the conditions i and ii together describe intrinsically the convergence of a sequence $\{\psi_m\}_{m=1}^\infty$ to zero in $H(\mathcal{D}(\mathbb{R}^n))$ as $m \to \infty$. $\square$

EXERCISES

1. (a) Prove that if $\varphi \in \mathcal{D}_{L^p}(\mathbb{R}^n)$, $p > 1$, then there exists a constant e_k depending upon $k = (k_1, k_2, \ldots, k_n)$ such that $|\varphi^{(k)}(t)| \leq e_k$ uniformly for all $t \in \mathbb{R}^n$.

   (b) Prove that $\varphi^{(k)}(t) \to 0$ as $|t| \to \infty$ for each fixed $k \in \mathbb{R}^n$.

   (c) Prove that the convergence of a sequence $\{\varphi_\nu(t)\}^\infty$ to zero in $\mathcal{D}_{L^p}(\mathbb{R}^n)$ implies the uniform convergence of the sequence $\{\varphi_\nu^{(k)}(t)\}_{\nu-1}^\infty$ to zero on $\mathbb{R}^n$.

   (d) Prove that a distribution of compact support on $\mathbb{R}^n$ is a subset of $(\mathcal{D}_{L^p}(\mathbb{R}^n))'$.

2. (a) Prove that the Schwartz testing function space $S(\mathbb{R}^n)$ is a proper subspace of the Schwartz testing function space $\mathcal{D}_{L^p}(\mathbb{R}^n)$, $p > 1$. *Hint:* Prove that the function $1 / \prod_{i=1}^n (1 + x_i^2)$ belongs to $\mathcal{D}_{L^p}(\mathbb{R}^n)$, $p > 1$ but not to the space $S(\mathbb{R}^n)$.

   (b) Prove that the space $\mathcal{D}(\mathbb{R}^n)$ is dense in the space $S(\mathbb{R}^n)$.

   (c) Prove that the space $S(\mathbb{R}^n)$ is dense in $\mathcal{D}_{L^p}(\mathbb{R}^n)$, $p > 1$.

   (d) Prove that the space $(\mathcal{D}_{L^p}(\mathbb{R}^n))' \subset S'(\mathbb{R}^n)$.

3. Solve the following boundary-value problem:

$$u_{xx} + u_{yy} + u_{zz} = 0, \qquad z > 0$$

$$\lim_{z \to 0^+} u(x, y, z) = p.v. \left(\frac{1}{xy} \right) \qquad \text{in } \left(\mathcal{D}_{L^p}(\mathbb{R}^3) \right)'$$

Hint: The Hilbert transform of $\pi \delta(t_1, t_2) = p.v. \left(\frac{1}{xy} \right)$.

6

FURTHER APPLICATIONS OF THE HILBERT TRANSFORM, THE HILBERT PROBLEM—A DISTRIBUTIONAL APPROACH

6.1. INTRODUCTION

In Chapter 2 we saw that a general Riemann-Hilbert problem can be reduced to finding a regular analytic function $F(z)$ in a given region, which is compliment of the curve C such that

$$F_+(x) - F_-(x) = f(x) \tag{6.1}$$

where $F_+(x)$, and $F_-(x)$ are defined as limits of $F(z)$ as $z \to x^+$ and $z \to x^-$, respectively, and x is a point on C. The concept of limits $z \to x^+$ and $z \to x^-$ are associated with the given curve C as discussed in Chapter 2. There is another equation,

$$F_+(x) + F_-(x) = \frac{1}{\pi i}(P) \int_C \frac{f(t)}{t - x} dt \tag{6.2}$$

that is associated with (6.1). The solution to (6.1) can also be obtained by solving (6.2), and vice versa. In fact (6.1) and (6.2) have common solutions for $F(z)$ under certain set of condition on $F(z)$. Writing $g(x) = \frac{1}{\pi i}(P) \int_C \frac{f(t)}{t-x} dt = i(Hf)(x)$, we can write the Riemann-Hilbert problem and the associated Riemann-Hilbert problem together as

$$F_+(x) - F_-(x) = f(x) \tag{6.3}$$

$$F_+(x) + F_-(x) = g(x) \tag{6.4}$$

where $g(x) = i(Hf)(x)$. So $f(x) = i(Hg)(x)$, a.e., where

$$(Hf)(x) = \frac{1}{\pi}(P)\int_C \frac{f(t)}{x - t}\, dt$$

More explicitly we have

$$g(x) = \frac{1}{\pi i}(P)\int_{-\infty}^{\infty} \frac{f(t)}{t - x}\, dt \qquad (6.5)$$

and

$$f(x) = \frac{1}{\pi i}(P)\int_{-\infty}^{\infty} \frac{g(t)}{t - x}\, dt \qquad (6.6)$$

Any one of the equations (6.3) and (6.4) can be called a *Riemann-Hilbert equation*, and the Riemann-Hilbert problem consists in solving any one of the equations (6.3) and (6.4) subject to a given set of conditions.

It is assumed that the given function f is such that its Hilbert transform exists a.e.; if $f \in L^p(R)$, $p > 1$, then

$$F(z) = \frac{1}{2\pi i}\int_{-\infty}^{\infty} \frac{f(t)}{t - z}\, dt, \qquad \text{Im } z \neq 0$$

where the limit $F_\pm(x)$ of $F(z)$ is interpreted in the $L^p(R)$ sense when $y \to 0^\pm$. When $f(t) \in \mathcal{D}'_{L^p}(R)$, then

$$F(z) = \frac{1}{2\pi i}\left\langle f(t), \frac{1}{t - z}\right\rangle$$

is solution to both (6.3) and (6.4). In this case $F_\pm(x)$ are interpreted as

$$F_\pm(x) = \lim_{y \to \pm 0} \frac{1}{2\pi i}\left\langle f(t), \frac{1}{t - x - iy}\right\rangle \qquad \text{in } \mathcal{D}'_{L^p}(R)$$

Also $g = iHf$ and $f = iHg$, $g, f, \in \mathcal{D}'_{L^p}(R)$. Now the problem arises to find an n-dimensional analogue of equations (6.3) and (6.4). That is, if $f \in L^p(R^n)$ or $\mathcal{D}'_{L^p}(R^n)$, we want to know what will be the corresponding analogues of equations (6.3) and (6.4).

For $n = 2$ we will show that an analogue of the equation (6.3) is

$$F_{++}(x_1, x_2) - F_{+-}(x_1, x_2) - F_{-+}(x_1, x_2) + F_{--}(x_1, x_2) = f(x_1, x_2) \qquad (6.7)$$

and that an analogue of (6.4) is

$$F_{++}(x_1, x_2) + F_{+-}(x_1, x_2) + F_{-+}(x_1, x_2) + F_{--}(x_1, x_2) = g(t_1, t_2) = i^2(Hf) \qquad (6.8)$$

The solution to (6.7) or (6.8) will be

$$F(z_1, z_2) = \left(\frac{1}{2\pi i}\right)^2 (P)\int_{-\infty}^{\infty}\int_{-\infty}^{\infty} \frac{f(t_1, t_2)}{(t_1 - z)(t_2 - z_2)}\, dt_1\, dt_2$$

if $f \in L^p(\mathbb{R}), p > 1$;

$$F_+(x) = \lim_{y \to 0^+} F(x + iy) \qquad \text{in } \mathcal{D}'_{L^p}(\mathbb{R})$$

If $f \in \mathcal{D}'_{L^p}(\mathbb{R}^2)$, then

$$F(z_1, z_2) = \left(\frac{1}{2\pi i}\right)^2 \left\langle f(t_1, t_2), \frac{1}{(t_1 - z_1)(t_2 - z_2)} \right\rangle$$

$\operatorname{Im} z_1 \neq 0$ and $\operatorname{Im} z_2 \neq 0$.

When $n = 3$, an analogue of (6.3) is the equation

$$F_{+++}(x_1 x_2 x_3) - F_{++-}(x_1, x_2, x_3)$$
$$+ F_{+--}(x_1 x_2 x_3) - F_{+-+}(x_1, x_2, x_3)$$
$$- F_{-++}(x_1, x_2, x_3) + F_{-+-}(x_1, x_2, x_3)$$
$$+ F_{--+}(x_1, x_2, x_3) - F_{---}(x_1, x_2, x_3)$$
$$= f \tag{6.9}$$

and that of (6.4) is

$$F_{+++}(x_1, x_2, x_3) + F_{++-}(x_1, x_2, x_3)$$
$$+ F_{+--}(x_1 x_2, x_3) + F_{+-+}(x_1, x_2, x_3)$$
$$+ F_{-++}(x_1, x_2, x_3) + F_{-+-}(x_1, x_2, x_3)$$
$$+ F_{-++}(x_1, x_2, x_3) + F_{-++}(x_1, x_2, x_3)$$
$$= g(x_1, x_2, x_3) \equiv i^3 Hf \tag{6.10}$$

Here

$$Hf = \frac{1}{\pi^3} (P) \int_{\mathbb{R}^3} \frac{f(t_1, t_2, t_3)}{(x_1 - t_1)(x_2 - t_2)(x_3 - t_3)} \, dt_1 \, dt_2 \, dt_3$$

If $f \in \mathcal{D}'_{L^p}(\mathbb{R}^3)$, then

$$F(z_1, z_2, z_3) = \frac{1}{(2\pi i)^3} \left\langle f(t_1, t_2, t_3), \frac{1}{(t_1 - z_1)(t_2 - z_2)(t_3 - z_3)} \right\rangle$$

$\operatorname{Im} z_i \neq 0, i = 1, 2, 3$.

The sign rule for the left-hand-side expression in (6.9) or (6.7) is as follows: The sign depends upon whether the number of $-$ signs attached in the subscript of F is odd or even. With odd many $-$ signs in the lower subscript of F, we attach negative sign, and with even many $-$ signs in the lower subscript, we attach $+$ signs. But in the equation (6.10) and (6.8) we have all positive signs attached. These results will be discussed in detail in Section 6.5.

In brief we can state that an analogue of the Hilbert problem,

$$F_+(x) + F_-(x) = g(x) \tag{6.11}$$

when $f \in L^p(\mathbb{R}^n)$ or $\mathcal{D}'_{L^p}(\mathbb{R}^n)$, is

$$\sum_{k=1}^{2^n} F_{\sigma_k}(x_1, x_2, \ldots, x_n) = g(x_1, x_2, \ldots, x_n) \tag{6.12}$$

In this expression the σ_k are n-dimensional vectors consisting of plus and minus signs. Singh [90] points out that the n-dimensional analogue (or extension) of the equation (6.11) is

$$\sum_{k=1}^{2^n} F_{\sigma_k}(-1)^k(x_1, x_2, \ldots, x_n) = g(x_1, x_2, \ldots, x_n) \tag{6.13}$$

In my view it is more appropriate to say that (6.12) is an extension or analogue of

$$F_+(x) + F_-(x) = g(x)$$

and (6.13) is an extension or analogue of

$$F_+(x) - F_-(x) = g(x)$$

However, our objective in this section is, as in [18], to discuss the solutions to the Hilbert problems of the type

$$n = 1 \quad F_+(x) + F_-(x) = f(x) \tag{6.14}$$

$$n = 2 \quad F_{++}(x_1, x_2) + F_{--}(x_1, x_2) = f(x_1, x_2) \tag{6.15}$$

$$n = 3 \quad F_{+++}(x_1, x_2, x_3) + F_{---}(x_1, x_2, x_3) = f(x_1, x_2, x_3) \tag{6.16}$$

which is quite different from the Hilbert problems.

It is this problem that was discussed in my paper [18], which I have here named a Hilbert problem. It is actually a misnomer to name problems described by (6.15) and (6.16) and [18] as Hilbert problems, but for our purposes it will suffice to follow this nomenclature.

Let $F(s)$ be a holomorphic function in the region $\text{Im } z \neq 0$ of the n-dimensional complex space $\mathbb{C}^n$. Assume that

$$F_+(x) = \lim_{y \to 0_+} F(z) \quad \text{in } \mathcal{D}'_{L^p}(\mathbb{R}^n) \tag{6.17}$$

and that

$$F_-(x) = \lim_{y \to 0_-} F(z) \quad \text{in } \mathcal{D}'_{L^p}(\mathbb{R}^n) \tag{6.18}$$

with

$$z = (z_1, z_2, \ldots, z_n) = (x_1 + iy_1, x_2 + iy_2, \ldots, x_n + iy_n)$$

Here $y \to 0_+$ means that $y_1 \to 0_+, y_2 \to 0_+, \ldots, y_n \to 0_+$ simultaneously, with a similar interpretation for $y \to 0_-$. $\operatorname{Im} z \neq 0$ means $\operatorname{Im} z_i \neq 0$ for $i = 1, 2, 3, \ldots, n$. We will consider the following Hilbert problem or Riemann-Hilbert problem: Let $f \in \mathcal{D}'_{L^p}(\mathbb{R}^n)$. We wish to find a function $F(z) = F(z_1, z_2, \ldots, z_n)$ holomorphic and regular in the region $\operatorname{Im} z_i \neq 0$ for all $i = 1, 2, \ldots, n$ such that

$$F_+(x) + F_-(x) = f(x) \tag{6.19}$$

where $F_+(x), F_-(x)$ are as defined in (6.17) and (6.18), respectively. The convergence in (6.17), (6.18), and the equality (6.19) are interpreted in the sense of $\mathcal{D}'_{L^p}(\mathbb{R}^n)$. We will show that in one dimension this Hilbert problem can always be solved while in higher dimensions a number of compatibility conditions must be satisfied by $f(x)$.

6.2. THE HILBERT PROBLEM

Given a function f on the real line satisfying certain prescribed conditions, we wish to find a holomorphic function $F(z)$ in the complex plane such that

$$F_+(x) + F_-(x) = f(x) \tag{6.20}$$

where

$$F_+(x) = \lim_{y \to 0_+} F(z), \qquad z = x + iy$$

and

$$F_-(x) = \lim_{y \to 0_-} F(z) \tag{6.21}$$

The mode of convergence may be suitably chosen. The solution to the problem in the classical sense is discussed in Chapter 2 and in the distributional sense is given in [77]. We will attempt to solve the n-dimensional Hilbert problem for the distribution space $\mathcal{D}'_{L^p}(\mathbb{R}^n)$.

Let $F(z)$ be a function defined on the one-dimensional complex plane which is holomorphic in the upper half-plane $\operatorname{Im} z > 0$ and also in the lower half-plane $\operatorname{Im} z < 0$ satisfying the following conditions:

1. $F(z) = o(1)$, $|y| \to \infty$ uniformly for every $x \in \mathbb{R}$.
2. $\sup_{x \in \mathbb{R}, \, y \geq \delta} |F(z)| \leq A_\delta < \infty$.
3. $F_+(x) = \lim_{z \to x_+} F(z) = \lim_{y \to 0^+} F(x + iy)$ in $\mathcal{D}'_{L^p}(\mathbb{R})$.
4. $F_-(x) = \lim_{z \to x_-} F(z) = \lim_{y \to 0^-} F(x + iy)$ in $\mathcal{D}'_{L^p}(\mathbb{R})$.

Then we have

$$F(z) = \frac{1}{(2\pi i)} \left\langle F_+(t) - F_-(t), \frac{1}{t - z} \right\rangle, \qquad \operatorname{Im} z \neq 0 \tag{6.22}$$

If we consider the convergence in $D'(\mathbb{R})$, then

$$F(z) = \frac{1}{(2\pi i)} \left\langle F_+(t) - F_-(t), \frac{1}{t-z} \right\rangle + P(z), \qquad \text{Im } z \neq 0$$

where $P(z)$ is a polynomial in z. From now on, we will consider the convergence in the space $L^p(\mathbb{R})$ only, for $p > 1$. Writing $g = F_+ - F_-$, we have

$$F(z) = \frac{1}{2\pi i} \left\langle g(t), \frac{t - x + iy}{(t-x)^2 + y^2} \right\rangle$$

Then we have

$$\lim_{y \to 0_+} F(z) = F_+(x) = \frac{1}{2i}[-Hg + iIg] \qquad (6.23)$$

and

$$\lim_{y \to 0_-} F(z) = F_-(x) = \frac{1}{2i}[-Hg - iIg] \qquad (6.24)$$

where I is the identity operator. A detailed proof of the identities (6.23) and (6.24) is given in [77]. Adding (6.23) and (6.24), we obtain

$$F_+(x) + F_-(x) = -\frac{1}{i}Hg = f \qquad (6.25)$$

Hence, using the inversion formula (3.4), we deduce

$$g = iHf$$

So the required function $F(z)$, holomorphic for $\text{Im } z \neq 0$, is given by

$$F(z) = \frac{1}{2\pi} \left\langle Hf, \frac{1}{t-z} \right\rangle, \qquad \text{Im } z \neq 0 \qquad (6.26)$$

We now extend the problem to $\mathcal{D}'_{L^p}(\mathbb{R}^2)$. Let $f \in \mathcal{D}'_{L^p}(\mathbb{R}^2)$ and let $F(z_1, z_2)$ be a function holomorphic in the region $\text{Im } z_1 \neq 0$, $\text{Im } z_2 \neq 0$ satisfying similar conditions as in the case of one dimension:

1. $F(z_1, z_2) = o(1)$ as $|y_1|, |y_2| \to \infty$.
2. $\sup_{\substack{|y_1| \geq \delta_1 > 0 \\ |y_2| \geq \delta_2 > 0}} |F(z_1, z_2) \leq A_\delta < \infty$, $\delta = (\delta_1, \delta_2)$.
3. $\lim_{y_1 \to 0_+, y_2 \to 0_+} F(z_1, z_2) = F_{++}(x_1, x_2)$,
   $\lim_{y_1 \to 0_+, y_2 \to 0_-} F(z_1, z_2) = F_{+-}(x_1, x_2)$,
   $\lim_{y_1 \to 0_-, y_2 \to 0_+} F(z_1, z_2) = F_{-+}(x_1, x_2)$, and
   $\lim_{y_1, y_2 \to 0_-} F(z_1, z_2) = F_{--}(x_1, x_2)$, in $\mathcal{D}'_{L^p}(\mathbb{R}^2)$, where

$$z_j = x_j + iy_j, \qquad j = 1, 2$$

Then we have

$$F(z_1, z_2) = \left(\frac{1}{2\pi i}\right)^2 \left\langle (F_{++} - F_{+-} - F_{-+} + F_{--})(t), \frac{1}{(t_1 - z_1)(t_2 - z_2)} \right\rangle \quad (6.27)$$

Writing $g = F_{++} - F_{+-} - F_{-+} + F_{--}$, we have

$$F(z_1, z_2) = \frac{1}{(2\pi i)^2} \left\langle g(t), \frac{1}{(t_1 - z_1)(t_2 - z_2)} \right\rangle$$

It was proved in [80] that

$$F_{++} = \frac{1}{(2i)^2}(-H_1 + iI_1)(-H_2 + iI_2)g,$$

where I_1, I_2 are the identity operators:

$$I_1 g(t_1, t_2) = g(x_1, t_2),$$
$$I_2 g(t_1, t_2) = g(t_1, x_2),$$

$$H_1(g(t_1, t_2)) = -\frac{1}{\pi} P \int_{\mathbb{R}} \frac{g(t_1, t_2)}{(t_1 - x_1)} dt_1$$

and

$$H_2(g(t_1, t_2)) = -\frac{1}{\pi} P \int_{\mathbb{R}} \frac{g(t_1, t_2)}{(t_2 - x_2)} dt_2$$

Similarly we have

$$F_{--} = \frac{1}{(2i)^2}(-H_1 - iI_1)(-H_2 - iI_2)g$$

Hence $f = F_{++} + F_{--}$ gives

$$-\frac{1}{2}[H_1 H_2 - I_1 I_2]g = f$$

That is,

$$(H - I)g = -2f \quad (6.28)$$

where $H = H_1 H_2$ and $I = I_1 I_2$ are the two-dimensional Hilbert transform and the identity operators on $\mathcal{D}'_{L^p}(\mathbb{R}^2)$, respectively. Using the inversion formula (5.46), we obtain

$$(I - H)g = -2Hf \quad (6.29)$$

Adding (6.28) and (6.29), we deduce that

$$f + Hf = 0 \quad (6.30)$$

Hence, if f does not satisfy (6.30), the solution of the aforesaid Hilbert problem does not exist. In [89] it was shown that there do exist functions satisfying (6.30). So let f satisfy (6.30), and let $g_1, g_2, \ldots, g_m$ in $L^p(\mathbb{R})$ be such that they satisfy

$$y - Hy = 0 \tag{6.31}$$

Then we have that

$$g = \sum_{j=1}^{m} c_j g_j + f \tag{6.32}$$

where c_j ($j = 1, \ldots, m$) are constants, satisfies (6.29). Substituting g for $F_{++} - F_{+-} + F_{-+} + F_{--}$ in (6.27), a class of solutions to the Hilbert problem is obtained.

Let us now consider the solution to the Hilbert problem in the next higher dimension. Let $F(z_1, z_2, z_3)$, where $z_j = x_j + iy_j$ ($j = 1, 2, 3$) be a function of z_1, z_2, z_3, which is analytic in the region

$$\{(z_1, z_2, z_3) : \operatorname{Im} z_1 \neq 0, \operatorname{Im} z_2 \neq 0, \operatorname{Im} z_3 \neq 0\}$$

of $\mathbb{C}^3$ and satisfies the following conditions:

1. $|F(z_1, z_2, z_3)| = o(1)$ as $|y_1|, |y_2|, |y_3| \to \infty$, the asymptotic order being valid uniformly $\forall x_1, x_2, x_3 \in \mathbb{R}^n$.
2. $\lim_{\substack{y_1 \to 0_\pm \\ y_2 \to \pm \\ y_3 \to \pm}} F(z_1, z_2, z_3) = F_{\pm\pm\pm}$ in $L^p(\mathbb{R})$.
3. $\sup_{\substack{|y_1| \geq \delta_1 > 0 \\ |y_2| \geq \delta_2 > 0 \\ |y_3| \geq \delta_3 > 0}} |F(z_1, z_2, z_3)| = A_\delta < \infty$, where $\delta = (\delta_1, \delta_2, \delta_3)$.

Now in view of the results proved in [89], there exists $g \in L^p(\mathbb{R})$ such that

$$F(z_1, z_2, z_3) = \frac{1}{(2\pi i)^3} \left\langle g(t), \frac{1}{\prod_{i=1}^{3}(t_1 - z_i)} \right\rangle$$

Therefore using results in [89], we obtain

$$F_{+++} = \frac{1}{(2i)^3}(-H_1 + il_1)(-H_2 + il_2)(-H_3 + il_3)g$$

and

$$F_{---} = \frac{-1}{(2i)^3}(H_1 + il_1)(H_2 + il_2)(H_3 + il_3)$$

so that

$$f = F_{+++} + F_{---} = -\frac{2}{(2i)^3}(H_1 H_2 H_3 + H_1 + H_2 + H_3)g$$

That is,

$$-4fi = (+H + H_1 + H_2 + H_3)g \tag{6.33}$$

Applying the operation $(H + H_1 + H_2 + H_3)$ to both sides of (6.33), we deduce

$$\begin{aligned}
-4i = (H + H_1 + H_2 + H_3)f &= [H^2 - (H_1 + H_2 + H_3)^2]g \\
&= [-1 - (H_1^2 + H_2^2 + H_3^2 + 2H_1H_2 + 2H_1H_3 + 2H_2H_3)]g \\
&= [-1 + 3 - 2(H_1H_2 + H_2H_3 + H_1H_3)]g \\
&= [2 + 2H(H_1 + H_2 + H_3)]g \tag{6.34}
\end{aligned}$$

Applying the operators $2H$ to both sides of (6.33) and adding the result to (6.34), we obtain

$$-4iHf - 4iH(H + H_1 + H_2 + H_3)f = 2H^2g + 2g = 0$$

or

$$f + (H + H_1 + H_2 + H_3)f = 0 \tag{6.35}$$

If the given f satisfies (6.35), then and only then a solution to the Hilbert problem exists. If f satisfies (6.35), then the solution to the Hilbert problem can be obtained by solving for g from (6.33) and substituting in the expression for $F(z_1, z_2, z_3)$. As we go to higher and higher dimensions, the problem becomes more and more difficult. We leave this as an open problem.

6.3. THE FOURIER TRANSFORM AND THE HILBERT TRANSFORM

It is known that $(\mathcal{F}Hf)(\xi) = i\,\text{sgn}(\xi)\mathcal{F}f(\xi)$, $\forall f \in L^2(\mathbb{R})$, where $\mathcal{F}$ and H are the operators of the classical Fourier and the Hilbert transformation, respectively. But such a result is not true in general $\forall f \in L^p(\mathbb{R})$, $p > 1$, $p \neq 2$, in the classical sense. The object of this section is to show that $\mathcal{F}Hf = i^n \prod_{i=1}^n \text{sgn}(\xi_i)(\mathcal{F}f)(\xi)$ for all Schwartz generalized functions f belonging to the dual space of $\mathcal{D}_{L^p}(\mathbb{R}^n)$, $p > 1$, with respect to the weak topology of the space $S_0'(\mathbb{R}^n)$. The space $S_0(\mathbb{R}^n)$ is a proper closed subspace of $S(\mathbb{R}^n)$, the Schwartz test function space of functions of rapid descent, equipped with the topology induced on it by that of $S(\mathbb{R}^n)$. In particular, the result is also true $\forall f \in L^p(\mathbb{R}^n)$, $p > 1$. These results are especially important because they are used in proving the fact that a bounded linear operator T from $L^p(\mathbb{R}^n)$ into itself, which commutes with both translations as well as dilatations, is a finite linear combination of the Hilbert-type transform and the identity operator.

In the sequel we will use the definition for the Fourier and the Hilbert transforms of functions f and g as

$$(\mathcal{F}f)(\xi) = \int_{-\infty}^{\infty} f(t)e^{it\xi}\, dt \tag{6.36}$$

and

$$(Hg)(x) = \frac{1}{\pi} P \int_{-\infty}^{\infty} \frac{g(t)}{x-t} \, dt \qquad (6.37)$$

respectively, provided that the integrals (6.36) and (6.37) exist. It is also known that

$$(\mathcal{F}(Hf))(\xi) = i \, \text{sgn}(\xi)(\mathcal{F}f)(\xi) \qquad (6.38)$$

$\forall f \in L^2(R)$, where

$$(\mathcal{F}f)(\xi) = \underset{N \to \infty}{\text{l.i.m}} \int_{-N}^{N} f(t)e^{i\xi t} \, dt$$

By using Fubini's theorem, we can easily show that

$$(\mathcal{F}(H\varphi))(\xi) = i \, \text{sgn}(\xi)(\mathcal{F}\varphi)(\xi) \qquad (6.39)$$

$\forall \varphi \in \mathcal{D}(\mathbb{R})$, where $\mathcal{D}(\mathbb{R})$ is the Schwartz test function space consisting of C^{∞} functions with compact supports. Since $\mathcal{D}(\mathbb{R})$ is dense in $L^2(\mathbb{R})$ and $\mathcal{F}$ and H are bounded linear operators from $L^2(\mathbb{R})$ into itself the result (6.38) follows from (6.39). But the result (6.38) is not true in general for $f \in L^p(\mathbb{R})$, $p > 1$, $p \neq 2$. In this note our object is to extend the result (6.38) for elements of $(\mathcal{D}_{L^p}(\mathbb{R}^n))'$ in the weak distributional sense:

$$\langle \mathcal{F}Hf, \varphi \rangle = \langle i \, \text{sgn}(x)(\mathcal{F}f)(x), \varphi(x) \rangle \qquad \forall \varphi \in S_0(\mathbb{R}^n)$$

where $S_0(\mathbb{R}^n)$ is a test function space of functions of rapid descent, namely those $C^{\infty}(\mathbb{R}^n)$ functions that vanish at the origin along with their derivative of all orders and for which

$$\sup_{t \in \mathbb{R}^n} \left| t_1^{m_1} t_2^{m_2} \cdots t_n^{m_n} \frac{\partial^{k_1 + k_2 + \cdots + k_n}}{\partial t_1^{k_1} \partial t_2^{k_2} \cdots \partial t_n^{k_n}} \varphi(t) \right| < \infty$$

for all nonnegative integers $m_1, m_2, \ldots, m_n, k_1, k_2, \ldots, k_n$.

Clearly $S_0(\mathbb{R}^n) \subset S(\mathbb{R}^n)$. For $x \in \mathbb{R}^n$ we define $\text{sgn}(x) = \prod_{i=1}^n \text{sgn}(x_i)$. The topology of $S_0(\mathbb{R}^n)$ is the same as that induced on $S_0(\mathbb{R}^n)$ by $S(\mathbb{R}^n)$ [110, p. 8]. We can see that the space $S_0(\mathbb{R}^n)$ is nonempty for a function φ defined by

$$\varphi(x) = \prod_{i=1}^{n} \varphi_i(x_i)$$

where

$$\varphi_i(x_i) = \begin{cases} e^{-(1/x_i^2) - x_i^2} & \text{when } x_i \neq 0 \\ 0 & \text{when } x_i = 0 \end{cases}$$

belongs to $S_0(\mathbb{R}^n)$.

6.4. DEFINITIONS AND PRELIMINARIES

Unless otherwise stated, $p > 1$ and the testing function space $\mathcal{D}_{L^p}(\mathbb{R}^n)$ consist of complex-valued C^∞ functions defined on $\mathbb{R}^n$ such that

$$\|\varphi^{(k)}\|_p = \left(\int_{\mathbb{R}^n} \left| \frac{\partial^{k_1}}{\partial x_1} \frac{\partial^{k_2}}{\partial x_2} \cdots \frac{\partial^{k_n}}{\partial x_n} \varphi(x) \right|^p dx \right)^{1/p} < \infty$$

for all nonnegative integers $k_1, k_2, \ldots, k_n$. We define $k = (k_1, k_2, \ldots, k_n)$ and $|k| = k_1 + k_2 + \cdots + k_n$.

The topology over $\mathcal{D}_{L^p}(\mathbb{R}^n)$ is defined by the sequence of seminorms $\{\gamma_m\}_{m=0}^\infty$ [67, pp. 169–170], where

$$\gamma_m(\varphi) = \left[\sum_{|k| \leq m} \|\varphi^{(k)}\|_p^p \right]^{1/p}$$

The test function space $S(\mathbb{R}^n)$ consists of C^∞ functions $\varphi(x)$ defined on $\mathbb{R}^n$ such that

$$\sup_t \left| t^m \varphi^{(k)}(t) \right| \equiv \sup_t \left| t_1^{m_1} y_2^{m_2} \cdots t_n^{m_n} \frac{\partial^{k_1}}{\partial t_1^{k_1}} \frac{\partial^{k_2}}{\partial t_2^{k_2}} \cdots \frac{\partial^{k_n}}{\partial t_n^{k_n}} \varphi(t) \right| < \infty$$

The topology on $S(\mathbb{R}^n)$ is defined by the sequence of seminorms $\{\gamma_{m,k}\}$, where

$$\gamma_{|m|,|k|}(\varphi) = \sup_t \left| t^m \varphi^{(k)}(t) \right|$$

Clearly $S_0(\mathbb{R}^n) \subset S(\mathbb{R}^n) \subset \mathcal{D}_{L^p}(\mathbb{R}^n)$ [67, pp. 106–107].

It is a well-known fact that $\mathcal{D}(\mathbb{R}^n)$ is dense in $\mathcal{D}_{L^p}(\mathbb{R}^n)$, and since $\mathcal{D}(\mathbb{R}^n) \subset S(\mathbb{R}^n) \subset \mathcal{D}_{L^p}(\mathbb{R}^n)$, it follows that $S(\mathbb{R}^n)$ is also dense in $\mathcal{D}_{L^p}(\mathbb{R}^n)$. If a sequence $\{\varphi_\nu\}$ in $S(\mathbb{R}^n)$ converges to zero as $\nu \to \infty$, then it also converges to zero in $\mathcal{D}_{L^p}(\mathbb{R}^n)$. As such

$$S'(\mathbb{R}^n) \supset (\mathcal{D}_{L^p}(\mathbb{R}^n))'$$

Each element of $(\mathcal{D}_{L^p}(\mathbb{R}^n))'$ can be identified by a unique element of $S'(\mathbb{R}^n)$. Therefore the Fourier transform $\mathcal{F}f$ of $f \in (\mathcal{D}_{L^p}(\mathbb{R}^n))'$ can be defined as a functional on $S(\mathbb{R}^n)$ by the relation

$$\langle \mathcal{F}f, \varphi \rangle = \langle f, \mathcal{F}\varphi \rangle \qquad \forall \varphi \in S(\mathbb{R}^n) \tag{6.40}$$

Clearly $\mathcal{F}f \in S'(\mathbb{R}^n)$ [67, p. 118].

The Hilbert transform of $f \in (\mathcal{D}_{L^p}(\mathbb{R}^n))'$ is defined as a functional on $\mathcal{D}_{L^p}(\mathbb{R}^n)$ by the relation

$$\langle Hf, \varphi \rangle = \langle f, (-1)^n H\varphi \rangle \qquad \forall \varphi \in \mathcal{D}_{L^p}(\mathbb{R}^n) \tag{6.41}$$

Clearly $Hf \in (\mathcal{D}_{L^p}(\mathbb{R}^n))'$ [89, p. 248].

6.5. THE ACTION OF THE FOURIER TRANSFORM ON THE HILBERT TRANSFORM, AND VICE VERSA

For $f \in (\mathcal{D}_{L^p}(\mathbb{R}^n))'$ we can see that $Hf \in (\mathcal{D}_{L^p}(\mathbb{R}^n))'$ as defined by (6.41) and that $\mathcal{F}(Hf) \in S'(\mathbb{R}^n)$ as defined by (6.40). Consequently

$$\langle \mathcal{F}(Hf), \varphi \rangle = \langle Hf, \mathcal{F}\varphi \rangle \qquad \forall \varphi \in S(\mathbb{R}^n) \tag{6.42}$$

Therefore

$$\langle \mathcal{F}Hf, \varphi, \rangle = \langle f, (-1)^n H \mathcal{F}\varphi \rangle \qquad \forall \varphi \in S(\mathbb{R}^n) \tag{6.43}$$

The right-hand-side expression of (6.42) is meaningful as

$$\mathcal{F}\varphi \in S \subset \mathcal{D}_{L^p}(\mathbb{R}^n) \qquad \forall \varphi \in S(\mathbb{R}^n)$$

Step (6.42) is now justified. Step (6.43) is also justified as $H\mathcal{F}\varphi \in \mathcal{D}_{L^p}(\mathbb{R}^n) \,\forall \varphi \in S(\mathbb{R}^n)$.

The result

$$H\mathcal{F}\varphi = (-i)^n \mathcal{F}\left(\prod_{i=1}^n \mathrm{sgn}(x_i)\varphi(x) \right) \qquad \forall \varphi \in S(\mathbb{R}^n) \tag{6.44}$$

$$= (-i)^n \mathcal{F}(\mathrm{sgn}(x)\varphi(x)) \qquad \forall \varphi \in S(\mathbb{R}^n)$$

can be justified by using Fubini's theorem. Therefore

$$\langle \mathcal{F}Hf, \varphi \rangle = \langle f, (-i)^n \, \mathrm{sgn}(x)\varphi(x) \rangle \qquad \forall \varphi \in S(\mathbb{R}^n)$$

The space $S_0(\mathbb{R}^n)$ is closed with respect to multiplication by $\mathrm{sgn}(x)$. Hence

$$\mathcal{F}(\mathrm{sgn}(x)\varphi(x)) \in S(\mathbb{R}^n) \qquad \forall \varphi \in S_0(\mathbb{R}^n)$$

Now we have

$$\langle \mathcal{F}Hf, \varphi \rangle = \langle i^n \, \mathrm{sgn}\, x \mathcal{F}f, \varphi \rangle \qquad \forall \varphi \in S_0(\mathbb{R}^n)$$

That is, $\mathcal{F}Hf = i^n \, \mathrm{sgn}(x)\mathcal{F}f$ in the weak topology of $S_0'(\mathbb{R}^n)$. In the same way it can be shown that

$$H\mathcal{F}f = (-i)^n \mathcal{F}(\mathrm{sgn}(x)f)$$

in the weak topology of $(\mathcal{F}(S_0(\mathbb{R}^n)))'$.

In particular, for $f \in L^p(\mathbb{R}^n)$ we have

$$\mathcal{F}Hf = i^n \, \mathrm{sgn}(x)\mathcal{F}f \qquad \text{in } S_0'(\mathbb{R}^n)$$

$$H\mathcal{F}f = (-i)^n \mathcal{F}(\mathrm{sgn}(x)f) \qquad \text{in } (\mathcal{F}(S_0(\mathbb{R}^n)))$$

We have therefore proved the following:

Theorem 1. Let $f \in (\mathcal{D}_{L^p}(\mathbb{R}^n))'$ and $\mathcal{F}$ and H be the operator of the Fourier and the Hilbert transformation. Then

   i. $\mathcal{F}Hf = i^n \operatorname{sgn}(x)\mathcal{F}f$ in the weak topology of the space $S_0'(\mathbb{R}^n)$.
   ii. $H\mathcal{F}f = (-i)^n \mathcal{F}(\operatorname{sgn}(x)f) \ \forall f \in (\mathcal{D}_{L^p}(\mathbb{R}^n))'$ in the weak topology of the space $(\mathcal{F}(S_0(\mathbb{R}^n)))'$.

6.6. CHARACTERIZATION OF THE SPACE $\mathcal{F}(S_0(\mathbb{R}^n))$

Proposition. $\varphi(t) \in \mathcal{F}(S_0(\mathbb{R}^n)) \iff \int_{\mathbb{R}_n} t^k \varphi(t) \, dt = 0$, and $\varphi(t) \in S(\mathbb{R}^n)$ $\forall |k| \geq 0$.

Proof. If $\varphi(t) \in \mathcal{F}(S_0(\mathbb{R}^n))$, there exists a function $\psi \in S_0(\mathbb{R}^n)$ satisfying

$$\varphi(t) = \int_{\mathbb{R}^n} \psi(x) e^{ix \cdot t} \, dt$$

Therefore

$$\int_{\mathbb{R}^n} t^k \, \varphi(t) \, dt = \int_{\mathbb{R}^n} \int_{\mathbb{R}^n} \psi(x) t^k e^{ix \cdot t} \, dt \, dx$$

$$= \int_{\mathbb{R}^n} \int_{\mathbb{R}^n} \psi(x) t^k e^{ix \cdot t} \, dx \, dt$$

$$= i^k \int_{\mathbb{R}^n} \int_{\mathbb{R}^n} \psi^{(k)}(x) e^{ix \cdot t} \, dx \, dt$$

$$= i^k \int_{\mathbb{R}_n} \left(\int_{\mathbb{R}^n} \psi^{(k)}(x) e^{ix \cdot t} \, dx \right) e^{it \cdot 0} \, dt$$

Hence

$$\int_{\mathbb{R}^n} t^k \varphi(t) \, dt = i^k (2\pi)^n \psi^{(k)}(0) \qquad \forall |k| \geq 0$$

The result follows. $\square$

   Let $B(L^p(\mathbb{R}^n))$ stand for the space of all bounded linear mappings from $L^p(\mathbb{R}^n)$ into itself. We say that the operator $T \in B(L^p(\mathbb{R}^n))$ commutes with translations if $\tau_a T = T \tau_a$ for all $a \in \mathbb{R}^n$. We define the operator of translation τ_a by

$$\tau_a f(x) = f(x_1 - a_1, x_2 - a_2, \ldots, x_n - a_n)$$

The operator $T \in B(L^p(\mathbb{R}^n))$ is said to commute with dilatations D_m if

$$T D_m = D_m T$$

where

$$D_m f(x) = \left(\prod_{i=1}^{n} m_i\right)^{-1/p} f\left(\frac{x_1}{m_1}, \frac{x_2}{m_2}, \ldots, \frac{x_n}{m_n}\right)$$

In the above expression $m_1, m_2, \ldots, m_n$ are all > 0. One can also see that

$$\|T_a f\|_p = \|f\|_p \quad \text{and} \quad \|D_m f\|_p = \|f\|_p \qquad \forall f \in L^p(\mathbb{R}^n)$$

Using the part i of Theorem 1, the following theorem is proved in [46, 69].

Theorem 2. Let $1 < p < \infty$, and let $T \in B(L^p(\mathbb{R}^n))$. Suppose that T commutes both with translations and with dilatations. Then there exist constants $a, a_1, a_2, \ldots,$ $a_{ij}, \ldots, b$ such that

$$T = aI + \sum_{i=1}^{n} a_i + H_i + \sum_{\substack{i,j=1 \\ i<j}}^{n} a_{ij} H_i H_j + \cdots + bH$$

where I is the identity operator on $L^p(\mathbb{R}^n)$. Here

$$(H_i f)(t_1, \ldots, t_2, \ldots, x_i, \ldots, t_n) = \frac{1}{\pi} P \int_{-\infty}^{\infty} \frac{f(t_1, t_2, \ldots, t_i, \ldots, t_n)}{x_i - t_i} dt$$

and

$$Hf(x) = \frac{1}{\pi^n} P \int_{\mathbb{R}^n} \frac{f(t)dt}{(x_1 - t_1)(x_2 - t_2) \ldots (x_n - t_n)}$$
$$= (H_1 H_2 \ldots H_n f)(x)$$

6.7. THE p-NORM OF THE TRUNCATED HILBERT TRANSFORM

It is shown that a bounded linear operator T from $L^p(\mathbb{R}^n)$ to itself that commutes both with translations and dilatations is a finite linear combination of Hilbert-type transforms. Using this, we show that the p-norm of the Hilbert transform is the same as the p-norm of its truncation to any Lebesgue measurable subset of $\mathbb{R}^n$ with nonzero measure.

Preliminaries: Recall that for a function $f(x)$ defined on the real line, the Hilbert transform $(Hf)(x)$ is given by the Cauchy principal value

$$(Hf)(x) = \frac{1}{\pi}(P) \int_{\mathbb{R}} \frac{f(t)}{x - t} dt \tag{6.45}$$

One of the fundamental results in the subject is that $(Hf)(x)$ exists for almost every x if $f \in L^p(\mathbb{R})$, $1 \le p < \infty$, $H : L^p(\mathbb{R}) \to L^p(\mathbb{R})$ is both continuous and linear, and that

$$\|Hf\|_p \le C_p\|f\|_p \qquad \text{for } 1 < p < \infty \qquad (6.46)$$

where C_p is a constant independent of f [99].

An n-dimensional Hilbert transform $(Hf)(x)$ for $f \in L^p(\mathbb{R}^n)$, $p > 1$, may also be defined as

$$(Hf)(x) = \frac{1}{\pi^n}(P) \int_{\mathbb{R}^n} \frac{f(t)}{\prod_{i=1}^n (x_i - t_i)} \, dt \qquad (6.47)$$

$$= \lim_{\varepsilon \to 0} \frac{1}{\pi^n} \int_{\substack{|x_i - t_i| > \varepsilon_i > 0 \\ i=1,2,\dots,n}} \frac{f(t)}{\prod_{i=1}^n (x_i - t_i)} \, dt$$

where $\varepsilon = \sqrt{\varepsilon_1^2 + \varepsilon_2^2 + \cdots + \varepsilon_n^2}$, $t = (t_1, t_2, \dots, t_n)$ and $dt = dt_1 dt_2 \cdots dt_n$. The existence of the singular integral in (6.1.3) and its boundedness property

$$\|Hf\|_p \le C_p^n \|f\|_p \qquad (6.48)$$

were proved by Kokilashvili [57]. Singh and Pandey [89] extended the n-dimensional Hilbert transform to the Schwartz distribution space $\mathcal{D}'(\mathbb{R}^n)$ [87] and proved that H is an automorphism on the distribution space $\mathcal{D}'_{L_p}(\mathbb{R}^n)$, $p > 1$ [87]. They also obtained the following inversion formula:

$$(H^2 f)(x) = (-1)^n f(x) \qquad \text{almost everywhere} \qquad (6.49)$$

for $f \in L^p(\mathbb{R}^n)$. The inversion formula (6.49) is a generalization of the corresponding one-dimensional result proved by Riesz; see Titchmarsh [89].

In 1972 Fefferman showed the iterative nature of the double Hilbert transform [41]. In 1989 Singh and Pandey [89] showed the iterative nature of the n-dimensional Hilbert transform over the spaces $L^p(\mathbb{R}^n)$ and $\mathcal{D}'_{L_p}(\mathbb{R}^n)$, $p > 1$. It was shown that

$$H = \prod_{i=1}^n H_i \qquad (6.50)$$

where

$$(H_i f)(t_1, \dots, t_{i-1}, x_i, t_{i+1}, \dots, t_n) = \frac{1}{\pi} \int_{\mathbb{R}} \frac{f(t_1, \dots, t_i, \dots, t_n)}{x_i - t_i} \, dt_i$$

The operation H_i and H_j, $i, j = 1, 2, \dots, n$, commute with each other.

In the 1960s Gohberg and Krupnik [46] and O'Neil and Weiss [69] had tried to obtain the best possible value C_p^* ($= \|H\|_p$) of C_p in (6.30). They gave the following upper and lower bounds for C_p^*:

$$\nu(p) \le C_p^* \le \frac{q}{\pi^{3/2}} \Gamma\left(\frac{1}{2p}\right) \Gamma\left(\frac{1}{p}\right)$$

where

$$
v(p) = \begin{cases} \tan\left(\dfrac{\pi}{2p}\right), & 1 < p \le 2 \\[2mm] \cot\left(\dfrac{\pi}{2p}\right), & 2 \le p < \infty \end{cases}
$$

and $1/p + 1/q = 1$. Later Pichorides [82] proved that $C_p^* = v(p)$ for $1 < p < \infty$. Recently McLean and Elliott [60] found the best possible constant $C_{p,E}^*$ $(= \|H_E\|_p)$, $1 < p < \infty$, for the truncated Hilbert transform H_E, defined by

$$
(H_E f)(x) = \frac{1}{\pi i}(P) \int_E \frac{f(t)}{x - t}\, dt, \qquad x \in E \tag{6.51}
$$

where E is a measurable subset of R. It is obvious that there exists a constant $C_{P,E} < \infty$ such that

$$
\|H_E f\|_p \le C_{p,E}\|f\|_p
$$

for every $f \in L^p(\mathbb{R}^n)$ and moreover the best constant $C_{p,E}^* \le C_p^*$. McLean and Elliott [60] proved that

$$
C_{p,E}^* = C_p^* = v(p) \qquad \text{for } 1 < p < \infty, \tag{6.52}
$$

provided that the Lebsegue measure of E is not zero.

Here we will extend the result (6.52) to n-dimensions. More precisely, we will show that for the n-dimensional Hilbert transform H defined in (6.47),

$$
C_{p,E}^{*n} = \|H_E\|_p = \|H\|_p = C_p^{*n} = \left[v(p)\right]^n \tag{6.53}
$$

for every measurable subset E of $\mathbb{R}^n$ with nonzero Lebesgue measure. The n-dimensional truncated Hilbert transform H_E is defined by

$$
(H_E f)(x) = \frac{1}{\pi^n}P \int_E \frac{f(t)}{\prod_{i=1}^n (x_i - t_i)}\, dt, \qquad x \in E \tag{6.54}
$$

In view of (6.50) and the fact that

$$
\|H_i\|_p = C_p^* = v(p), \qquad 1 \le i \le n
$$

it is easy to see that

$$
\|H\|_p = C_p^{*n} = \left[v(p)\right]^n
$$

Now we have proved the latter half of (6.53).

6.8. OPERATORS ON $L^p(\mathbb{R}^n)$ THAT COMMUTE WITH TRANSLATIONS AND DILATATIONS

Let $a = (a_1, a_2, \ldots, a_n)$, and let $m = (m_1, m_2, \ldots, m_n) \in \mathbb{R}^n$ with $m_i > 0$ for each i. We define the translation operator

$$\tau_a : L^p(\mathbb{R}^n) \to L^p(\mathbb{R}^n)$$

and the dilatation operators

$$D_m, D_{m*} : L^p(\mathbb{R}^n) \to L^p(\mathbb{R}^n)$$

by $\tau_a f(x) = f(x - a) = f((x_1 - a_1), (x_2 - a_2), \ldots, (x_n - a_n))$,

$$D_m f(x) = \left(\prod_{i=1}^{n} m_i \right)^{-1/p} f\left(\frac{x_1}{m_1}, \frac{x_2}{m_2}, \ldots, \frac{x_n}{m_n} \right)$$

$$D_{m*} f(x) = \left(\prod_{i=1}^{n} m_i \right)^{-1/p} f(m_1 x_1, m_2 x_2, \ldots, m_n x_n)$$

Both τ_a and D_m are isometric isomorphisms, since

$$(\tau_a)^{-1} = \tau_{-a}, \quad (D_m)^{-1} = D_{m*}$$

and

$$\|\tau_a f\|_p = \|f\|_p, \quad \|D_m f\|_p = \|f\|_p \qquad \text{for every } f \in L^p(\mathbb{R}^n)$$

Let β denote the space of all bounded linear operators from $L^p(\mathbb{R}^n)$ into itself. Then $T \in \beta$ is said to commute with translations if $\tau_a T = T\tau_a$ for all $a \in \mathbb{R}$, and similarly it commutes with dilatations if $D_m T = T D_m$ for all $m \in \mathbb{R}$ with $m_i > 0$ for $1 \leq i \leq n$. The following lemma, the proof of which is trivial, characterizes an integral operator commuting with translations or dilatations.

Lemma 1. Let K in β be an integral operator given by

$$Kf(x) = \int_{\mathbb{R}^n} K(x, y) f(y) \, dy, \qquad x \in \mathbb{R}^n$$

Then

     i. K commutes with translations if and only if K is a difference kernel:

$$K(x, y) = K(x - y, 0) = K(0, y - x)$$

     ii. K commutes with dilatations if and only if K is a Hardy kernel:

$$K(mx, my) = \left(\prod_{i=1}^{n} m_i \right)^{-1} K(x, y)$$

where by mx and my we mean $(m_1 x_1, m_2 x_2, \ldots, m_n x_n)$ and $(m_1 y_1, m_2 y_2, \ldots, m_n y_n)$, respectively.

Note that the n-dimensional Hilbert transform H commutes with both translations and dilatations, since

$$H = H_1 H_2 \ldots H_n$$

and each H_i commutes both with translations and dilatations. Actually H is essentially the only integral operator having this property. To prove this we need the following two lemmas.

Lemma 2. Let $T \in \beta, p > 1$ commute with translations. Then there exists a unique bounded complex-valued Borel measurable function $\sigma(\xi)$ satisfying

$$\left(\widehat{T\varphi}\right)(\xi) = \widehat{\varphi}(\xi)\sigma(\xi)$$

where $\sigma(\xi) \in L^\infty(\mathbb{R}^n)$.

Proof. If $T \in \beta$, then $\tau_a T (= T\tau_a) \in \beta$ for each $a \in \mathbb{R}^n$. The Schwartz testing functions space $\mathcal{D}(\mathbb{R}^n)$ is dense in $L^p(\mathbb{R}^n)$. Let $\varphi \in \mathcal{D}(\mathbb{R}^n)$, and let g_m be sequence of C^∞ functions with bounded support such that $\|g_m\|_p = 1$, and $g_m * \varphi \to \varphi$ as $m \to \infty$, in sup norm as well as in $L^p(\mathbb{R}^n)$ norm [7, pp. 6–8]. Since φ and g_m are of compact supports, $g_m * \varphi$ are also C^∞ functions with compact supports for all m. Therefore in view of the Riesz representation theorem [84, p. 131], there exists a bounded complex regular Borel measure μ on $\mathbb{R}^n$ such that

$$[T((g_m * \varphi)(y))](0) = \int_{\mathbb{R}^n} \left(\int_{\mathbb{R}^n} g_m(x)\varphi(y - x)\, dx \right) d\mu(y)$$

$$= \int_{\mathbb{R}^n} dx\, g_m(x) dy(y) \varphi(y - x)$$

$$= \int_{\mathbb{R}^n} g_m(-x)(T\varphi)(x)\, dx \qquad \text{(by Fubini's theorem)}$$

Hence

$$(g_m * T(\cdot))(0) : \mathcal{D}(\mathbb{R}^n) \to \mathbf{C}$$

is a bounded linear functional. The Riesz representations theorem asserts the existence of a regular Borel measure μ_m (depending on g_m) bounded on $\mathbb{R}^n$ such that

$$(g_m * T\varphi)(0) = \int_{\mathbb{R}^n} \varphi(-x)\, d\mu_m(x), \qquad \varphi \in \mathcal{D}(\mathbb{R}^n)\ [84, p. 131].$$

Hence

$$(g_m * T\varphi)(y) = \int_{\mathbb{R}^n} \varphi(y - x)\, d\mu_m \tag{6.55}$$

$$\tau_{-y}T(g_m * \varphi)(0) = (g_m * \tau_{-y}T\varphi)(0)$$

$$= (g_m * T\tau_{-y}\varphi)(0)$$

Since $|\mu_m(\mathbb{R}^n)| \leq \|T\|$, we can select a sequence g_m in such a way that

$$\lim_{m \to \infty} (g_m * T\varphi)(y) = (T\varphi)(y) \tag{6.56}$$

in $L^p(\mathbb{R}^n)$ norm as well as in sup norm. Hence from (3.2), and by selecting an appropriate subsequence $\{m_j\}$ of $\{m\}$ and letting $m_i \to \infty$, we have $\lim_{m_i \to \infty} \widehat{\mu}_{m_j} = \sigma(\xi)$, a bounded complex-valued measurable function

$$\left(\widehat{R\varphi}\right)(\xi) = \widehat{\varphi}(\xi)\sigma(\xi)\varphi \in \mathcal{D}(R^n) \tag{6.57}$$

[5, pp. 132, 133]. □

Corollary 1. For $T \in \beta$ commuting with translations, there exists $\sigma \in L^\infty(\mathbb{R}^n)$ such that

$$\widehat{Tf}(\xi) = \sigma(\xi)\widehat{f}(\xi), \qquad \xi \in \mathbb{R}^n, \, f \in L^p(\mathbb{R}^n) \tag{6.58}$$

where $\widehat{}$ denotes the operator of Fourier transform.

Proof. Using the definition of the Fourier transform of f in $L^p(\mathbb{R}^n)$, where f is treated as a regular tempered distribution in $S'(\mathbb{R}^n)$ [7, pp. 131–132], it follows that

$$\widehat{f}(\xi) = \lim_{\substack{\min N_j \to \infty \\ 1 \leq j \leq n}} \int_{|x_j| < N_j} f(x)e^{+ix \cdot \xi} \, dx$$

where the above limit is interpreted in the sense of $S'(\mathbb{R}^n)$ and $x \cdot \xi$ is the inner product of x and ξ in $\mathbb{R}^n$. Since $\mathcal{D}(\mathbb{R}^n)$ is dense in $L^p(\mathbb{R}^n)$ the result (6.58) follows from Lemma 2, Bergh and Löfström [7, pp. 132–133], and Stein [94, p. 28]. □

Theorem 3. Let $1 < p < \infty$ and $T \in \beta$. Suppose that T commutes both with translations and with dilatations. Then there exist constant $a, a_i, a_{i,j}, \ldots, b$ such that

$$T = aI + \sum_{i=1}^{n} a_iH_i + \sum_{\substack{i,j=1 \\ i<j}} a_{ij}H_iH_j + \cdots + bH \tag{6.59}$$

where I is the identity operator on $L^p(\mathbb{R}^n)$.

Proof. Let $T \in \beta$, $1 < p < \infty$, commuting with translations and dilatations. Then from (6.58) we have

$$\widehat{Tf}(\xi) = \sigma(\xi)\widehat{f}(\xi), \qquad \xi \in \mathbb{R}^n, \, f \in L^p(\mathbb{R}^n)$$

for some $\sigma \in L^\infty(\mathbb{R}^n)$. Since

$$\widehat{D_m f}(\xi) = \left(\prod_{i=1}^{n} m_i\right)^{1-(1/p)} \widehat{f}(m_1\xi_1, m_2\xi_2, \ldots, m_n\xi_n)$$

and T commutes with dilatations, we have $\sigma(\xi) = \sigma(m_1\xi_1, m_2\xi_2, \ldots, m_n\xi_n)$, for $\xi = (\xi_1, \xi_2, \ldots, \xi_n) \in (\mathbb{R}^n)$ and $m_1, \ldots, m_n > 0$.

Hence

$$\sigma(\xi) = \sigma(\operatorname{sgn}\xi_1, \ldots, \operatorname{sgn}\xi_n),$$

where

$$\operatorname{sgn}\xi_j = \begin{cases} +1, & \text{if } \xi_j > 0 \\ -1, & \text{if } \xi_j < 0 \end{cases}$$

When $n = 2$, it is easy to see that

$$\begin{aligned}
\sigma(\xi_1, \xi_2) = \frac{1}{2^2}\Big[&[\sigma(1,1) + \sigma(1,-1) + \sigma(-1,1) + \sigma(-1,-1)] \\
&+ [\sigma(1,1) + \sigma(1,-1) - \sigma(-1,1) - \sigma(-1,-1)]\operatorname{sgn}\xi_1 \\
&+ [\sigma(1,1) - \sigma(1,-1) + \sigma(-1,1) - \sigma(-1,-1)]\operatorname{sgn}\xi_2 \\
&+ [\sigma(1,1) - \sigma(1,-1) - \sigma(-1,1) + \sigma(-1,-1)]\operatorname{sgn}\xi_1\operatorname{sgn}\xi_2\Big]
\end{aligned}$$

Generalizing this expression we obtain the n-dimensional case

$$\begin{aligned}
\sigma(\xi) = \frac{1}{2^2}\Big[&\sum_{i=1}^{2^n}\sigma(i_1, i_1, \ldots, i_n) + \sum_{j=1}^{n}\left(\sum_{i=1}^{2^n} i_j\sigma(i_1, i_2, \ldots, i_n)\right)\operatorname{sgn}\xi_j \\
&+ \sum_{j,k=1}^{n}\left(\sum_{i=1}^{2^n} i_j i_k \sigma(i_1, \ldots, i_n)\right)\operatorname{sgn}\xi_j \cdot \operatorname{sgn}\xi_k + \cdots \\
&+ \left(\sum_{i=1}^{2^n}\left(\prod_{j=1}^{n} i_j\right)\sigma(i_1, \ldots, i_n)\right)\prod_{j=1}^{n}\operatorname{sgn}\xi_j\Big] \\
= a + &\sum_{j=1}^{n} a_j\operatorname{sgn}\xi_j + \sum_{\substack{j,k=1 \\ j<k}}^{n} a_{jk}\operatorname{sgn}\xi_j\operatorname{sgn}\xi_k + \cdots + b\prod_{j=1}^{n}\operatorname{sgn}\xi_j
\end{aligned}$$

where $i_j = +1$ or -1 for $j = 1, 2, \ldots, n$. Since $\widehat{Hf}(\xi) = \prod_{j=1}^{n}\operatorname{sgn}\xi_j\widehat{f}(\xi)$ and $\widehat{H_jf}(\xi) = \operatorname{sgn}\xi_j\widehat{f}(\xi)$, we have the desired result (6.59); see [74]. $\square$

Remark. The n Riesz transforms $R_1, R_2, \ldots, R_n$ are defined as

$$(R_jf)(x) = \lim_{\varepsilon\to 0} c_n \int_{|y|>\varepsilon} \frac{y_j}{|y|^{n+1}} f(x-y)\, dy, \qquad j = 1, \ldots, n$$

with

$$c_n = \frac{\Gamma((n+1)/2)}{\pi^{(n+1)/2}}, \qquad \text{for } f \in L^p(\mathbb{R}^n),\ 1 \le p < \infty \ [95, p.\ 57]$$

It is easy to see that, in general, they do not commute with dilatations D_m for $m = (m_1, \ldots, m_n) \in (\mathbb{R}^n)$, $m_1, \ldots, m_n > 0$. Hence none of the R_j's can be written in the form (6.59), despite the fact that in the particular case when $m = (m_1, m_1, \ldots, m_1)$ with $m_1 > 0$, the n Riesz transforms commute with dilatations. But only when $n = 1$, does the Riesz transform $\mathbb{R}$ commute both with translations and with dilatations so that it can be written in the form (6.59).

For a measurable set $E \subset (\mathbb{R}^n)$, define

$$\chi_E : L^p(\mathbb{R}^n) \to L^p(\mathbb{R}^n)$$

by

$$\chi_E f(x) = \begin{cases} f(x), & \text{if } x \in E \\ 0, & \text{otherwise} \end{cases}$$

Since any $f \in L^p(\mathbb{R}^n)$ can be written as

$$f = \chi_E f + (1 - \chi_E) f$$

the space $L^p(\mathbb{R}^n)$ is the direct sum

$$L^p(\mathbb{R}^n) = L^p(E) \oplus L^p(\mathbb{R}^n - E)$$

(The vector space W is said to be the direct sum of the vectors spaces U and V iff every $w \in W$ can be written as $w = u + v$, $u \in U$, $v \in V$.)

Thus the space $L^p(E)$ can be treated as closed subspace of $L^p(\mathbb{R}^n)$, and for any bounded linear operator T on $L^p(\mathbb{R}^n)$, we define the truncated operator

$$T_E = \chi_E T \chi_E$$

For $E \subset (\mathbb{R}^n)$ and $m, a \in \mathbb{R}^n$,

$$a + E = \{a + x : x \in E\}$$
$$mE = \{(m_1 x_1, \ldots, m_n x_n) : x \in E\}$$
$$mE = \{(mx_1, \ldots, mx_n) : x \in E\} \qquad m \in \mathbb{R}$$

Then we have the following theorem.

Theorem 4. Let E be any measurable subset of $\mathbb{R}^n$.

  i. If T commutes with translations, then

$$\|T_{a+E}\|_p = \|T_E\|_p, \qquad \forall a \in \mathbb{R}^n$$

  ii. If T commutes with dilatations, then

$$\|T_{mE}\|_p = \|T_E\|_p \qquad \forall m \in \mathbb{R}^n, \, m_1, \ldots, m_n > 0$$

Proof. The proof is similar to the one given by McLean and Elliott [60, thm. 2.2] for the one-dimensional case. $\square$

Let μ be the Lebesgue measure on $\mathbb{R}^n$. Denote by $J_\delta(x)$ the open box centered at x:

$$J_\delta(x) = \prod_{i=1}^{n}(x_i - \delta_i, x_i + \delta_i), x = (x_1, \ldots, x_n) \in \mathbb{R}^n$$

$$\delta = (\delta_1, \ldots, \delta_n) \in \mathbb{R}^n \qquad \text{with each } \delta_i > 0$$

The density of E at x is defined by

$$d_E(x) = \lim_{\delta \to 0^+} \frac{\mu(E \cap J_\delta(x))}{\mu(J_\delta(x))} \tag{6.60}$$

provided that the limit exists. Clearly $0 \leq d_E(x) \leq 1$. On the one hand, $d_E(x) = 0$ when $x \notin \overline{E}$ (the closure of E); on the other hand, $d_E(x) = 1$ when $x \in E^0$ (the interior of E). The Lebesgue density theorem [20, p. 184] asserts that

$$d_E(x) = 1 \qquad \text{for almost every } x \in E \tag{6.61}$$

Lemma 3. If J is a bounded box centered at 0 and $m > 0$, then

$$\lim_{m \to \infty} \mu(J \cap mE) = d_E(0)\mu(J)$$

Proof. Let E be a measurable subset of $\mathbb{R}^n$. Then for $m > 0$, we have

$$\mu(mE) = \mu\{(mx_1, \ldots, mx_n) : x = (x_1, \ldots, x_n) \in E\} = m\mu(E)$$

and $m(E_1 \cap E_2) = (mE_1) \cap (mE_2)$, for E_1, E_2 measurable subset of $\mathbb{R}^n$. Suppose that $J = (-M, M) \times \cdots \times (-M, M)$ (n factors), and let $m = M/\delta$, $\delta > 0$. Then $mJ_\delta(0) = J$, and hence

$$d_E(0) = \lim_{\delta \to 0^+} \frac{\mu(E \cap J_\delta(0))}{\mu(J_\delta(0))} = \lim_{m \to \infty} \frac{\mu(mE \cap J)}{\mu(J)}$$

which proves the lemma. Then, following Lemma 3 and Theorem 3 in Chapter 3, we have proofs similar to that of Lemma 2 and Theorem 3 in Chapter 3 of McLean and Elliott [60], so we state them without proof. $\square$

Lemma 4. For $1 \leq p < \infty$, the following are equivalent:

i. $d_E(0) = 1$.

ii. $\lim_{m \to \infty} \|\chi_{mE} f\|_p = \|f\|_p \; \forall f \in L^p(\mathbb{R}^n), m > 0$.

iii. $\lim_{m \to \infty} \|(1 - \chi_{mE})f\|_p = 0 \; \forall f \in L^p(\mathbb{R}^n), m > 0$.

Theorem 5. Suppose that $d_E(0) = 1$. If $T \in \beta$ commutes with dilatations, then

$$\|T_E\|_p = \|T\|_p.$$

Since the n-dimensional Hilbert transform H commutes both with translations and with dilatations, Theorems 3, 4, and 5 are true for H. Let E be a subset of $\mathbb{R}^n$ such that $\mu(E) \neq 0$. Then there exists an $x \in E$ such that $d_E(x) = 1$, by (6.61). Hence $d_{-x+E}(0) = 1$. Therefore

$$\|H_E\|_p = \|H_{-x+E}\|_p = \|H\|_p$$

Thus we have proved the following theorem:

Theorem 6. If $\mu(E) \neq 0$, then $\|H_E\|_p = \|H\|_p$.

6.9. FUNCTIONS WHOSE FOURIER TRANSFORMS ARE SUPPORTED ON ORTHANTS

We characterize the functions in $L^p(\mathbb{R}^n)$ and generalized functions in $\mathcal{D}'_{L^p}(\mathbb{R}^n)$, $1 < p < \infty$, whose Fourier transforms vanish on one or more orthants of $\mathbb{R}^n$. It is a fairly difficult problem to characterize the functions in $L^p(\mathbb{R}^n)$ whose Fourier transform vanishes in some orthants of $\mathbb{R}^n$. Very little is known concerning this problem except the classical Paley-Wiener theorem in one dimension that characterizes the functions in $L^2(\mathbb{R})$ having their Fourier transforms vanish for negative values of the variable [54, p. 175]. Later some results for the space $L^2(\mathbb{R}^n)$ were obtained by Stein and Weiss [93, p. 112].

Concerning the Fourier transform of a distribution with compact support, it was shown that the Fourier transform of a distribution f with bounded support is a function $F(z) = f(\exp(-2\pi i z \cdot x))$, which may be continued to all complex numbers z as an entire function of exponential growth. The converse is also true [96, p. 15]. For further references, see [96, 94, 97]. But none of those give the explicit characterization of functions in $L^p(\mathbb{R}^n)$ and distributions in the Schwartz space $\mathcal{D}'_{L^p}(\mathbb{R}^n)$, whose Fourier transforms are supported on a given number of orthants in $\mathbb{R}^n$. The aim of the present section is to give a complete answer to the problem for functions in $L^p(\mathbb{R}^n)$ and distributions in $\mathcal{D}'_{L^p}(\mathbb{R}^n)$, $1 < p < \infty$.

For $f \in L^p(\mathbb{R}^n)$, $1 < p < \infty$, we construct the following holomorphic function $F(z)$, $z \in \mathbb{C}^n$, as

$$F(z) = \frac{1}{(2\pi i)^n} \int_{\mathbb{R}^n} f(t) \frac{1}{\prod_{j=1}^n (t_j - z_j)} \, dt \qquad (6.62)$$

where $z_j = x_j + i y_j$ and $y_j \neq 0 \; \forall j = 1, 2, \ldots, n$. For a distribution $f \in \mathcal{D}'_{L^p}(\mathbb{R}^n)$ the corresponding function $F(z)$ is defined as

$$F(z) = \frac{1}{(2\pi i)^n} \left\langle f(t), \frac{1}{\prod_{j=1}^n (t_j - z_j)} \right\rangle, \qquad y_j \neq 0 \; \forall j \qquad (6.63)$$

There are 2^n different ways in which $y \to 0$ depending upon the way the various components y_j of y tend to either 0_+ or 0_-. Thus we get 2^n different boundary values of $F(z)$ as $y \to 0$. To denote them, we adopt the following notation:

Let $\sigma_k = \{\sigma_k(1), \sigma_k(2), \dots, \sigma_k(n)\}$ be a sequence of length n whose elements are $+$ and $-$ for $1 \le k \le 2^n$. Then 2^n orthants of $\mathbb{R}^n$ are denoted by S_{σ_k}, $1 \le k \le 2^n$, where

$$S_{\sigma_k} = \left\{x \in \mathbb{R}^n \mid x_j > 0 \quad \text{if } \sigma_k(j) = + \text{ and } x_j < 0 \right\} \quad (6.64)$$

if $\sigma_k(j) = -$, $j = 1, 2, \dots, n$. For example, when $n = 2$ the various quadrants of $\mathbb{R}^2$ are denoted by $S_{++}, S_{-+}, S_{+-},$ and S_{--}, where

$$S_{+-} = \left\{x \in \mathbb{R}^2 \mid x_1 > 0 \text{ and } x_2 < 0\right\}$$

and so on. Similarly the various limits of $F(z)$ as $y \to 0$ are denoted by

$$F_{\sigma_k}(x) = \lim_{y_1 \to 0_{\sigma_k(1)}, \dots, y_n \to 0_{\sigma_k(n)}} F(z) \quad (6.65)$$

where $0_{\sigma_k(j)} = 0^+$ if $\sigma_k(j) = +$; otherwise, it is 0^-.

For $f \in \mathcal{D}'_{L^p}(\mathbb{R}^n)$ or $L^p(\mathbb{R}^n)$, with the limits taken in respective spaces, we have proved that

$$f = \sum_{k=1}^{2^n} (-1)^{m_k} F_{\sigma_k} \quad \text{in } \mathcal{D}'_{L^p}(\mathbb{R}^n) \text{ (or } L^p(\mathbb{R}^n)), \ 1 < p < \infty \quad (6.66)$$

and

$$(-1)^{m_k} \widehat{F}_{\sigma_k}(\xi) = \begin{cases} \hat{f}(\xi) & \text{for } \xi \in S_{\sigma_k} \\ 0, & \text{elsewhere} \end{cases} \quad (6.67)$$

where m_k is the number of minus signs in the sequence σ_k and $\hat{f}$ is the Fourier transform of f in the following sense:

$$\langle \hat{f}, \varphi \rangle = \langle f, \hat{\varphi} \rangle$$

$$= \left(\int_{\mathbb{R}^n} f \hat{\varphi} \text{ if } f \in L^p(\mathbb{R}^n)\right) \quad \forall \varphi \in S(\mathbb{R}^n) \quad (6.68)$$

where $\hat{\varphi}$ is the classical Fourier transform of φ defined as

$$\hat{\varphi} = \int_{\mathbb{R}^n} \varphi(t) e^{it \cdot x} \, dt$$

The space $S(\mathbb{R}^n)$ is the testing function space of rapid descent [1, 8, 67, 88, 109].

From (6.67) we are able to prove the following Paley-Wiener theorem for $\mathcal{D}'_{L^p}(\mathbb{R}^n)$ (or $L^p(\mathbb{R}^n)$) $(1 < p < \infty)$:

Theorem 7. For $f \in \mathcal{D}'_{L^p}(\mathbb{R}^n)$,

$$\hat{f}(\xi) = 0 \quad \text{for } \xi \in \bigcup_{k=1}^{\ell} S_{\sigma_k}, \ 1 \leq \ell \leq 2^n$$

$$\text{iff } \sum_{k=1}^{\ell} (-1)^{m_k} F_{\sigma_k}(\xi) = 0 \quad \text{for } \xi \in \bigcup_{k=1}^{\ell} S_{\sigma_k}$$

in some space $S'_0(\mathbb{R}^n)$.

The space $S_0(\mathbb{R}^n)$ is a subspace of $S(\mathbb{R}^n)$ that is closed with respect to the multiplication by the function sgn x. The "sgn" is defined as

$$\text{sgn}(x) = \prod_{i=1}^{n} \text{sgn}(x_i) \tag{6.69}$$

In the process we prove the M. Riesz and Titchmarsh inequality and many related classical results for $L^p(\mathbb{R}^n)$.

As an application of our theory, we characterize the solution space of the following Dirichlet boundary-value problem:

$$\Delta u = 0 \tag{6.70}$$

where

$$\Delta = \prod_{j=1}^{n} \left(\frac{\partial^2}{\partial x_j^2} + \frac{\partial^2}{\partial y_j^2} \right)$$

with boundary conditions

$$\lim_{y \to 0_{\sigma_k}} u = F_{\sigma_k} \quad \text{in } \mathcal{D}'_{L^p}(\mathbb{R}^n) \text{ (or } L^p(\mathbb{R}^n)), \ 1 \leq k \leq 2^n \tag{6.71}$$

Here F_{σ_k}, $1 \leq k \leq 2^n$, are arbitrary elements of $\mathcal{D}'_{L^p}(\mathbb{R}^n)$. Incidently for fixed F_{σ_k} $(1 \leq k \leq 2^n)$ in $\mathcal{D}'_{L^p}(\mathbb{R}^n)$ (or $L^p(\mathbb{R}^n)$), the system (6.70) and (6.71) has

$$F(z) = \frac{1}{(2\pi i)^n} \left\langle \sum_{k=1}^{2^n} (-1)^{m_k} F_{\sigma_k}(t), \frac{1}{\prod_{j=1}^{n}(t_j - z_j)} \right\rangle, \quad \text{Im } z_j \neq 0 \ \forall j \tag{6.72}$$

as a unique solution.

6.9.1. The Schwartz Distribution Space $\mathcal{D}'_{L^p}(\mathbb{R}^n)$

A C^∞ complex-valued function $\varphi(x)$ on $\mathbb{R}^n$ belongs to the space $\mathcal{D}_{L^p}(\mathbb{R}^n)$ iff $\partial^\alpha \varphi(x)$ belongs to $L^p(\mathbb{R}^n)$ for each $|\alpha| = 0, 1, 2, \ldots$, where $\alpha = (\alpha_1, \alpha_2, \ldots, \alpha_n)$, α_i's are nonnegative integers and $|\alpha| = \sum_{i=1}^{n} \alpha_i$. The topology over $\mathcal{D}_{L^p}(\mathbb{R}^n)$ is generated by the countable family of separating seminorms [67, 87, 110]

$$\gamma_\alpha(\varphi) = \left[\int_{\mathbb{R}^n} |\partial^\alpha \varphi(x)|^p \, dx \right]^{1/p}$$

The space $\mathcal{D}_{L^p}(\mathbb{R}^n)$ is a sequentially complete, locally convex, Hausdorff topological linear space.

In conformity with the notation used by Laurent Schwartz [87], we will denote $\mathcal{D}'_{L^p}(\mathbb{R}^n)$, $p > 1$, as the dual space of $\mathcal{D}_{L^q}(\mathbb{R}^n)$ where $\frac{1}{p} + \frac{1}{q} = 1$. It can be shown [67, p. 173] that for $f \in \mathcal{D}'_{L^p}(\mathbb{R}^n)$, there exist measurable functions f_α in $L^p(\mathbb{R}^n)$ and a $k \in \mathbb{N}$ such that

$$f = \sum_{|\alpha| \le k} \partial^\alpha f_\alpha \tag{6.73}$$

Let $S'(\mathbb{R}^n)$ denote the space of tempered distributions and $S(\mathbb{R}^n)$ the corresponding testing function space of rapid descent [67]. One can see that $S(\mathbb{R}^n) \subset \mathcal{D}_{L^p}(\mathbb{R}^n)$ and is dense in $\mathcal{D}_{L^p}(\mathbb{R}^n)$ [67]. Therefore the restriction of $f \in \mathcal{D}'_{L^p}(\mathbb{R}^n)$ to $S(\mathbb{R}^n)$ is in $S'(\mathbb{R}^n)$, and each element of $\mathcal{D}'_{L^p}(\mathbb{R}^n)$ can be identified with an element of $S'(\mathbb{R}^n)$ in a one-to-one way. Hence with this kind of identification, $\mathcal{D}'_{L^p}(\mathbb{R}^n) \subset S'(\mathbb{R}^n)$. Therefore the Fourier transform $\hat{f}$ of f in $\mathcal{D}'_{L^p}(\mathbb{R}^n)$ can be defined by

$$\langle \hat{f}, \varphi \rangle = \langle f, \hat{\varphi} \rangle \qquad \forall \varphi \in S(\mathbb{R}^n)$$

Theorem 8. Let $f \in L^p(\mathbb{R}^n)$, $1 < p < \infty$. Define

$$F(x, y) = \int_{\mathbb{R}^n} f(t) \prod_{j=1}^{n} \frac{t_j - x_j}{(t_j - x_j)^2 + y_j^2} \, dt \tag{6.74}$$

where $x = (x_1, x_2, \ldots, x_n)$, $y = (y_1, y_2, \ldots, y_n)$, $t = (t_1, t_2, \ldots, t_n)$ are in $\mathbb{R}^n$ and $y_j \ne 0$ $(j = 1, 2, \ldots, n)$. Then we have

$$\partial_x^\alpha \partial_y^\beta F(x, y) = \int_{\mathbb{R}^n} f(t) \partial_x^\alpha \partial_y^\beta \left[\prod_{j}^{n} \frac{t_j - x_j}{(t_j - x_j)^2 + y_j^2} \right] dt \tag{6.75}$$

where

$$|\alpha|, |\beta| = 0, 1, 2, \ldots$$

$$\partial_x^\alpha \equiv \frac{\partial^{\alpha_1}}{\partial x_1^{\alpha_1}} \frac{\partial^{\alpha_2}}{\partial x_2^{\alpha_2}} \cdots \frac{\partial^{\alpha_n}}{\partial x_n^{\alpha_n}}, \quad \partial_y^\beta = \frac{\partial^{\beta_1}}{\partial y_1^{\beta_1}} \cdots \frac{\partial^{\beta}}{\partial y_n^{\beta_n}}$$

and the α_j's and β_j's are non-negative integers. Also $F(x, y)$ and $\partial_x^\alpha \partial_y^\beta F(x, y)$ are continuous functions of $x, y \in \mathbb{R}^n$. Thus $F(x, y) \in C^\infty(\mathbb{R}^{2n})$.

Proof. Set

$$u(x, y) = \int_{\mathbb{R}^n} f(t) \partial_x^\alpha \prod_{j=1}^n \left[\frac{t_j - x_j}{(t_j - x_j)^2 + y_j^2} \right] dt$$

$$= \int_{\mathbb{R}^n} f(t + x) \partial_t^\alpha \prod_{j=1}^n \left(\frac{t_j}{t_j^2 + y_j^2} \right) dt$$

Thus by Holder's inequality we have

$$|u(x, y| \le \|f\|_p \left\| \prod_{j=1}^n \partial_{t_j}^{\alpha_j} \left(\frac{t_j}{t_j^2 + y_j^2} \right) \right\|_q$$

Hence the integral representing $u(x, y)$ is uniformly convergent $\forall\, x$ in $\mathbb{R}^n$ and a fixed $y \in \mathbb{R}^n$ having all nonzero components. By using the mean value theorem, we can prove the continuity of $F(x, y)$ and $u(x, y)$ with respect to both x and y. These results are true for arbitrary α. Hence, using a standard classical theorem [100, p. 59], it follows that

$$\partial_x^\alpha F(x, y) = \int_{\mathbb{R}^n} f(t) \partial_x^\alpha \left[\prod_{j=1}^n \frac{t_j - x_j}{(t_j - x_j)^2 + y_j^2} \right] dt$$

Also we have

$$|\partial_y^\beta F(x, y)| \le \|f\|_p \left\| \prod_{j=1}^n \partial_{y_j}^{\beta_j} \left(\frac{t_j}{t_j^2 + y_j^2} \right) \right\|_q$$

Using the fact that

$$\left| \frac{y_j}{t_j^2 + y_j^2} \right|^q \le \left| \frac{a_j + \delta_j}{t_j^2 + (a_j - \delta_j)^2} \right|^q$$

$\forall y_j \in (a_j - \delta_j, a_j + \delta_j)$, we can see that for arbitrary β the integral representing $\partial_y^\beta F(x, y)$ is uniformly convergent in an appropriately chosen rectangle lying in the region

$$\{ y \in \mathbb{R}^n \mid |y_j| > 0,\ j = 1, 2, \ldots, n \}.$$

Therefore we have

$$\partial_y^\beta F(x, y) = \int_{\mathbb{R}^n} f(t) \partial_y^\beta \prod_{j=1}^n \frac{t_j - x_j}{(t_j - x_j)^2 + y_j^2} dt$$

See [100, p. 59]. □

Lemma 5. Let $x, y, t \in \mathbb{R}^n$ be such that $y_j \neq 0 \; \forall j = 1, 2, \ldots, n$. For $f \in \mathcal{D}'_{L^p}(\mathbb{R}^n)$ $(1 < p < \infty)$, define a function

$$F(x, y) = \left\langle f(t), \prod_{j=1}^{n} [(t_j - x_j) \mid ((t_j - x_j)^2 + y_j^2)] \right\rangle \tag{6.76}$$

Then

$$\partial_y^\beta \partial_x^\alpha F(x, y) = \left\langle f(t), \partial_y^\beta \partial_x^\alpha \prod_{j=1}^{n} [(t_j - x_j) \mid ((t_j - x_j)^2 + y_j^2)] \right\rangle \tag{6.77}$$

Proof. Since

$$\prod_{j=1}^{n} \frac{t_j - x_j}{(t_j - x_j)^2 + y_j^2} \in L^q(\mathbb{R}^n)$$

as a function of t for a fixed x and y, and $f \in \mathcal{D}'_{L^p}(\mathbb{R}^n)$, the dual of $L^q(\mathbb{R}^n)$ $\left(\frac{1}{p} + \frac{1}{q} = 1\right)$, $F(x, y)$ is well defined for each $x, y \in \mathbb{R}^n$ with y having all nonzero components. Using the structure formula (6.73) for $f \in \mathcal{D}'_{L^p}(\mathbb{R}^n)$, we see that

$$F(x, y) = \sum_{|\gamma| \le k} \left\langle f_\gamma(t), (-1)^{|\gamma|} \partial_t^\gamma \prod_{j=1}^{n} \frac{t_j - x_j}{(t_j - x_j)^2 + y_j^2} \right\rangle$$

$$= \sum_{|\gamma| \le k} \left\langle f_\gamma(t), \partial_x^\gamma \prod_{j=1}^{n} \frac{t_j - x_j}{(t_j - x_j)^2 + y_j^2} \right\rangle, \qquad f_\gamma \in L^p(\mathbb{R}^n)$$

Then, using relation (6.72), we obtain

$$\partial_y^\beta \partial_x^\alpha F(x, y) = \sum_{|\gamma| \le k} \left\langle f_\gamma(t), \partial_y^\beta \partial_x^{\alpha + \gamma} \prod_{j=1}^{n} \left[\frac{t_j - x_j}{(t_j - x_j)^2 + y_j^2} \right] \right\rangle$$

$$= \sum_{|\gamma| \le k} \left\langle f_\gamma(t), (-1)^{|\gamma|} \partial_t^\gamma \partial_y^\beta \partial_x^\alpha \prod_{j=1}^{n} \left[\frac{t_j - x_j}{(t_j - x_j)^2 + y_j^2} \right] \right\rangle$$

$$= \left\langle f(t), \partial_y^\beta \partial_x^\alpha \prod_{j=1}^{n} \left[\frac{t_j - x_j}{(t_j - x_j)^2 + y_j^2} \right] \right\rangle. \qquad \square$$

6.9.2. An Approximate Hilbert Transform and Its Limit in $L^p(\mathbb{R}^n)$

Some of the results proved in Sections 6.3, 6.4, and 6.5 are proved by Tillmann [97, 98] and Vladimirov [104, ch. 5]. So the results proved in Section 6.3, 6.4, and 6.5 are not entirely new. However, our techniques are different in that we make an extended use of the results proved by Riesz and Titchmarsh [99], thereby making our treatment simpler. Our main results proved in Section 6.6 are new and are not proved anywhere else. In our analysis we heavily rely upon the result that

$$\mathcal{F}(Hf) = i^n \prod_{j=1}^n \mathrm{sgn}(x_j)(\mathcal{F}f) \qquad \forall f \in (D_{L^p}(\mathbb{R}^n))', \; p > 1$$

in the weak topology of $S_0(\mathbb{R}^n)$. The space $S_0(\mathbb{R}^n)$ is a subspace of the Schwartz testing function space $S(\mathbb{R}^n)$ such that every element of $S_0(\mathbb{R}^n)$ vanishes at the origin along with all its derivatives. The topology of $S_0(\mathbb{R}^n)$ is the same as that induced on $S_0(\mathbb{R}^n)$ by $S(\mathbb{R}^n)$.

Let H be the operator of the classical Hilbert transform from $L^p(\mathbb{R}^n)$, $p > 1$, into itself defined by

$$(Hf)(x) = \lim_{\substack{\max \epsilon_j \to 0 \\ 1 \le j \le n}} \frac{1}{\pi^n} \int_{|t_j - x_j| > \epsilon_j} \frac{f(t)}{\prod_{j=1}^n (x_j - t_j)} \, dt$$

$$= \frac{1}{\pi^n} P \int_{\mathbb{R}^n} \frac{f(t)}{\prod_{j=1}^n (x_j - t_j)} \, dt \qquad (6.78)$$

It is a known fact that the limit exists a.e. [57] and that $(Hf)(x) \in L^p(\mathbb{R}^n)$. Also

$$\|Hf\|_p \le C_p \|f\|_p \qquad [34, 57, 89, 111] \qquad (6.79)$$

where C_p is a constant independent of f [57, 93].

Titchmarsh [99] proved that if $f \in L^p(\mathbb{R})$, $p > 1$, then its approximate Hilbert transform

$$(H_y f)(x) = \frac{1}{\pi} \int_{\mathbb{R}} \frac{f(t) \cdot (x - t)}{(t - x)^2 + y^2} \, dt, \qquad y \ne 0 \qquad (6.80)$$

exists a.e., and

$$\lim_{y \to 0} \frac{1}{\pi} \int_{\mathbb{R}} \frac{x - t}{(t - x)^2 + y^2} f(t) \, dt = (Hf)(x) \qquad \text{in } L^p(\mathbb{R})$$

It is also known that

$$\|(H_y f)(x)\|_p \le C_p \|f\|_p \qquad (6.81)$$

where C_p is a constant independent of f and y. Stein and Weiss [94, p. 218] proved similar result for $L^p(\mathbb{R})$ over the Lebesgue set of f. We extend the above results to n-dimensions.

Definition. The n-dimensional approximate Hilbert transform $(H_y f)(x)$ of $f \in L^p(\mathbb{R}^n)$, $p > 1$, is defined by

$$(H_y f)(x) = \frac{1}{\pi^n} \int_{\mathbb{R}^n} \prod_{j=1}^n \frac{x_j - t_j}{(t_j - x_j)^2 + y_j^2} f(t) \, dt, \qquad y_j \ne 0, \; \forall j = 1, 2, 3, \dots, n$$

$$(6.82)$$

Theorem 9. The operator H_y as defined by (6.82) is a bounded linear operator from $L^p(\mathbb{R}^n)$ into itself.

Proof. We will first prove the result for $n = 2$. Let $f \in L^p(\mathbb{R}^2)$. Then we have

$$\|f\|_p = \left(\int_{\mathbb{R}^2} |f|^p \, dx \, dy\right)^{1/p} = \left(\int_{-\infty}^{\infty} dy \int_{-\infty}^{\infty} dx \, |f(x,y)|^p\right)^{1/p}$$

$$= \left(\int_{-\infty}^{\infty} dx \int_{-\infty}^{\infty} dy \, |f(x,y)|^p\right)^{1/p} \qquad \text{(by Fubini's theorem [48])}$$

so that

$$\|f\|_p = \|f(\cdot, y)\|_{1,p;2,p} = \|f(x, \cdot)\|_{2,p;1,p} \tag{6.83}$$

where

$$\|f(x, \cdot)\|_{2,p} = \left(\int_{\mathbb{R}} |f(x,y)|^p \, dy\right)^{1/p}$$

$$\|f(\cdot, y)\|_{1,p} = \left(\int_{\mathbb{R}} |f(x,y)|^p \, dx\right)^{1/p}$$

Now

$$\|f(x,y)\|_{1,p;2,p} = L^p \qquad \text{norm of } \|f(\cdot, y)\|_{1,p} \text{ as a function of } y$$

and

$$\|f(x,y)\|_{2,p;1,p} = L^p \qquad \text{norm of } \|f(x, \cdot)\|_{2,p} \text{ as a function of } x$$

If one of the expressions in (6.83) exists, the remaining two also exist. Since

$$(H_y f)(x_1, x_2) = \frac{1}{\pi^2} \int_{\mathbb{R}^2} f(t) \prod_{j=1}^{2} \left[\frac{x_j - t_j}{(t_j - x_j)^2 + y_j^2}\right] dt$$

by (6.81) we have

$$\|(H_y f)(x_1, x_2)\|_p = \|(H_y f)(x_1, x_2)\|_{1,p;2,p}$$

$$\leq C_p \left\| \int_{\mathbb{R}} \frac{x_2 - t_2}{(t_2 - x_2)^2 + y_2^2} f(\cdot, t_2) \, dt_2 \right\|_{1,p;2,p}$$

where C_p is a constant independent of f and y [101]. But

$$\left\| \int_{\mathbb{R}} \frac{t_2 - x_2}{(t_2 - x_2)^2 + y_2^2} f(\cdot, t_2) \, dt_2 \right\|_{2,p;1,p} \leq C_p \|f(t_1, \cdot)\|_{2,p;1,p}$$

$$\leq C_p^2 \|f\|_p$$

In view of Fubini's theorem [43],

$$\|H_y f\|_p \leq C_p^2 \|f\|_p$$

Thus the theorem is proved for $n = 2$. Using similar techniques and induction on n, it can be shown that for $f \in L^p(\mathbb{R}^n)$,

$$\|H_y f\|_p \leq C_p^n \|f\|_p \qquad \square \tag{6.84}$$

Recall that the space $X(\mathbb{R}^n)$ is defined to be the collection of $\varphi \in D(\mathbb{R}^n)$ which are finite sums of the form

$$\varphi(x) = \Sigma \, \varphi_{m1}(x_1)\varphi_{m2}(x_2) \ldots \varphi_{mn}(x_n)$$

where

$$\varphi_{mj}(x_j) \in \mathcal{D}(\mathbb{R}), \qquad 1 \leq j \leq n$$

The space $X(\mathbb{R}^n)$ is dense in $L^p(\mathbb{R}^n)$ [106, p. 71].

Theorem 10. For $f \in L^p(\mathbb{R}^n)$, define $(Hf)(x)$ (the Hilbert transform of f) and $(H_y f)(x)$ (the approximate Hilbert transform of f) as in (6.78) and (6.82), respectively. Then

$$\lim_{y_1, y_2, \ldots, y_n \to 0} (H_y f)(x) = (Hf)(x) \qquad \text{in } L^p(\mathbb{R}^n) \text{ norm}$$

Proof. Let φ_m be a sequence in $X(\mathbb{R}^n)$ converging to f in $L^p(\mathbb{R}^n)$. Then

$$\lim_{m \to \infty} \|f(x) - \varphi_m(x)\|_p = 0$$

Now

$$H_y f - Hf = H_y f - H_y \varphi_m + H_y \varphi_m - H\varphi_m + (H\varphi_m - Hf)$$

So

$$\|H_y f - Hf\|_p \leq \|H_y(f - \varphi_m)\|_p + \|H_y \varphi_m - H\varphi_m\|_p + \|H(\varphi_m - f)\|_p$$
$$\leq C_p^n \|f - \varphi_m\|_p + \|H_y \varphi_m - H\varphi_m\|_p + C_p^n \|\varphi_m - f\|_p$$

It is a simple exercise to show that $\|H_y \varphi_m - H\varphi_m\|_p \to 0$ as $y \to 0$. Letting $y \to 0$, we deduce

$$\lim_{y \to 0} \|H_y f - Hf\|_p \leq 2C_p^n \|f - \varphi_m\|_p$$

Now letting $m \to \infty$, we obtain

$$\lim_{y \to 0} \|H_y f - Hf\|_p = 0 \qquad \square$$

Theorem 11. Let $f \in L^p(\mathbb{R}^n)$, $1 < p < \infty$, and let $y_1, y_2, \ldots, y_n$ be nonzero real numbers. Then

i. $\displaystyle (I_y f)(x) = \frac{1}{\pi^n} \int_{\mathbb{R}^n} f(t) \prod_{j=1}^{n} \left[\frac{y_j}{(t_j - x_j)^2 + y_j^2} \right] dt$ (6.85)

    which, as a function of x, belongs to $L^p(\mathbb{R}^n)$.

ii. $\|I_y f\|_p \le C_p^n \|f\|_p$, where C_p is a constant independent of f and y.

iii. $\|I_y f - f\|_p \to 0$ as $y \to 0_+$; that is, $y_1, y_2, \ldots, y_n \to 0_+$.

Proof. The proof is very similar to that given for Theorem 9. We can use the fact that for $g \in L^p(\mathbb{R})$,

$$(I_y g)(x) = \frac{1}{\pi} \int_{\mathbb{R}} \frac{y g(t)}{(t - x)^2 + y^2} dt \in L^p(\mathbb{R}), \qquad y \ne 0$$

$$\lim_{y \to 0_+} (I_y g)(x) = g(x) \text{ in } L^p(\mathbb{R}) \qquad [99]$$

$$\|I_y g\| \le C_p \|g\|_p$$

The result i can also be proved by using [48, p. 400]. It is easy to see that if $f \in L^p(\mathbb{R}^n)$,

$$\|I_y f\|_p \le C_p^n \|f\|_p \qquad \forall f \in L^p(\mathbb{R}^n) \qquad \square$$

Theorem 12. For $f \in L^p(\mathbb{R}^n)$, $p > 1$, and $x, y \in \mathbb{R}^n$, define

$$(Tf)(x) = \frac{1}{\pi^n} \int_{\mathbb{R}^n} f(t) \left[\prod_{j=1}^{m} \frac{x_j - t_j}{(t_j - x_j)^2 + y_j^2} \right] \left[\prod_{k=m+1}^{n} \frac{y_k}{(t_k - x_k)^2 + y_k^2} \right] dt$$

$$(6.86)$$

where $0 \le m \le n$. Then T is a bounded linear operator from $L^p(\mathbb{R}^n)$ into itself,

$$\|Tf\|_p \le C_p^n \|f\|_p \qquad (6.87)$$

and

$$\lim_{y_1, y_2, \ldots, y_n \to 0^+} (Tf)(x) = (H_1 H_2 \ldots H_m I_{m+1} \ldots I_n f)(x)$$

$$= (H_1 H_2 \ldots H_m f)(x) \qquad (6.88)$$

where $I_1, I_2, \ldots, I_n$ are all one-dimensional identity operators and $H_1, H_2, \ldots, H_n$ are all one-dimensional Hilbert transform operators.

Proof. The proof of (6.87) can be given by using the technique followed in Theorems 2 and 3 in Chapter 4, and then (6.88) can be proved by using (6.87) and the density of $X(\mathbb{R}^n)$ in $L^p(\mathbb{R}^n)$ [106, p. 71]. $\square$

6.9.3. Complex Hilbert Transform

Let $f \in L^p(\mathbb{R}^n)$, $1 < p < \infty$, and $z = (z_1, z_2, \ldots, z_n) \in \mathbb{C}^n$ such that $\mathrm{Im}\, z_j = y_j \neq 0$ $\forall j = 1, 2, \ldots, n$. We define the n-dimensional complex Hilbert transform $(Hf)(z)$ of f by

$$(Hf)(z) = \frac{1}{\pi^n} \int_{\mathbb{R}^n} \frac{f(t)}{\prod_{j=1}^n (z_j - t_j)}\, dt$$

$$= \frac{1}{\pi^n} \int_{\mathbb{R}^n} f(t) \prod_{j=1}^n \frac{(x_j - t_j) - iy_j}{(t_j - x_j)^2 + y_j^2}\, dt \tag{6.89}$$

Then we have the following:

Theorem 13. For $f \in L^p(\mathbb{R}^n)$, $1 < p < \infty$, its complex Hilbert transform $(Hf)(z)$ as a function of x belongs to $L^p(\mathbb{R}^n)$ for a fixed y with all nonzero components. Also

$$\|Hf\|_p \leq (2C_p)^n \|f\|_p, \qquad \text{(Titchmarsh and Riesz inequality)} \tag{6.90}$$

and

$$\lim_{y_1, y_2, \ldots, y_n \to 0+} (Hf)(z) = \left(\prod_{j=1}^n (H_j - iI_j) \right) f(x) \qquad \text{in } L^p(\mathbb{R}^n) \tag{6.91}$$

where

$$(H_j f)(x) = \frac{1}{\pi} P \int_{\mathbb{R}} \frac{f(x_1, \ldots, x_{j-1}, t_j, x_{j+1}, \ldots, x_n)}{x_j - t_j}\, dt_j \tag{6.92}$$

and

$$(I_j f)(x) = I_j f(x_1, \ldots, x_{j-1}, t_j, x_{j+1}, \ldots, x_n) = f(x) \tag{6.93}$$

Similarly

$$\lim_{\ldots, y_j \to 0+, \ldots, y_k \to 0-, \ldots} (Hf)(z) = (\ldots (H_j - iI_j) \ldots (H_k + iI_k) \ldots) f(x) \tag{6.94}$$

Proof. The proof can be given by using the technique followed in Theorems 9, 10, and 11. $\square$

Theorem 14. For $f \in L^p(\mathbb{R}^n)$, $p > 1$, define

$$F(x) = \frac{1}{\pi^n} \int_{\mathbb{R}^n} f(t) \prod_{j=1}^n \frac{t_j - x_j}{(t_j - x_j)^2 + y_j^2}\, dt, \qquad y_j \neq 0\, \forall j \tag{6.95}$$

Then

$$\partial^\alpha F(x) \in L^p(\mathbb{R}^n)$$

Proof. We will prove the result for the simple case where $\partial^\alpha = \frac{\partial}{\partial x_1}$, and the general result will follow by induction. Now

$$\frac{\partial}{\partial x_1} F(x_1, x_2, \ldots, x_n) = \frac{1}{\pi^n} \int_{\mathbb{R}} \frac{(t_1 - x_1)^2 - y_1^2}{[(t_1 - x_1)^2 + y_1^2]^2} \, dt_1 \int_{\mathbb{R}^{n-1}} f(t) \prod_{j=2}^{n} \frac{(t_j - x_j) \, dt_j}{(t_j - x_j)^2 + y_j^2}$$

Therefore

$$\left\| \frac{\partial}{\partial x_1} F(x) \right\|_p \leq C_p^{n-1} \left\| \frac{t_1^2 - y_1^2}{(t_1^2 + y_1^2)^2} \right\|_{L^1(\mathbb{R})} \|f(t_1, \ldots)\|_{L^p(\mathbb{R}^{n-1})} \qquad [48, p. 401]$$

$$\leq C_p^n \frac{\pi}{2y_1} \|f\|_p \qquad \square$$

Corollary. For $f \in \mathcal{D}_{L^p}'(\mathbb{R}^n)$, $p > 1$, and fixed real numbers $y_1, y_2, \ldots, y_n$ different from zero, define

$$F(x) = \frac{1}{\pi^n} \left\langle f(t), \prod_{j=1}^{n} \left[\frac{x_j - t_j}{(t_j - x_j)^2 + y_j^2} \right] \right\rangle \qquad (6.96)$$

Then $F(x) \in L^p(\mathbb{R}^n)$.

Proof. Using the structure formula (6.12)

$$f = \sum_{|\alpha| \leq k} \partial^\alpha f_\alpha$$

where each $f_\alpha \in L^p(\mathbb{R}^n)$, and Lemma 5, we have

$$F(x) = \sum_{|\alpha|=0}^{k} \partial_x^\alpha \left\langle f_\alpha(t), \prod_{j=1}^{n} \left[\frac{x_j - t_j}{(t_j - x_j)^2 + y_j^2} \right] \right\rangle$$

The result now follows in view of Theorems 3 and 9. $\square$

Theorem 15. For $f \in \mathcal{D}_{L^p}'(\mathbb{R}^n)$, $p > 1$, and $y_j \neq 0$, $1 \leq j \leq n$, define

$$(H_y f)(x) = F(x)$$

as defined in (6.96). Then

$$\lim_{|y| \to 0} (H_y f)(x) = (H_1 \ldots H_n f)(x) = (Hf)(x) \qquad (6.97)$$

where the limit is interpreted as the weak limit on $\mathcal{D}_{L^p}'(\mathbb{R}^n)$.

Proof. In view of Theorem 14 and its corollary, $(H_y f)(x)$ can be interpreted as a regular distribution on $\mathcal{D}_{L^q}(\mathbb{R}^n)$. Therefore, for each $\varphi \in \mathcal{D}_{L^q}(\mathbb{R}^n)$,

$$\left\langle \left\langle f(t), \prod_{j=1}^{n} \left[\frac{x_j - t_j}{(t_j - x_j)^2 + y_j^2} \right] \right\rangle, \varphi(x) \right\rangle$$

$$= \left\langle \left\langle \sum_{|\alpha|=0}^{m} \partial_t^\alpha f_\alpha(t), \prod_{j=1}^{n} \left[\frac{x_j - t_j}{(t_j - x_j)^2 + y_j^2} \right] \right\rangle, \varphi(x) \right\rangle \qquad \text{[67, p. 175]}$$

$$= \sum_{|\alpha|=0}^{m} \left\langle \left\langle f_\alpha(t), \partial_x^\alpha \prod_{j=1}^{n} \left[\frac{x_j - t_j}{(t_j - x_j)^2 + y_j^2} \right] \right\rangle, \varphi(x) \right\rangle$$

$$= \sum_{|\alpha|=0}^{m} \left\langle \partial_x^\alpha \left\langle f_\alpha(t), \prod_{j=1}^{n} \left[\frac{x_j - t_j}{(t_j - x_j)^2 + y_j^2} \right] \right\rangle, \varphi(x) \right\rangle \qquad \text{[Lemma 5]}$$

$$= \sum_{|\alpha|=0}^{m} \left\langle \left\langle f_\alpha(t), \prod_{j=1}^{n} \left[\frac{x_j - t_j}{(t_j - x_j)^2 + y_j^2} \right] \right\rangle, (-\partial_x)^\alpha \varphi(x) \right\rangle$$

$$= \sum_{|\alpha|=0}^{m} (-1)^{|\alpha|} \left\langle f_\alpha(t), \int_{\mathbb{R}^n} (\partial_x^\alpha \varphi(x)) \prod_{j=1}^{n} \left[\frac{x_j - t_j}{(t_j - x_j)^2 + y_j^2} \right] dx \right\rangle \qquad (6.98)$$

Since

$$f_\alpha \in L^p(\mathbb{R}^n) \quad \text{and} \quad \partial_x^\alpha \varphi(x) \in \mathcal{D}_{L^q}(\mathbb{R}^n)$$

by using the duality theorems and the limiting processes, the switch in the order of integration is justified. Now letting $|y| \to 0$ in (6.98), we obtain

$$\lim_{|y| \to 0} \langle (H_y f)(x), \varphi(x) \rangle = \sum_{|\alpha|=0}^{m} (-1)^{|\alpha|} \langle f_\alpha(t), (-1)^n H(\partial^\alpha \varphi(t)) \rangle \qquad (6.99)$$

The steps in (6.99) can easily be justified in view of Theorem 4. Now using the commutativity of the distributional differentiation ∂^α and H [77, 89], we deduce

$$\lim_{|y| \to 0} \langle (H_y f)(x), \varphi(x) \rangle = \left\langle H \sum_{|\alpha|=0}^{m} \partial_t^\alpha f_\alpha(t), \varphi(t) \right\rangle$$

$$= \langle Hf, \varphi \rangle$$

Therefore

$$\lim_{|y| \to 0} H_y f = Hf \quad \text{in } \mathcal{D}'_{L^p}(\mathbb{R}^n) \qquad \square$$

Corollary. For $f \in \mathcal{D}'_{L^p}(\mathbb{R}^n)$, define the complex Hilbert transform of f by

$$F(z) = \frac{1}{\pi^n} \left\langle f(t), \frac{1}{\prod_{j=1}^{n}(z_j - t_j)} \right\rangle, \qquad \text{Im } z_j = y_j \neq 0 \; \forall j \qquad (6.100)$$

Then

$$\lim_{y_1,\ldots,y_n \to 0^+} F(z) = \prod_{j=1}^{n} (H_j - iI_j) f \tag{6.101}$$

Proof. The proof is similar to the proof of Theorem 15. $\square$

6.9.4. Distributional Representation of Holomorphic Functions

The holomorphic function $F(z)$ given by (6.100) satisfies the uniform asymptotic orders (uniformity with respect to x is assumed here)

$$|F(z)| = O\left(\frac{1}{(y_1 y_2 \cdots y_n)^{(p-1)/p}}\right) \qquad \text{as } y_1, y_2, \ldots, y_n \to \infty$$

Let us now reverse the problem. Let $F(z)$ be holomorphic in $y_j > 0$ ($j = 1, 2, \ldots, n$), that is, on $S_{++\cdots+}$, and satisfies the relation

$$\sup_{\substack{x_j \in \mathbb{R}, y_j \geq \delta > 0 \\ 1 \leq j \leq n}} |F(x + iy)| < A_\delta < \infty \tag{6.102}$$

and the uniform asymptotic order (with respect to x)

$$|F(x + iy)| = o(1), \qquad y \to \infty \tag{6.103}$$

Assume also that

$$\lim_{y_1, y_2, \ldots, y_n \to 0^+} F(z) = F_{++\cdots+}(x) \qquad \text{in } \mathcal{D}'_{L^p}(\mathbb{R}^n) \tag{6.104}$$

Then by using the technique of [27], it can be shown that

$$\frac{1}{(2\pi i)^n} \left\langle F_{++\cdots+}(t), \frac{1}{\prod_{j=1}^{n}(z_j - t_j)} \right\rangle = \begin{cases} F(z), & \text{for } y \in S_{++\cdots+} \\ 0, & \text{elsewhere} \end{cases} \tag{6.105}$$

where the positive orthant $S_{++\cdots+} = \{y \in \mathbb{R}^n \mid y_j > 0, j = 1, 2, \ldots, n\}$. Results similar to (6.105) can be obtained by taking $F(z)$ holomorphic in other of the $2^n - 1$ orthants and evaluating the corresponding limits of $F(z)$. Let

$$\Omega = \{z \in \mathbb{C}^n \mid \operatorname{Im} z_j = y_j \neq 0 \, \forall j = 1, 2, \ldots, n\} \tag{6.106}$$

For $F(z) \in Hol(\Omega)$, there are 2^n different ways of evaluating $\lim_{y \to 0} F(z)$ depending upon the various components of y going to either 0^+ or 0^-. These limits are denoted by $F_{\sigma_k}(x)$.

Example 1. When $n = 2$, there are four quadrants S_{++}, S_{-+}, S_{+-}, and S_{--} and four different limits $F_{++}, F_{-+}, F_{+-}, F_{--}$, where, for example,

$$S_{-+} = \{y \in \mathbb{R}^2 \mid y_1 < 0 \text{ and } y_2 > 0\}$$

and

$$F_{-+}(x) = \lim_{y_1 \to 0_-, y_2 \to 0_+} F(z)$$

Let $Hol(\Omega)$ be the space of homorphic functions defined on Ω. Then

$$M = \{F(z) \in Hol(\Omega) \mid F(z)$$

$$\text{satisfies the following conditions } (A), (B), \&(C)\} \qquad (6.107)$$

$$\sup_{\substack{x_j \in \mathbb{R}, |y_j| \geq \delta > 0 \\ 1 \leq j \leq n}} |F(x + iy)| < A_\delta < \infty \qquad (A)$$

$$|F(x + iy)| = o(1) \qquad \text{as } |y_1|, \ldots, |y_n| \to \infty \qquad (B)$$

independently of each other and the asymptotic order is valid uniformly $\forall\, x \in \mathbb{R}^n$ and

$$\lim_{y \to 0_{\sigma_k}} F(z) = F_{\sigma_k}(x) \text{ in } \mathcal{D}'_{L^p}(\mathbb{R}^n), \qquad k = 1, 2, \ldots, 2^n \qquad (C)$$

where $y \to 0_{\sigma_k}$ means $y_j \to 0_{\sigma_k(j)}$, $1 \leq j \leq n$. Then we have the following theorem:

Theorem 16. For any $F(z) \in M$, we have

$$F(z) = \frac{1}{(2\pi i)^n} \left\langle \sum_{k=1}^{2^n} (-1)^{m_k} F_{\sigma_k}(t), \frac{1}{\prod_{j=1}^{n} (z_j - t_j)} \right\rangle \qquad (6.108)$$

where $m_k = $ the number of minus signs present in the sequences σ_k. For example, when $n = 2$,

$$F(z) = \frac{1}{(2\pi i)^2} \left\langle F_{++} + (-1)F_{-+} + (-1)F_{+-} + (-1)^2 F_{--}, \frac{1}{(z_1 - t_1)(z_2 - t_2)} \right\rangle$$

$$= -\frac{1}{4\pi^2} \left\langle (F_{++} - F_{-+} - F_{+-} + F_{--})(t), \frac{1}{(z_1 - t_1)(z_2 - t_2)} \right\rangle$$

6.9.5. Action of the Fourier Transform on the Hilbert Transform

If $f \in L^2(\mathbb{R})$, then

$$(\widehat{Hf})(x) = i\,\mathrm{sgn}(x)\hat{f}(x) \qquad \text{a.e. [94, p. 219]} \qquad (6.109)$$

where the Fourier transform $\hat{f}$ of f is defined by

$$\hat{f}(x) = \int_{\mathbb{R}} f(t) \exp(t \cdot x)\, dt \qquad (6.110)$$

Note that in the right-hand-side expression of (6.109) Stein and Weiss use $(-i)$ in place of i as their Hilbert transform differs from ours by a constant factor only. The result (6.109) can easily be extended to $L^2(\mathbb{R}^n)$ as follows:

Recall that the space $X(\mathbb{R}^n)$ consisting of finite linear combinations of functions of the type $\varphi_1(x_1)\varphi_2(x_2)\ldots\varphi_n(x_n)$, where each $\varphi_j(x_j) \in \mathcal{D}(\mathbb{R})$, is dense in $L^2(\mathbb{R}^n)$ [106, p. 71]. Therefore for $f \in L^2(\mathbb{R}^n)$, we can find a sequence ψ_m in $X(\mathbb{R}^n)$ such that $\psi_m(x) \to f(x)$ in $L^2(\mathbb{R}^n)$ as $m \to \infty$. Denoting by $\mathcal{F}$ the Fourier transform operator, we have

$$(\mathcal{F}(H(\psi_m)))(x) = i^n \operatorname{sgn}(x)(\mathcal{F}\psi_m)(x) \tag{6.111}$$

where

$$\operatorname{sgn}(x) = \prod_{j=1}^{n} \operatorname{sgn}(x_j)$$

Now letting $m \to \infty$ in (6.111) and interpreting the convergence in $L^2(\mathbb{R}^n)$, we deduce

$$(\mathcal{F}Hf)(x) = i^n \operatorname{sgn}(x)(\mathcal{F}f)(x) \tag{6.112}$$

The question now arises whether or not such a result can be proved for the space $L^p(\mathbb{R}^n)$, $p > 1$. We are able to prove the result (6.112) for $p = 2$ because of the fact that the Fourier transform maps $L^2(\mathbb{R}^n)$ into itself. But such a result is not true in general for $p > 1$, $p \neq 2$. If $f \in L^p(\mathbb{R}^n)$, $1 < p < \infty$, its Fourier transform can be defined, treating f as a regular tempered distribution, as follows:

$$\langle \hat{f}, \varphi \rangle = \langle f, \hat{\varphi} \rangle = \int_{\mathbb{R}^n} f\hat{\varphi}\, dx \qquad \forall \varphi \in S(\mathbb{R}^n)$$

where $\hat{\varphi}(x)$ is the classical Fourier transform of $\varphi(t)$ given by

$$\hat{\varphi}(x) = \int_{\mathbb{R}^n} \varphi(t) \exp(t \cdot x)\, dt$$

where $t \cdot x$ is now the inner product of t and x. For $f \in L^p(\mathbb{R}^n)$, let ψ_m be a sequence in $X(\mathbb{R}^n)$ tending to f in $L^p(\mathbb{R}^n)$, as $m \to \infty$. Then we have

$$\lim_{m \to \infty} \hat{\psi}_m = \hat{f} \text{ in } S'(\mathbb{R}^n)$$

Since the Hilbert transform H is a bounded linear operator from $L^p(\mathbb{R}^n)$ into itself [99], it follows that

$$\lim_{m \to \infty} \mathcal{F}(H(\psi_m)) = \mathcal{F}(Hf)$$

As $\psi_m \in L^2(\mathbb{R}^n)$, from (6.39) we conclude that

$$\lim_{m \to \infty} i^n \prod_{j=1}^{n} \operatorname{sgn}(x_j)\hat{\psi}_m = \mathcal{F}(H(f)) \tag{6.113}$$

That is,

$$\langle \mathcal{F}Hf, \varphi \rangle = \lim_{m \to \infty} \langle i^n \operatorname{sgn}(x)\hat{\psi}_m(x), \varphi(x) \rangle \tag{6.114}$$

$$= \lim_{m \to \infty} i^n \int_{\mathbb{R}^n} \operatorname{sgn}(x)\hat{\psi}_m \varphi(x)\, dx \qquad \forall \varphi \in S(\mathbb{R}^n)$$

But still we cannot, in general, say that the limit in (6.114) equals $i^n \operatorname{sgn}(x)\hat{f} \ \forall f \in L^p(\mathbb{R}^n)$, $1 < p < \infty$.

We now construct a testing function space $S_0(\mathbb{R}^n)$ that is a subspace of $S(\mathbb{R}^n)$ closed with respect to multiplication by $\prod_{j=1}^{n} \operatorname{sgn}(x_j)$. The topology of $S_0(\mathbb{R}^n)$ is the same as that induced on it by $S(\mathbb{R}^n)$. $S_0(\mathbb{R}^n)$ is a nonempty subspace of $S(\mathbb{R}^n)$. All functions in $S(\mathbb{R}^n)$ that vanish at the origin along with all of their derivatives are in $S_0(\mathbb{R}^n)$. For example,

$$\varphi(x) = \begin{cases} \prod_{j=1}^{n} \exp(-x_j^2 - x_j^{-2}), & x_j \neq 0, \ \forall j = 1, 2, \ldots, n, \\ 0, & \text{otherwise} \end{cases}$$

and

$$\psi(x) = \begin{cases} \exp\left(-\dfrac{1}{1 - |x|^2}\right) \exp(-|x|^2), & |x| > 1, \\ 0, & |x| \leq 1 \end{cases}$$

are members of $S_0(\mathbb{R}^n)$. The convergence of a sequence to zero in $S_0(\mathbb{R}^n)$ implies its convergence to zero in $\mathcal{D}_{L^p}(\mathbb{R}^n)$. Therefore the restriction of $f \in \mathcal{D}'_{L^p}(\mathbb{R}^n)$ to $S_0(\mathbb{R}^n)$ is in $S'_0(\mathbb{R}^n)$. We express this fact by saying that $\mathcal{D}'_{L^p}(\mathbb{R}^n) \subset S'_0(\mathbb{R}^n)$. Elements of $\mathcal{D}'_{L^p}(\mathbb{R}^n)$ cannot be identified with the elements of $S'_0(\mathbb{R}^n)$ in a one-to-one manner, since $S_0(\mathbb{R}^n)$ is not dense in $\mathcal{D}_{L^p}(\mathbb{R}^n)$. Therefore

$$\mathcal{F}(Hf) = i^n \prod_{j=1}^{n} \operatorname{sgn}(x_j)\mathcal{F}f \qquad \text{on } S_0(\mathbb{R}^n), \ \forall f \in L^p(\mathbb{R}^n) \tag{6.115}$$

Because

$$\langle \mathcal{F}(H\psi_m), \varphi \rangle = \langle i^n \operatorname{sgn}(x)(\mathcal{F}\psi_m)(x), \varphi(x) \rangle \qquad \text{(from 6.112)}$$
$$= \langle i^n \hat{\psi}_m(x), \operatorname{sgn}(x)\,\varphi(x) \rangle \qquad \forall \varphi \in S_0(\mathbb{R}^n)$$

Now taking the limit $m \to \infty$, we obtain

$$\langle \mathcal{F}(Hf), \varphi \rangle = \langle i^n \hat{f}(x), \operatorname{sgn}(x)\,\varphi(x) \rangle$$
$$= \langle i^n \operatorname{sgn}(x)\hat{f}, \varphi(x) \rangle \qquad \forall \varphi(x) \in S_0(\mathbb{R}^n)$$

Definition. The Hilbert transform Hf of $f \in \mathcal{D}'_{L^p}(\mathbb{R}^n)$ is defined by

$$\langle Hf, \varphi \rangle = \langle f, (-1)^n H\varphi \rangle \qquad \forall \varphi \in \mathcal{D}_{L^q}(\mathbb{R}^n)$$

where $(H\varphi)(x)$ is the Hilbert transform of $\varphi \in \mathcal{D}_{L^q}(\mathbb{R}^n)$, given by (6.78).

Definition. The Fourier transform $\mathcal{F}f \ (= \hat{f})$ of $f \in \mathcal{D}'_{L^p}(\mathbb{R}^n)$ is defined by

$$\langle \hat{f}, \varphi \rangle = \langle f, \hat{\varphi} \rangle \qquad \forall \varphi \in S(\mathbb{R}^n) \tag{6.116}$$

Then we have the following:

Theorem 17. Let $f \in \mathcal{D}'_{L^p}(\mathbb{R}^n)$, $1 < p < \infty$. Then

$$(\mathcal{F}(Hf))(x) = i^n \prod_{j=1}^{n} \text{sgn}(x_j)\hat{f}(x) \qquad \text{on } S_0(\mathbb{R}^n) \qquad (6.117)$$

Proof. Let $f \in \mathcal{D}'_{L^p}(\mathbb{R}^n)$, then for every $\varphi \in S_0(\mathbb{R}^n)$, we have

$$\langle \mathcal{F}Hf, \varphi \rangle = \left\langle \mathcal{F}H \sum_{|\alpha| \le m} \partial_t^\alpha f_\alpha, \varphi \right\rangle \qquad \text{[from (6.73)]}$$

$$= \left\langle \sum_{|\alpha| \le m} \partial_t^\alpha Hf_\alpha, \hat{\varphi} \right\rangle \qquad \text{[78]}$$

$$= \sum_{|\alpha| \le m} \langle Hf_\alpha, (-1)^{|\alpha|} \partial_t^\alpha \hat{\varphi} \rangle$$

$$= \sum_{|\alpha| \le m} \langle \mathcal{F}Hf_\alpha, (-1)^{|\alpha|} (ix)^\alpha \varphi \rangle \qquad \text{[96, p. 9]}$$

$$= \sum_{|\alpha| \le m} \langle i^n \, \text{sgn}(x) \hat{f}_\alpha(x), (-1)^{|\alpha|} (ix)^\alpha \varphi(x) \rangle$$

$$= \sum_{|\alpha| \le m} \langle i^n f_\alpha, (-1)^{|\alpha|} \partial_x^\alpha (\mathcal{F}(\text{sgn}(t)\varphi(t)))(x) \rangle$$

$$= \left\langle \sum_{|\alpha| \le m} i^n \partial_x^\alpha f_\alpha(x), (\mathcal{F}(\text{sgn}(t)\varphi(t)))(x) \right\rangle$$

$$= \langle i^n \, \text{sgn}(x) \hat{f}(x), \varphi(x) \rangle \qquad \text{[from (6.73)]} \qquad \square$$

Another proof of this theorem is given in [78]. In [78] the result

$$\mathcal{F}(Hf)(\xi) = i^n \, \text{sgn}(\xi)(Ff) \qquad \text{in } S'_0(\mathbb{R}^n)$$

is made use of to prove the fact that a bounded linear operator T from $L^p(\mathbb{R}^n)$ into itself that commutes with the operators of translation as well as dilatation is a finite linear combination of the identity operator I and the Hilbert transform-type operator $H_1, H_2, \ldots, H_n, H_iH_j, H_iH_jH_k, \ldots, H$.

For $f \in \mathcal{D}'_{L^p}(\mathbb{R}^n)$, define a holomorphic function

$$F(z) = \frac{1}{(2\pi i)^n} \left\langle f(t), \frac{1}{\prod_{j=1}^{n}(z_j - t_j)} \right\rangle, \qquad y_j \ne 0, \; j = 1, 2, \ldots, n \qquad (6.118)$$

where y_j is the imaginary part of z_j. Then we have the following decomposition theorem:

Theorem 18. For $f \in \mathcal{D}'_{L^p}(\mathbb{R}^n)$, $1 < p < \infty$, define $F(z)$ as in (6.118). Then

$$f = \sum_{k=1}^{2^n}(-1)^{m_k}F_{\sigma_k} \qquad \text{in } \mathcal{D}'_{L^p}(\mathbb{R}^n) \tag{6.119}$$

and

$$(-1)^{m_k}\widehat{F}_{\sigma_k}(\xi) = \begin{cases} \hat{f}(\xi), & \text{for } \xi \in S_{\sigma_k} \\ 0, & \text{elsewhere} \end{cases} \tag{6.120}$$

on $S_0(\mathbb{R}^n)$, where

$$F_{\sigma_k} = \lim_{y_1 \to 0_{\sigma_k(1)},\dots,y_n \to 0_{\sigma_k(n)}} F(z) \tag{6.121}$$

Here m_k stands for the number of negative signs in the sequence σ_k.

Proof. Without loss of generality, we can take $n = 2$. Then

$$F(z) = -\frac{1}{4\pi^2}\left\langle f(t), \frac{1}{(z_1 - t_1)(z_2 - t_2)} \right\rangle, \qquad y_1, y_2 \neq 0$$

Now

$$F_{++}(x) = \lim_{y_1,y_2 \to 0^+} F(z) = -\frac{1}{4}\left((H_1 - iI_1)(H_2 - iI_2)f\right)(x)$$

[Cor. 5.2]

$$F_{--}(x) = \lim_{y_1,y_2 \to 0^-} F(z) = -\frac{1}{4}\left((H_1 + iI_1)(H_2 + iI_2)f\right)(x)$$

Similarly we have

$$F_{+-}(x) = -\frac{1}{4}\left((H_1 - iI_1)(H_2 + iI_2)f\right)(x)$$

and

$$F_{-+}(x) = -\frac{1}{4}\left((H_1 + iI_1)(H_2 - iI_2)f\right)(x).$$

Then

$$[F_{++} - F_{+-} - F_{-+} + F_{--}](x) = -\frac{1}{4}[-4I]f(x) = f(x)$$

and

$$[F_{++} + F_{+-} + F_{-+} + F_{--}](x) = i^2(Hf)(x).$$

Using the method of induction, these results can be generalized to any positive integer n. Also

$$F_{++}(x) = -\frac{1}{4}[H - i(H_1 I_2 + H_2 I_1) - I]f(x) \qquad (6.122)$$

where $H = H_1 H_2$ and $I = I_1 I_2$. Taking the Fourier transform of equation (6.122), we get

$$\widehat{F}_{++}(\xi) = -\frac{1}{4}\left[i^2 \operatorname{sgn}(\xi_1) \operatorname{sgn}(\xi_2) - i^2(\operatorname{sgn} \xi_1 + \operatorname{sgn} \xi_2) - 1\right] \hat{f}(\xi)$$

$$= \frac{1}{4}\left[\operatorname{sgn}(\xi_1) \operatorname{sgn}(\xi_2) + \operatorname{sgn}(\xi_1) + \operatorname{sgn}(\xi_2) + 1\right] \hat{f}(\xi)$$

Case 1. $\xi_1, \xi_2 > 0$:

$$\widehat{F}_{++}(\xi) = \frac{1}{4}[1 - 1 - 1 + 1]\hat{f}(\xi) = 0$$

Case 2. $\xi_1, \xi_2 < 0$:

$$\widehat{F}_{++}(\xi) = \frac{1}{4}[1 + 1 + 1 + 1]\hat{f}(\xi) = \hat{f}(\xi)$$

Case 3. $\xi_1 > 0, \xi_2 < 0$:

$$\widehat{F}_{++}(\xi) = \frac{1}{4}[-1 - 1 + 1 + 1]\hat{f}(\xi) = 0$$

Case 4. $\xi_1 < 0, \xi_2 > 0$:

$$\widehat{F}_{++}(\xi) = \frac{1}{4}[-1 + 1 - 1 + 1]\hat{f}(\xi) = 0$$

Hence

$$\widehat{F}_{++} = \begin{cases} \hat{f}, & \text{for } \xi \in S_{--} = \{\xi \in \mathbb{R}^2 \mid \xi_1, \xi_2 < 0\} \\ 0, & \text{elsewhere} \end{cases}$$

Thus we have proved the theorem for $n = 2$. Using the induction, the proof can be given for any $n > 1$. $\square$

Note that

$$\mathcal{F}(F_{++} + F_{--})(\xi) = \begin{cases} \hat{f}(\xi), & \text{for } \xi \in S_{++} \cup S_{--} \\ 0, & \text{elsewhere} \end{cases}$$

Likewise

$$\left(\widehat{F}_{++} - \widehat{F}_{+-} + \widehat{F}_{--}\right)(\xi) = \begin{cases} \hat{f}(\xi), & \text{for } \xi \in S_{++} \cup S_{+-} \cup S_{--} \\ 0, & \text{elsewhere} \end{cases}$$

Similar results hold for all the other possible combinations of $\widehat{F}_{\sigma_k}$. Theorem 13 is analogous to the result proved by Tillmann [97, p. 19] for the space $H'(\mathbb{R}^n)$. However, our techniques is operator theoretic. In other words, it is based upon properties of the complex Hilbert transform and its limit, whereas the techniques used by Tillmann are essentially an outcome of complex integration in $\mathbb{C}^n$ on appropriately chosen Jordan arcs. The space $H(\overline{\mathbb{R}^n})$ chosen by Tillman [97] is a subspace of $\mathcal{D}_{L^p}(\mathbb{R}^n)$. The convergence of a sequence in $H(\overline{\mathbb{R}^n})$ to zero necessarily implies its convergence to zero in the space $\mathcal{D}_{L^p}(\mathbb{R}^n)$, and as such, the restriction of any $t \in (\mathcal{D}_{L^p}\mathbb{R}^n))'$ to $H(\overline{\mathbb{R}^n})$ is in $H'(\mathbb{R}^n)$. That is, in Tillman [97, p. 19] $(\mathcal{D}_{L^p}(\mathbb{R}^n))' \subset H'(\mathbb{R}^n)$. However, the advantage of our space $(\mathcal{D}_{L^p}(\mathbb{R}^n))'$ is that it is a Fourier as well as Hilbert transformable space. By Theorem 17 we are thus able to prove a Paley-Wiener-type Theorem 19. Some special cases of our representation formulas are also proved by Vladimirov [104, ch. 5].

Analyzing in the same manner yields the following results for $f \in \mathcal{D}'_{L^p}(\mathbb{R}^n)$:

Lemma 6. For $f \in \mathcal{D}'_{L^p}(\mathbb{R}^n)$, $1 < p < \infty$, and $F(z)$ defined as in (6.123), we have

$$\sum_{k=1}^{\ell} (-1)^{m_k} \widehat{F}_{\sigma_k}(\xi) = \begin{cases} \hat{f}(\xi), & \text{if } \xi \in \bigcup_{k=1}^{\ell} S_{\sigma_k} \\ 0, & \text{elsewhere,} \end{cases} \tag{6.123}$$

for $1 \le \ell \le 2^n$, equality in the sense of $S'_0(\mathbb{R}^n)$.

Suppose that one of the summands, say, $F_{\sigma_{k_0}}(\xi)$, for some $1 \le k_0 \le 2^n$, is zero $\forall \xi \in S_{\sigma_{k_0}}$. Then, since $\hat{f} = \sum_{k=1}^{2^n}(-1)^{m_k}\widehat{F}_{\sigma_k}$, equation (6.120) implies that $\hat{f}(\xi) = 0$ $\forall \xi \in S_{\sigma_{k_0}}$. Conversely, suppose that $\hat{f} = 0$ $\forall \xi \in S_{\sigma_{k_0}}$. Then again equation (6.120) gives $\widehat{F}_{\sigma_{k_0}} = 0$:

$$\langle \widehat{F}_{\sigma_{k_0}}, \varphi \rangle = 0 \qquad \forall \varphi \in S_0(\mathbb{R}^n)$$

So

$$\langle F_{\sigma_{k_0}}, \hat{\varphi} \rangle = 0, \qquad \forall \varphi \in S_0(\mathbb{R}^n)$$

We can generalize the above argument to obtain the following:

Theorem 19. Paley-Wiener Theorem for $\mathcal{D}'_{L^p}(\mathbb{R}^n)$. Let $f \in \mathcal{D}'_{L^p}(\mathbb{R}^n)$, $1 < p < \infty$. Define $F(z)$ by

$$F(z) = \frac{1}{(2\pi i)^n} \left\langle f(t), \frac{1}{\prod_{j=1}^{n}(z_j - t_j)} \right\rangle, \qquad \text{Im } z_j \ne 0, \ j = 1, 2, \dots, n$$

Then we have

$$\hat{f}(\xi) = 0 \qquad \text{for } \xi \in \bigcup_{k=1}^{\ell} S_{\sigma_k} \text{ in } S'_0(\mathbb{R}^n)$$

$$\text{iff } \sum_{k=1}^{\ell} (-1)^{m_k} F_{\sigma_k} = 0 \; \forall \, \xi \in \bigcup_{k=1}^{\ell} S_{\sigma_k} \text{ in } \mathcal{F}(S_0(\mathbb{R}^n))'$$

That is,

$$\left\langle \sum_{k=1}^{\ell} (-1)^{m_k} F_{\sigma_k}, \hat{\varphi} \right\rangle = 0 \qquad \forall \, \varphi \in S_0(\mathbb{R}^n)$$

with support contained in $\bigcup_{k=1}^{\ell} S_{\sigma_k}$.

Remark. Lemma 1 (Chapter 6), Theorems 13 and 14 are also true when we replace $\mathcal{D}'_{L^p}(\mathbb{R}^n)$ by $L^p(\mathbb{R}^n)$ and $F(z)$ by

$$\frac{1}{(2\pi i)^n} \int_{\mathbb{R}^n} f(t) \frac{1}{\prod_{j=1}^{n}(z_j - t_j)} \, dt, \qquad \operatorname{Im} z_j \neq 0, \; j = 1, 2, \ldots, n$$

and treating $L^p(\mathbb{R}^n)$ as a subspace of $\mathcal{D}'_{L^p}(\mathbb{R}^n)$.

6.10. THE DIRICHLET BOUNDARY-VALUE PROBLEM

Let $F(z) \in M$ as defined by (6.26). Then, by (6.7), we have

$$\Delta F(z) = \Delta \left(\frac{1}{2\pi i}\right)^n \left\langle \sum_{k=1}^{2^n} (-1)^{m_k} F_{\sigma_k}(t), \frac{1}{\prod_{j=1}^{n}(t_j - z_j)} \right\rangle \tag{6.124}$$

where

$$\Delta = \prod_{j=1}^{n} \left(\frac{\partial^2}{\partial x_j^2} + \frac{\partial^2}{\partial y_j^2} \right) \tag{6.125}$$

Using a method similar to that used in proving Lemma 1 in Chapter 3, we see that

$$\Delta F(z) = 0$$

So we have proved the next theorem.

Theorem 20. $\Delta F(z) = 0 \; \forall F(z) \in M$.

Consider the operator equation

$$\Delta u = 0 \tag{6.126}$$

with the following boundary conditions:

$$\lim_{y \to 0_{\sigma_k}} u = F_{\sigma_k} \qquad \text{in } \mathcal{D}'_{L^p}(\mathbb{R}^n), \; 1 \leq k \leq 2^n \tag{6.127}$$

Then

$$F(z) = \frac{1}{(2\pi i)^n} \left\langle \sum_{k=1}^{2^n} (-1)^{m_k} F_{\sigma_k}(t), \frac{1}{\prod_{j=1}^{n}(t_j - z_j)} \right\rangle, \qquad \text{Im } z_j \neq 0, \ j = 1, 2, \ldots, n$$

(6.128)

is in M (from Theorem 11), and it is also a solution of (6.126) with (6.127) as the boundary condition. The fact that $F(z)$ given by (6.128) is a unique solution in M of (6.126) and (6.127) follows from the representation formula (6.128).

6.11. EIGENVALUES AND EIGENFUNCTIONS OF THE OPERATOR H

Let $f \in \mathcal{D}'_{L^p}(\mathbb{R}^n)$, $p > 1$, be an eigenfunction of the operator of the generalized Hilbert transformation as defined in Chapters 3 and 4, and let λ be the corresponding eigenvalue. Then

$$Hf = \lambda \qquad (6.129)$$

Operating both sides of (6.127) by H, we have

$$H^2 f = \lambda H f$$

$$(-1)^n f = \lambda H f = \lambda(\lambda f) \qquad \text{(by Case 1)}$$

Therefore

$$\lambda^2 f = (-1)^n f$$

or

$$\lambda^2 = (-1)^n$$

If n is odd, we have

$$\lambda = \pm i$$

and if n is even, we have

$$\lambda = \pm 1$$

We now show that there do exist eigenfunctions corresponding to the possible eigenvalues ± 1, $\pm i$.

Case 1. n is even: (a) $\lambda = 1$ and $Hf = f$; take $f = g + Hg$ where $g \in \mathcal{D}'_{L^p}(\mathbb{R}^n)$. Then

$$H(g + Hg) = Hg + H^2 g = Hg + (-1)^n g = Hg + g$$

since n is even. So $g + Hg$ is an eigenfunction corresponding to the eigenvalue $\lambda = 1$.

(b) $\lambda = -1$; take $f = g - Hg$. We have

$$Hf = Hg - H^2g = Hg - (-1)^n g = Hg - g$$
$$= (-1)[g - Hg] = (-1)f$$

So $g - Hg$ is an eigenfunction corresponding to the eigenvalue $\lambda = -1$.

Case 2. n is odd: (a) $\lambda = i$; take $f = g - Hg$,

$$H[g - iHg] = Hg - iH^2g = Hg - i(-1)^n g = i[g - iHg]$$
$$Hf = if$$

Then $g - iHg$ is an eigenfunction corresponding to the eigenvalue $\lambda = i$.

(b) $\lambda = -i$; take $f = g + iHg$. Then $Hf = Hg + iH^2g = Hg - ig = (-i)f$.

Therefore $g + iHg$ is an eigenfunction corresponding to the eigenvalue $\lambda = -i$.

It is an open problem to find the class of all eigenvalues and eigenfunctions of the operator H and to find whether or not an eigenfunction expansion of Hf can be done interpreting the convergence of the series appropriately.

EXERCISES

1. (a) Find the distributional Fourier transform of $\sin t$.

   (b) Find the distributional Hilbert transform of $\cos t$.

   (c) Find $(\mathcal{F}H) \sin t$ and $(H\mathcal{F}) \sin t$ where $\mathcal{F}, H$ stand for Fourier and the Hilbert transform, respectively, taken in the distributional sense. You must specify the space of generalized functions over which your results are valid.

2. Construct a function $\varphi(t) \in S$ defined on the real line such that

$$\int_{-\infty}^{\infty} t^m \varphi(t)\, dt = 0 \qquad \forall m = 0, 1, 2, 3, \ldots$$

3. Solve the integral equation

$$f(t) + \frac{1}{\pi} \int_{-\infty}^{\infty} \frac{f(t)}{x - t}\, dt = \delta(t)$$

   in the space $(\mathcal{D}'_{L^p})$, $p > 1$.

4. Solve the Hilbert problem

$$F_+(x) - F_-(x) = \delta(t)$$

Stating various conditions satisfied by $F(z)$, interpreting convergence (a) in $\mathcal{D}'_{L^p}$, $p > 1$, and (b) $\mathcal{D}'$.

5. 5. Solve the following Hilbert problem in R^2,

$$F_{++}(x_1, x_2) - F_{-+}(x_1 x_2) - F_{+-}(x_1, x_2) + F_{--}(x_1, x_2) = \delta(x_1, x_2)$$

where

$$\langle \delta(x_1, x_2), \varphi(x_1, x_2) \rangle = \varphi(0, 0) \qquad \forall \varphi \in \mathcal{D}_{L^p}(R^2)$$

6. Generalize the problem given in Exercises 4 and 5, and find its solution.

7

PERIODIC DISTRIBUTIONS, THEIR HILBERT TRANSFORM AND APPLICATIONS

7.1. THE HILBERT TRANSFORM OF PERIODIC DISTRIBUTIONS

Let $f(t)$ be a periodic function with period 2τ defined on the real line R. A new definition of the Hilbert transform $(Hf)(x)$ of this function is given by

$$(Hf)(x) = \lim_{N \to \infty} \frac{1}{\pi} (P) \int_{-N}^{N} \frac{f(t)}{x - t} \, dt \qquad (7.1)$$

provided this limit exists; the integral in (7.1) is taken in the sense of Cauchy principal value. It is shown that the definition (7.1) is equivalent to the conventional definition

$$(Hf)(x) = \frac{1}{2\tau} (P) \int_{-\tau}^{\tau} f(x - t) \cot\left(\frac{t\pi}{2\tau}\right) dt \qquad (7.2)$$

where the integral in (7.2) is also taken in the sense of Cauchy principal value. We then use our results to extend the Hilbert transform to periodic distributions of period 2τ defined on the real line, that is, to $P'_{2\tau}$ where $P_{2\tau}$ is the space of infinitely differentiable functions with period 2τ defined on the real line. An inversion formula over the space $P'_{2\tau}$ is proved. Many other related results are proved, and our results are used in solving some singular integral equations in the space $P'_{2\tau}$.

7.1.1. Introduction

Assume that $f(z)$ is analytic in the region $\operatorname{Im} z \geq 0$, and let $u(x)$ and $v(x)$ be its real and imaginary parts on the real axis. Let us further assume that $f(z) = O(\frac{1}{|z|})$, $|z| \to \infty$ uniformly $\forall 0 \leq \theta \leq \pi$.

Using the contour integration technique over the semicircle in the upper half-plane, we can get

$$f(x, 0) = \frac{1}{\pi i}(P) \int_{-\infty}^{\infty} \frac{f(t, o)\,dt}{t - x}$$

Thus

$$u(x) + iv(x) = +i\frac{1}{\pi}(P) \int_{-\infty}^{\infty} \frac{u(t)\,dt}{x - t} - \frac{1}{\pi}(P) \int_{-\infty}^{\infty} \frac{v(t)\,dt}{x - t}$$

Therefore

$$u(x) = -(Hv)(x) \tag{7.3}$$

$$v(x) = (Hu)(x) \tag{7.4}$$

where H is the Hilbert transform defined on the real line. Relations (7.3) and (7.4) are called *reciprocity relations*. Now let us assume that $H_R(e^{iw})$ and $H_I(e^{iw})$ are the real and imaginary parts of the z-transform of a causal sequence $h(n)$, $n \geq 0$. Then over the unit circle we have

$$H_I(e^{iw}) = \frac{1}{2\pi}(P) \int_{-\pi}^{\pi} H_R(e^{i\theta}) \cot\left(\frac{\theta - \omega}{2}\right) d\theta \tag{7.5}$$

$$H_R(e^{iw}) = h(o) - \frac{1}{2\pi}(P) \int_{-\pi}^{\pi} H_I(e^{i\theta}) \cot\left(\frac{\theta - \omega}{2}\right) d\theta \tag{7.6}$$

[71, p. 344]. Motivated by the Hilbert reciprocity relation given by (7.3) and (7.4), we say that right-hand-side integral in (7.5) is the Hilbert transform of $H_R(e^{\theta i})$ and equals $H_I(e^{iw})$, and that (7.6) is an inversion formula for the Hilbert transform. Consequently the Hilbert transform of a periodic function $f(t)$ with period 2π is defined as

$$\frac{1}{2\pi}(P) \int_{-\pi}^{\pi} f(t) \cot\left(\frac{x - t}{2}\right) dt \tag{7.7}$$

provided that this integral exists in the sense of Cauchy principal value. By a simple transformation in (7.7), we can show that

$$(Hf)(x) = \frac{1}{2\pi}(P) \int_{-\pi}^{\pi} f(x - t) \cot\frac{t}{2}\, dt \tag{7.8}$$

This transform appears in the discussion of the convergence and the divergence of the Fourier series of a periodic function and the associated conjugate series [112, pp. 20–22; pp. 1–2; pp. 145–147]. The convergence a.e. of (7.8) implies the convergence of the conjugate Fourier series, and vice versa, provided that f is of bounded variation. There is, however, another good reason to choose (7.7) or (7.8) as the definition for the Hilbert transform of a periodic function with period 2π. In [65, pp. 67–69] it is shown that if t_o is an arbitrary point on a closed contour L and $\psi(t)$ is a given function satisfying H condition (i.e., $|\psi(t_1) - \psi(t_2)| \leq A |t_1 - t_2|^{\mu}$, $\mu > 0$, A is independent of t_1, t_2) and $\varphi(t)$ is an unknown function satisfying

$$\frac{1}{\pi i} \int_L \frac{\varphi(t) \, dt}{t - t_o} = \psi(t_o) \qquad (7.9)$$

then the only solution of (7.9) in the space of functions satisfying H condition on L is

$$\frac{1}{\pi i} \int_L \frac{\psi(t) \, dt}{t - t_o} = -\varphi(t_0) \qquad (7.10)$$

Using the transformation $t = e^{i\theta}$, $t_o = e^{i\theta_o}$, we have

$$\frac{dt}{t - t_o} = \frac{1}{2} \cot\left(\frac{\theta - \theta_o}{2}\right) d\theta + \frac{i}{2} d\theta \qquad \text{[65, p. 69]}$$

So using further restrictions on φ and ψ that they are periodic functions with period 2π, (7.9) and (7.10) are transformed into

$$\frac{1}{2\pi} \int_o^{2\pi} \varphi(\theta) \cot\left(\frac{\theta - \theta_o}{2}\right) d\theta + \frac{i}{2\pi} \int_o^{2\pi} \varphi(\theta) \, d\theta = \psi(\theta_o) \qquad (7.11)$$

$$\frac{1}{2\pi} \int_o^{2\pi} \psi(\theta) \cot\left(\frac{\theta - \theta_o}{2}\right) d\theta + \frac{i}{2\pi} \int_o^{2\pi} \psi(\theta) \, d\theta = -\varphi(\theta_o) \qquad (7.12)$$

With an additional restriction on φ and ψ,

$$\int_o^{2\pi} \varphi(\theta) \, d\theta = \int_o^{2\pi} \psi(\theta) \, d\theta = 0$$

we have

$$-\frac{1}{2\pi} \int_o^{2\pi} \varphi(\theta) \cot\left(\frac{\theta_o - \theta}{2}\right) d\theta = \psi(\theta_o) \qquad (7.13)$$

$$\frac{1}{2\pi} \int_o^{2\pi} \psi(\theta) \cot\left(\frac{\theta_o - \theta}{2}\right) d\theta = \varphi(\theta_o) \qquad (7.14)$$

Thus (7.13) is taken as the Hilbert transform of a periodic function with period 2π, and (7.14) gives its inversion formula. Now using the definition (7.1) which is similar to the definition of the Hilbert transform of a function on the real line, we will arrive at a similar definition of the Hilbert transform of periodic functions. Butzer and Nessel [12] have extended the definition (7.13) and its inversion formula to the class $L_{2\pi}^p$ (periodic functions with period 2π satisfying $\int_{-\pi}^{\pi} |f(x)|^p \, dx < \infty$, $p \geq 1$), and we will be using many of their results proved in [2]. We will see that the uniqueness theorem in general is not valid for the class of functions $L_{2\tau}^p$, $\tau > 0$, and $p \geq 1$.

We now state without proof a lemma that will be used in the sequel.

Lemma 1. Let $\tau > 0$ and $t \in \mathbb{R}$. Then

$$\frac{\pi}{2\tau} \cot\left(\frac{t\pi}{2\tau}\right) = \frac{1}{t} + \sum_{\substack{k=-\infty \\ k\neq o}}^{\infty} \left[\frac{1}{t - 2k\tau} + \frac{1}{2k\tau}\right], \qquad t \neq 0 \qquad (7.15)$$

$$\frac{\pi}{2\tau} \cot\left(\frac{t\pi}{2\tau}\right) = \frac{1}{t} + \lim_{n\to\infty} \left[\sum_{\substack{k=-n \\ k\neq 0}}^{n} \frac{1}{t - 2k\tau}\right], \qquad t \neq 0 \qquad (7.16)$$

where $n \in N_+$. The equation (7.16) is an immediate consequence of (7.15), and (7.15) is an immediate consequence of the results [12, p. 335]. Note that the series in (7.15) and (7.16) converge uniformly over any compact subset of the real line not containing the set of points $2k\tau \; \forall k = 0, \pm 1, \pm 2, \pm 3, \ldots$.

We now prove

Theorem 1. Let $f \in L_{2\tau}^p$, $p \geq 1$. That is, $f(t)$ defined on the real line is a periodic function of t with period 2τ and $\int_{-\tau}^{\tau} |f(t)|^p \, dt < \infty$.
Then the Hilbert transform $(Hf)(x)$ of $f(t)$ as defined by (7.1) is

$$\frac{1}{2\tau}(P) \int_{-\tau}^{\tau} f(x - t) \cot\left(\frac{t\pi}{2\tau}\right) dt$$

provided that the integral exists.

Proof. If the limit in (7.1) exists, then

$$(Hf)(x) = \lim_{n\to\infty} \frac{1}{\pi}(P) \int_{-n\tau}^{n\tau} \frac{f(t)}{x - t} \, dt, \qquad n \in N_+$$

$$= \lim_{n\to\infty} \frac{1}{\pi}(P) \int_{-(2n+1)\tau}^{(2n+1)\tau} \frac{f(t)}{x - t} \, dt, \qquad n \in N_+$$

$$= \lim_{n\to\infty} \frac{1}{\pi}(P) \int_{-(2n+1)\tau}^{(2n+1)\tau} \frac{f(x - t)}{t} \, dt$$

$$= \frac{1}{\pi}\left[\cdots(P) \int_{-5\tau}^{-3\tau} \frac{f(x - t)}{t} \, dt + (P) \int_{-3\tau}^{-\tau} \frac{f(x - t)}{t} \, dt + (P) \int_{-\tau}^{\tau} \frac{f(x - t)}{t} \, dt\right.$$

$$\left. + (P) \int_{\tau}^{3\tau} \frac{f(x - t)}{t} \, dt + (P) \int_{3\tau}^{5\tau} \frac{f(x - t)}{t} \, dt + \cdots\right]$$

$$= \lim_{N\to\infty} \frac{1}{\pi}(P) \int_{-\tau}^{\tau} f(x - t)\left[\frac{1}{t} + \sum_{\substack{n=-N \\ n\neq 0}}^{N} \frac{1}{t - 2n\tau}\right] dt \qquad (7.17)$$

$$= \lim_{N\to\infty} \frac{1}{\pi}(P) \int_{-\tau}^{\tau} f(x - t)\left[\frac{1}{t} + \sum_{\substack{n=-N \\ n\neq 0}}^{N} \left(\frac{1}{t - 2n\tau} + \frac{1}{2n\tau}\right)\right] dt \qquad (7.18)$$

The equality of (7.17) and (7.18) follows from Lemma 1. The series in (7.18) converges absolutely and uniformly to $\frac{\pi}{2\tau}\cot(\frac{t\pi}{2\tau})$ in a compact subset of the real line obtained by deleting the origin and the points $\pm 2\tau, \pm 4\tau, \pm 6\tau, \ldots$. Therefore

$$(Hf)(x) = \frac{1}{2\tau}(P) \int_{-\tau}^{\tau} f(x-t) \cot\left(\frac{t\pi}{2\tau}\right) dt \qquad (7.19)$$

provided that the integral in the right-hand side of (7.19) exists. It is a fact that if $f \in L_{2\tau}^p$, $p \geq 1$ the integral in (7.19) exists a.e. [12, p. 335]. Using the fact that f is periodic with period 2τ, we can also show that

$$(Hf)(x) = \frac{1}{2\tau}(P) \int_{-\tau}^{\tau} f(t) \cot\left(\frac{(x-t)\pi}{2\tau}\right) dt \qquad (7.20)$$

We now prove a theorem that will be used in the sequel. $\square$

Theorem 2. If $\varphi(t)$ is an infinitely differentiable periodic function with period 2τ, then its Hilbert transform $(H\varphi)(x)$ is a periodic function with period 2τ, and

$$\frac{d^k}{dx^k}(H\varphi)(x) = (H\varphi^{(k)})(x) \qquad (7.21)$$

Proof.

$$F(x) = (H\varphi)(x) = \frac{1}{2\tau}(P) \int_{-\tau}^{\tau} \varphi(x-t) \cot\left[\frac{t\pi}{2\tau}\right] dt$$

$$= \frac{1}{2\tau}(P) \int_{0}^{\tau} [\varphi(x-t) - \varphi(x+t)] \cot\left(\frac{t\pi}{2\tau}\right) dt$$

$$= \frac{1}{2\tau}(P) \int_{0}^{\tau} \frac{\varphi(x-t) - \varphi(x+t)}{t} \left(t \cot\frac{t\pi}{2\tau}\right) dt$$

Define $\psi(x,t)$ as a function of two variables as follows:

$$\psi(x,t) = \frac{\varphi(x-t) - \varphi(x+t)}{t}, \qquad t \neq 0$$

$$= -2\varphi^{(1)}(x) \qquad \text{when } t = 0$$

$\psi(x,t)$ is an infinitely differentiable function with respect to t and x. Therefore

$$F(x) = \frac{1}{2\tau} \int_{0}^{\tau} \psi(x,t) \left\{t \cot\left(\frac{t\pi}{2\tau}\right)\right\} dt$$

or

$$F^{(k)}(x) = \frac{1}{2\tau} \int_{0}^{\tau} \left(\frac{\partial}{\partial x}\right)^k \psi(x,t) t \cot\frac{t\pi}{2\tau} dt$$

$$= \frac{1}{2\tau} \int_{0}^{\tau} \frac{\varphi^{(k)}(x-t) - \varphi^{(k)}(x+t)}{t} t \cot\frac{t\pi}{2\tau} dt$$

or

$$F^{(k)}(x) = \frac{1}{2\tau}(P)\int_o^\tau [\varphi^{(k)}(x-t) - \varphi^{(k)}(x+t)]\cot\left(\frac{t\pi}{2\tau}\right)dt$$

$$= \frac{1}{2\tau}(P)\int_{-\tau}^\tau \varphi^{(k)}(x-t)\cot\left(\frac{t\pi}{2\tau}\right)dt$$

The fact that $F(x)$ is a periodic function with period 2τ is obvious. ☐

7.2. DEFINITIONS AND PRELIMINARIES

7.2.1. Testing Function Space $P_{2\tau}$

A function $\varphi(t)$ defined on the real line is said to belong to the space $P_{2\tau}$ iff $\varphi(t)$ is an infinitely differentiable function with period 2τ. Let us assume that $\mathcal{D}$ is the Schwartz testing function space consisting of infinitely differentiable functions defined on $\mathbb{R}$ with compact support [1, 87]. Any testing function φ in $\mathcal{D}$ generates a unique testing function θ in $P_{2\tau}$ through the expression

$$\theta(t) = \sum_{n=-\infty}^{\infty} \varphi(t - 2n\tau) \tag{7.22}$$

Over any bounded t interval there are only a finite number of nonzero terms in the summation in (7.22) because φ has bounded support. Thus we may differentiate (7.22) term by term to get

$$\theta^{(k)}(t) = \sum_{n=-\infty}^{\infty} \varphi^{(k)}(t - n2\tau), \qquad k = 1, 2, 3, \ldots \tag{7.23}$$

Convergence in the space $P_{2\tau}$ is defined as follows: A sequence $\{\theta_\nu(t)\}_{\nu=1}^\infty$ is said to converge in $P_{2\tau}$ to a limit θ if every θ_ν is in $P_{2\tau}$ and if for each nonnegative integer k the sequence $\{\theta_\nu^{(k)}(t)\}_{\nu=1}^\infty$ converges to $\theta^{(k)}(t)$ uniformly for all t. The limit function θ will also be in $P_{2\tau}$, and as such $P_{2\tau}$ is closed under convergence. A function $\xi(t)$ is said to be unitary if it is an element of $\mathcal{D}$ and if there exists a real number τ for which

$$\sum_{n=-\infty}^{\infty} \xi(t - 2n\tau) = 1 \tag{7.24}$$

Many such functions exist. An example of one of them for $2\tau = 1$ is

$$\xi(t) = \frac{\int_{|t|}^1 \exp[-1/x(1-x)]\,dx}{\int_0^1 \exp[-1/x(1-x)]\,dx} - 1 < t < 1, \qquad \xi(t) = o, \; |t| \geq 1 \tag{7.25}$$

It is easy to see that

$$\sum_{n=-\infty}^{\infty} \xi^{(k)}(t - n2\tau) = 0, \qquad k = 1, 2, 3, \ldots \tag{7.26}$$

[109, p. 315].

We denote the space of all functions that are unitary with respect to a fixed real number 2τ by the symbol $U_{2\tau}$. Clearly, if θ is in $P_{2\tau}$ and ξ is in $U_{2\tau}$, then $\xi\theta$ is in D. Also, if $\{\theta_\nu\}_{\nu=1}^\infty$ converges in $P_{2\tau}$ to θ, then $\{\xi\theta_\nu\}_{\nu=1}^\infty$ converges in D to $\xi\theta$. Every θ in $P_{2\tau}$ can be generated through (7.22) from some φ in D. This is because

$$\sum_{n=-\infty}^{\infty} \xi(t - 2n\tau)\theta(t - 2n\tau) = \theta(t) \sum_{n=-\infty}^{\infty} \xi(t - 2n\tau) = \theta(t) \tag{7.27}$$

7.2.2. The Space $P'_{2\tau}$ or Periodic Distributions

For $f \in D'$, f is said to be a periodic distribution with period 2τ if

$$f(t - 2\tau) = f$$

$$\langle f(t), \varphi(t) \rangle = \langle f(t - 2\tau), \varphi(t) \rangle \qquad \forall \varphi \in D$$

For example, $f(t) = \sin(\frac{t\pi}{\tau})$ is a regular distribution with period 2τ, and $g(t) = \sum_{n=-\infty}^{\infty} \delta(t - 2n\pi)$ is a periodic distribution with period 2τ. One can easily see that $g(t - 2\pi) = \sum_{n=-\infty}^{\infty} \delta(t - 2(n+1)\pi) = g(t)$. Also

$$\left\langle \sum_{n=-\infty}^{\infty} \delta(t - 2n\pi), \varphi \right\rangle = \sum_{n=-\infty}^{\infty} \varphi(2n\pi) \tag{7.28}$$

Clearly (7.28) defines g as a continuous linear functional over D, since only finitely many terms in (7.28) survive as φ is of compact support. So $g(t)$ is a periodic distribution with period 2π; that is, $g(t) \in P'_{2\tau}$.

Theorem 3. If f is a periodic distribution with period 2τ (i.e., $f \in P'_{2\tau}$), then it can be identified as a continuous linear functional on $P_{2\tau}$.

Proof. The complex number that f assigns to any θ in $P_{2\tau}$ will be denoted by the dot product $f \cdot \theta$ in order to avoid confusion with the number $\langle f, \varphi \rangle$ that f assigns to any φ in D. This number $f \cdot \theta$ is defined by

$$f \cdot \theta \overset{\triangle}{=} \langle f, \xi\theta \rangle \tag{7.29}$$

where ξ is any unitary function in $U_{2\tau}$. We first show that the right-hand-side expression in (7.29) is independent of the choice of $\xi \in P_{2\tau}$. It is a fact that

$$f(t) = \sum_{n=-\infty}^{\infty} f(t)\xi(t - 2n\tau) \tag{7.30}$$

The summation in (7.30) has only finitely many nonzero terms over any bounded open t interval. We may write for any $\varphi \in \mathcal{D}$,

$$\left\langle \sum_n f(t)\xi(t - 2n\tau), \varphi(t) \right\rangle = \left\langle f(t), \varphi(t) \sum_n \xi(t - 2n\tau) \right\rangle$$

$$= \langle f(t), \varphi(t) \rangle$$

Now let ξ and k be any two elements in $U_{2\tau}$. By using (7.30) and the fact that $f(t) = f(t + 2n\tau)$, we see that for every θ in $P_{2\tau}$,

$$\langle f, \xi\theta \rangle = \left\langle \sum_n f(t)k(t - 2n\tau), \xi(t)\theta(t) \right\rangle$$

$$= \sum_n \langle f(t)k(t - 2n\tau)\,\xi(t), \theta \rangle$$

$$= \sum_n \langle f(t + 2n\tau)k(t)\xi(t + 2n\tau), \theta(t) \rangle$$

$$= \left\langle \sum_n f(t)\xi(t + 2n\tau), k(t)\theta(t) \right\rangle$$

$$= \langle f, k\theta \rangle$$

Let $\theta_1, \theta_2 \in P_{2\tau}$ and α, β be any two numbers. We have

$$f \cdot (\alpha\theta_1 + \beta\theta_2) = \langle f, \xi(\alpha\theta_1 + \beta\theta_2) \rangle$$

$$= \alpha \langle f, \xi\theta_1 \rangle + \beta \langle f, \xi\theta_2 \rangle$$

$$= \alpha(f \cdot \theta_1) + \beta(f \cdot \theta_2)$$

Thus the linearity of f over $P_{2\tau}$ is proved. $\square$

Now we show that f is also continuous over $P_{2\tau}$. If $\{\theta_\nu\}_{\nu=1}^\infty$ converges in $P_{2\tau}$ to θ, then $\{\xi\theta_\nu\}_{\nu=1}^\infty$ converges in $\mathcal{D}$ to $\xi\theta$. Hence, as $\nu \to \infty$,

$$f \cdot \theta_\nu = \langle f, \xi\theta_\nu \rangle \to \langle f, \xi\theta \rangle = f \cdot \theta$$

Thus f is a continuous linear functional over $P_{2\tau}$.

Thus far we have identified a periodic distribution with period 2τ as a continuous linear functional over $P_{2\tau}$. Now we will show that continuous linear functional over $P_{2\tau}$ can also be identified as a periodic distribution with period 2τ in such a way that

$$f \cdot \theta = \langle f, \xi\theta \rangle \tag{7.31}$$

still holds true. This is done as follows: If φ in $\mathcal{D}$ will generate a θ in $P_{2\tau}$ through the expression

$$\theta(t) = \sum_{n=-\infty}^{\infty} \varphi(t - 2n\tau) \tag{7.32}$$

then, by virtue of the fact that f is a functional on $P_{2\tau}$, we define the number $\langle f, \varphi \rangle$ by the relation

$$\langle f, \varphi \rangle \overset{\triangle}{=} f \cdot \theta \tag{7.33}$$

This defines f as a functional on $\mathcal{D}$. Moreover, since both $\varphi(t)$ and $\varphi(t + 2\tau)$ generate through (7.31) the same $\theta(t)$, we have that

$$\langle f(t), \varphi(t) \rangle = f(t) \cdot \theta(t) = \langle f(t), \varphi(t + 2\tau) \rangle \tag{7.34}$$

Theorem 4. If f is a continuous linear functional on $P_{2\tau}$ and if θ and φ are related by (7.32), then (7.33) defines f as a periodic distribution with period 2τ.

Proof. We first prove that f defined on $\mathcal{D}$ by (7.31) is a linear functional

$$\langle f, \alpha\varphi_1 + \beta\varphi_2 \rangle = f \cdot (\alpha\theta_1 + \beta\theta_2) = \alpha(f \cdot \theta_1) + \beta(f \cdot \theta_2)$$
$$= \alpha\langle f, \varphi_1 \rangle + \beta\langle f, \varphi_2 \rangle$$

To show that f defined by (7.32) is continuous on $\mathcal{D}$, we assume that $\{\varphi_\nu\}_{\nu=1}^{\infty}$ converges in $\mathcal{D}$ to φ. Therefore $\{\theta_\nu\}_{\nu=1}^{\infty}$ converges in $P_{2\tau}$ to θ defined by (7.32). Since f is a continuous functional on $P_{2\tau}$, we have

$$\langle f, \varphi_\nu \rangle = f \cdot \theta_\nu \rightarrow f \cdot \theta = \langle f, \varphi \rangle$$

From (7.34) it follows that f is periodic with period 2τ. $\square$

We have assumed that the definitions (7.31) and (7.33) are consistent. In fact it is so. For by using (7.32) and the fact that $f(t) = f(t + 2n\tau)$, we have $\forall \xi \in U_{2\tau}$,

$$\langle f, \xi\theta \rangle = \left\langle f(t), \xi(t) \sum_n \varphi(t - 2n\tau) \right\rangle$$
$$= \sum_n \langle f(t), \xi(t)\varphi(t - 2n\tau) \rangle$$
$$= \sum_n \langle f(t + 2n\tau), \xi(t + 2n\tau)\varphi(t) \rangle$$
$$= \left\langle f(t), \varphi(t) \sum_n \xi(t + 2n\tau) \right\rangle = \langle f, \varphi \rangle$$

Corollary 1. Two distributions f and g in $P'_{2\tau}$ are equal if $f \cdot \theta = g \cdot \theta \; \forall \theta \in P_{2\tau}$. In view of the relations (7.32) and (7.33) we obtain for every φ in $\mathcal{D}$,

$$\langle f, \varphi \rangle = f \cdot \theta = g \cdot \theta = \langle g, \varphi \rangle$$

so that $f = g$.

Example 1. Let f be a regular distribution generated by locally integrable periodic function of period 2τ. Then

$$f \cdot \theta = \int_a^{a+2\tau} f(t)\theta(t)\,dt$$

since

$$f \cdot \theta = \langle f, \xi\theta \rangle = \int_{-\infty}^{\infty} f(t)\xi(t)\theta(t)\,dt$$

$$= \sum_{n=-\infty}^{\infty} \int_{a+2n\tau}^{a+2n\tau+2\tau} f(t)\xi(t)\theta(t)\,dt$$

$$= \sum_n \int_a^{a+2\tau} f(t + 2n\tau)\xi(t + 2n\tau)\theta(t + 2n\tau)\,dt$$

$$= \int_a^{a+2n\tau} f(t)\theta(t) \sum_n \xi(t + 2n\tau)\,dt = \int_a^{a+2\tau} f(t)\theta(t)\,dt$$

The interchange of the summation and the integration is again justified by virtue of the fact that there are only a finite number of nonzero terms in the summation.

Example 2. Let

$$\delta_{2\tau}(t) = \sum_{n=-\infty}^{\infty} \delta(t - 2n\tau), \qquad \tau > 0$$

Then

$$\delta_{2\tau} \cdot \theta = \theta(0)$$

$$\delta_{2\tau}(t - \tau) \cdot \theta(t) = \theta(\tau)$$

It can be readily checked that $\delta_{2\tau}(t) \in P'_{2\tau}$. Therefore

$$\delta_{2\tau} \cdot \theta = \langle \delta_{2\tau}, \xi\theta \rangle = \left\langle \sum_n \delta(t - 2n\tau), \xi(t)\theta(t) \right\rangle$$

$$= \sum_n \xi(2n\tau)\theta(2n\tau) = \theta(0) \sum_n \xi(2n\tau) = \theta(0)$$

since $\sum_{n=-\infty}^{\infty} \xi(t - 2n\tau) = 1$. Similarly we can show that

$$\delta_{2\tau}(t - a) \cdot \theta(t) = \theta(a)$$

Example 3. Consider

$$\delta_{2\tau}^{(k)}(t) = \sum_{n=-\infty}^{\infty} \delta^{(k)}(t - 2n\tau)$$

Then

$$\delta_{2\tau}^{(k)} \cdot \theta = (-1)^k \theta^{(k)}(0) \qquad \forall \theta \in P_{2\tau}$$

since

$$\delta_{2\tau}^{(k)} \cdot \theta = \sum_{n=-\infty}^{\infty} (-1)^k \frac{d^k}{dt^k}[\xi(t)\theta(t)]\Big|_{t=2n\tau}$$

$$= \sum_{n=-\infty}^{\infty} (-1)^k \sum_{p=0}^{k} \binom{k}{p} \xi^{(p)}(2n\tau)\theta^{(k-p)}(2n\tau)$$

$$= (-1)^k \sum_{p=9}^{k} \binom{k}{p} \theta^{(k-p)}(0) \sum_{n=-\infty}^{\infty} \xi^{(p)}(2n\tau) \qquad (7.35)$$

$$= (-1)^k \theta^{(k)}(0)$$

Note that the right-hand-side summation in (7.35) vanishes for $p \geq 1$. So the nonzero term in (7.35) is obtained only for $p = 0$.

Theorem 5. The space $P_{2\tau}'$ is complete. In other words, the limit f of every sequence $\{f_\nu\}_{\nu=1}^{\infty}$ that converges in $P_{2\tau}'$ is also in $P_{2\tau}'$.

Proof. For every φ in $\mathcal{D}$ there exists a θ in $P_{2\tau}$ related through (7.32). Therefore, from (7.33), and using the assumption that $P_{2\tau}'$ is closed under convergence, we have

$$\langle f_\nu, \varphi \rangle = f_\nu \cdot \theta$$

$\{f_\nu, \varphi >\}_{\nu=1}^{\infty}$ is convergent as is $\{f_\nu \cdot \theta\}_{\nu=1}^{\infty}$. Since $\mathcal{D}'$ is weakly complete, there exists $f \in \mathcal{D}'$ satisfying

$$\lim\langle f_\nu, \varphi \rangle = \langle f, \varphi \rangle = f \cdot \theta$$

that is, $\lim f_\nu \cdot \theta = f \cdot \theta$. Again, each $f_\nu \in P_{2\tau}'$, therefore $f \in P_{2\tau}'$. Also we have

$$\lim f_\nu \cdot \theta = f \cdot \theta \qquad \square$$

Theorem 6. The sequence $\{f_\nu\}_{\nu=1}^{\infty}$ converges in $P_{2\tau}'$ to f if and only if it converges in $\mathcal{D}'$ to f and each f_ν is in $P_{2\tau}'$.

Proof. In the preceding theorem it has already been shown that the convergence in $P'_{2\tau}$ implies convergence in $\mathcal{D}'$. Conversely assume that for each $f_\nu \in P'_{2\tau}$ the sequence $\{f_\nu\}$ converges in $\mathcal{D}'$. It follows that the limit f must also be in $P'_{2\tau}$.

Then, for ξ in $U_{2\tau}$ and θ in $P_{2\tau}$

$$f_\nu \cdot \theta = \langle f_\nu, \xi\theta \rangle \rightarrow \langle f, \xi\theta \rangle = f \cdot \theta \qquad \square$$

7.3. SOME WELL-KNOWN OPERATIONS ON $P'_{2\tau}$

1. Addition of periodic distributions on $P'_{2\tau}$:

$$(f + g) \cdot \theta = f \cdot \theta + g \cdot \theta$$

2. Multiplication of a periodic distribution by a constant α:

$$(\alpha f) \cdot \theta \overset{\Delta}{=} \alpha(f \cdot \theta)$$

So $P'_{2\tau}$ is a linear space.
3. The shifting property of a periodic distribution:

$$f(t - \tau) \cdot \theta(t) \overset{\Delta}{=} f(t) \cdot \theta(t)$$

4. The differentiation of a periodic distribution:

$$f^{(k)} \cdot \theta \overset{\Delta}{=} (-1)^k f \cdot \theta^{(k)}, \qquad k = 1, 2, 3, \ldots$$

One can also show that

$$\frac{d}{dt}(\alpha f + \beta g) = \alpha \frac{df}{dt} + \beta \frac{dg}{dt}$$

5. The multiplication of a distribution f in $P'_{2\tau}$ by a periodic testing function p in $P_{2\tau}$:

$$(pf) \cdot \theta \overset{\Delta}{=} f \cdot (p\theta)$$

That is, f is periodic with period 2τ.

7.4. THE FUNCTION SPACE $L^p_{2\tau}$ AND ITS HILBERT TRANSFORM, $p \geq 1$

A measurable function $f(t)$ defined on the real line is said to belong to the space $L^p_{2\tau}$ if and only if

1. $f(t)$ is periodic with period 2τ, that is, $f(t + 2\tau) = f(t)$ a.e.
2. $\int_{-\tau}^{\tau} |f(t)|^p \, dt < \infty$.

It is a fact that every element $f \in L_{2\tau}^p$ for $p > 1$ is Hilbert transformable in that $(Hf)(x)$ exists a.e. in the sense of (7.2) and that $(Hf)(x) \in L^p$ satisfying

$$\|Hf\| \leq C_p \|f\|_p \tag{7.36}$$

[12, p. 336][9, p. 147].

Since $(Hf)(x)$ is also periodic with period 2τ, it follows that $(Hf)(x) \in L_{2\tau}^p$. Thus we see that the function space $L_{2\tau}^p$ is closed with respect to the operation H defined by (7.1). If $\{\varphi_\nu(t)\}$ is a sequence of periodic functions tending to zero in $P_{2\tau}$, then $\varphi_\nu^{(k)}(t) \to o$ as $\nu \to \infty$ uniformly for a fixed $k = 0, 1, 2, 3, \ldots$.

Therefore

$$\|\varphi_\nu^{(k)}\|_p \to 0 \qquad \text{as } \nu \to \infty$$

Hence

$$\|(H\varphi_\nu)^{(k)}\|_p = \|H\varphi_\nu^{(k)}\|_p \leq C_p \|\varphi_\nu^{(k)}\|_p \to 0 \qquad \text{as } \nu \to \infty$$

where

$$\|\varphi\|_p = \left(\int_{-\tau}^{\tau} |f(t)|^p \, dt \right)^{1/p}, \qquad p > 1$$

It is a fact that if $f \in L_{2\tau}^p$, $p > 1$ and $g \in L_{2\tau}^q$, $q > 1$ such that $\frac{1}{p} + \frac{1}{q} = 1$, then $\int_{-\tau}^{\tau} (Hf)(x)g(x) \, dx = -\int_{-\tau}^{\tau} f(x)(Hg)(x) \, dx$ [12, p. 339]. In duality notation

$$\langle Hf, g \rangle = \langle f, -Hg \rangle \tag{7.37}$$

The Hilbert transform Hf of $f \in P_{2\tau}'$ in analogy with (7.37) can now be defined as

$$Hf \cdot \theta = -f \cdot (H\theta) \tag{7.38}$$

Therefore

$$Hf \cdot \theta = \langle f, -\xi H\theta \rangle \qquad \forall \theta \in P_{2\tau}, \ \xi \in U_{2\tau}$$

Since the space $P_{2\tau}$ is closed with respect to the operation H it follows that (7.38) defines a functional Hf over $P_{2\tau}$. The linearity of the functional over $P_{2\tau}$ is trivial. We prove the continuity of Hf as follows:

Let $\{\theta_\nu\}_{\nu=1}^{\infty}$ be a sequence $\to 0$ as $\nu \to \infty$ in $P_{2\tau}$. Then it also converges to zero in the space $D_{L^p[-\tau,\tau]}$, where $D_{L^p[-\tau,\tau]}$ consists of infinitely differentiable functions defined on $[-\tau, \tau]$, which along with all its derivative is L^p bounded, and $p > 1$. The topology in the space $D_{L^p[-\tau,\tau]}$ is defined by a separating collection of seminorms $\{\gamma_k\}_{k=0}^{\infty}$ where

$$\gamma_k(\theta) = \left[\int_{-\tau}^{\tau} |\theta^{(k)}(t)|^p \, dt \right]^{1/p}, \qquad p > 1$$

[110, p. 8] [87, p. 201].

Of course convergence in $P_{2\tau}$ implies convergence in $\mathcal{D}_{L^p[-\tau,\tau]}$. Thus, by the relation

$$\langle Hf, \theta_\nu \rangle = \langle f, -H\theta_\nu \rangle \tag{7.39}$$

and the fact that if the sequence $\theta_\nu \to 0$ in $\mathcal{D}_{L^p[-\tau,\tau]}$, there exists a constant C and a nonnegative integer $k \geq 0$ such that

$$|\langle Hf, \theta_\nu \rangle| = |\langle f, -H\theta_\nu \rangle| \leq C\gamma_k(\theta_\nu) \to 0 \qquad \text{as } \nu \to \infty$$

[110, p. 16].

Hf as defined by (7.38) is a continuous functional over $P_{2\tau}$. The continuity of the functional Hf over $P_{2\tau}$ can be proved more easily by using the fact that if $\theta_\nu \to 0$ in $P_{2\tau}$. Then $\xi H\theta_\nu \to 0$ in D as $\nu \to \infty$. The fact that Hf can be identified as a periodic distribution with period 2τ is proved by using the relation

$$\langle Hf, \varphi \rangle = Hf \cdot \theta \qquad \text{where } \theta = \sum_{n=-\infty}^{\infty} \varphi(t - 2n\tau)$$

$$= \langle Hf, \varphi(t + 2\tau) \rangle$$

7.5. THE INVERSION FORMULA

If $f \in L_{2\tau}^p$, $p > 1$, then using the technique in [12, p. 339], we can show that

$$H^2 f = -f + \frac{1}{2\tau} \int_{-\tau}^{\tau} f(t)\,dt \tag{7.40}$$

Therefore, if $f \in P_{2\tau}'$, we have $\forall \theta \in P_{2\tau}$,

$$\langle H^2 f, \xi\theta \rangle = \langle Hf, -\xi H\theta \rangle = \langle f, \xi(-H)^2\theta \rangle$$

$$= \langle f, \xi H^2\theta \rangle = \left\langle f, -\xi\left[\theta - \frac{1}{2\tau}\int_{-\tau}^{\tau}\theta(t)\,dt\right]\right\rangle$$

$$= \left\langle -f, \xi\left[2; \theta - \frac{1}{2\tau}\int_{-\tau}^{\tau}\theta(t)\,dt\right]\right\rangle$$

[12, p. 339]. The constant $\frac{1}{2\tau}\int_{-\tau}^{\tau}\theta(t)\,dt \in P_{2\tau}$.

Therefore (7.40) is meaningful. Hence we have derived the following inversion formula:

$$H^2 f = -f + (-f)\frac{1}{2\tau}$$

where $\langle -f(\frac{1}{2\tau}), \xi\theta \rangle = \langle -f, \xi\int_{-\tau}^{\tau}\frac{\theta}{2\tau}\,dt \rangle \;\forall \theta \in P_{2\tau}$. The uniqueness theorem clearly is not true for the Hilbert transform of periodic distributions of period 2τ in the space

$P'_{2\tau}$. We will now find a space of distributions $Q'_{2\tau}$ of period 2τ, where the uniqueness theorem for the operator H holds true.

7.6. THE TESTING FUNCTION SPACE $Q_{2\tau}$

An infinitely differentiable periodic function $\varphi(x)$ of period 2τ belongs to the space $Q_{2\tau}$ if and only if $\int_{-\tau}^{\tau} \varphi(x)\,dx = 0$. A sequence $\{\varphi_\nu\}_{\nu=1}^\infty$ in $Q_{2\tau}$ converges to zero if and only if $\varphi_\nu^{(k)} \to 0$ as $\nu \to \infty$ uniformly over the interval $[-\tau, \tau]$ for each $k = 0, 1, 2, 3, \ldots$. The Hilbert transform Hf of $f \in Q'_{2\tau}$ will be defined as

$$\langle Hf, \varphi \rangle = \langle f, -H\varphi \rangle \qquad \forall \varphi \in Q_{2\tau} \qquad (7.41)$$

Note now the difference in notation, as $Q'_{2\tau}$ stands here for the dual spaces of $Q_{2\tau}$. Clearly, since $\int_{-\tau}^{\tau} \varphi(x)\,dx = 0$, we have

$$H^2 f = -f \qquad \forall f \in Q'_{2\tau}$$

Note that the testing function space $Q_{2\tau}$ is closed with respect to differentiation as well the operator H of the Hilbert transformation defined by (7.1) or (7.2). The space $Q'_{2\tau}$ is clearly a Frechet space. We now show that the testing function space $Q_{2\tau}$ is closed with respect to the operator H of the Hilbert transformation.

Let

$$\psi(x) = \frac{1}{2\tau}(P)\int_{-\tau}^{\tau} \varphi(x-t)\cot\left(\frac{t\pi}{2\tau}\right)dt$$

where $\varphi \in Q_{2\tau}$. By Theorem 2 we can say that ψ is infinitely differentiable. To show that $\int_{-\tau}^{\tau} \psi(x)\,dx = 0$, we proceed as follows:

$$\psi(x) = \frac{1}{2\tau}(P)\int_0^\tau [\varphi(x-t) - \varphi(x+t)]\cot\left(\frac{t\pi}{2\tau}\right)dt$$

$$= \frac{1}{2\tau}\int_0^\tau \psi(x,t)\left\{t\cot\left(\frac{t\pi}{2\tau}\right)\right\}dt$$

where $\psi(x,t)$ is the function as defined in Theorem 2:

$$\int_{-\tau}^{\tau} \psi(x)\,dx = \frac{1}{2\tau}\int_{-\tau}^{\tau} dx \int_0^\tau \psi(x,t)t\cot\frac{t\pi}{2\tau}\,dt$$

$$= \frac{1}{2\tau}\int_0^\tau \int_{-\tau}^{\tau} \psi(x,t)\,dx\, t\cot\frac{t\pi}{2\tau}\,dt = 0$$

as the switch in the order of integration is justified. So

$$\int_{-\tau}^{\tau} \psi(x) = \frac{1}{2\tau}\int_{-\tau}^{\tau} \psi(x,t)\,dx\, t\cot\left(\frac{t\pi}{2\tau}\right)dt = 0$$

Note that $\int_{-\tau}^{\tau} \varphi(x \pm t)\,dx = 0$. Since the testing function space $Q_{2\tau}$ is closed with respect to differentiation and the Hilbert transformation, we define Df, the derivative of f and Hf on $Q'_{2\tau}$ as

$$\langle Df, \varphi \rangle = \langle f, -D\varphi \rangle \qquad \forall \varphi \in Q_{2\tau}$$

$$\langle DHf, \varphi \rangle = \langle f, -H\varphi' \rangle \qquad \forall \varphi \in Q_{2\tau}$$

$$= \langle Hf', \varphi \rangle$$

Now

$$\langle H^2 f, \varphi \rangle = \langle f, (-H)^2 \varphi \rangle$$

$$= \langle f, H^2 \varphi \rangle$$

$$= \left\langle f, -\varphi + \frac{1}{2\tau} \int_{-\tau}^{\tau} \varphi(x)\, dx \right\rangle$$

$$= \langle f, -\varphi \rangle$$

Hence

$$H^2 f = -f$$

that is, $H^{-1} = -H$. By using Theorem 2, one can easily show that

$$D^k Hf = H(D^k f) \qquad \forall f \in Q'_{2\tau}$$

Any periodic distribution f with period of 2τ can be identified as a continuous linear functional over $Q_{2\tau}$ by the relation

$$f \cdot \varphi = \langle f, \xi\varphi \rangle, \qquad \xi \in P_{2\tau}, \ \varphi \in Q_{2\tau}$$

Therefore the Hilbert transform Hf of the periodic distribution f, when identified as a continuous linear functional over $Q_{2\tau}$, can be defined by the relation. By using the technique of Section 7.7.2, it can also be shown that any element of $Q_{2\tau}$ can be identified as a distribution. Now

$$Hf \cdot \varphi = f \cdot (-H\varphi) \qquad \forall \varphi \in Q_{2\tau}$$

or

$$\langle Hf, \xi\varphi \rangle = \langle f, -\xi H\varphi \rangle \qquad \forall f \in P'_{2\tau}, \ \int_{-\tau}^{\tau} \in Q_{2\tau}$$

By the results proved in Section 7.1.7, we have

$$H^2 f = -f$$

$$D^k Hf = HD^k f$$

for all period distributions f when identified as a continuous linear functional over $Q_{2\tau}$.

7.7. THE HILBERT TRANSFORM OF LOCALLY INTEGRABLE AND PERIODIC FUNCTION OF PERIOD 2τ

Example 4. Let $f(t)$ be a locally integrable and periodic function of period 2τ defined on the real line. Let f be the distribution represented by this locally integrable function. Then by definition the Hilbert transform Hf of the distribution f is defined by

$$\langle Hf, \xi\varphi \rangle = \langle f, -\xi H\varphi \rangle \qquad \forall \varphi \in P_{2\tau}$$

$$= \left\langle f(x), -\xi(x)\frac{1}{2\tau}(P)\int_{-\tau}^{\tau} \varphi(x-t)\cot\left(\frac{t\pi}{2\tau}\right)dt \right\rangle$$

$$= \left\langle f(x), -\xi(x)\frac{1}{2\tau}(P)\int_{0}^{\tau} \left\{\frac{\varphi(x-t)-\varphi(x+t)}{t}\right\}t\cot\left(\frac{t\pi}{2\tau}\right)dt \right\rangle$$

Define the function $\psi(x,t)$ as in Theorem 2. Now $\psi(x,t)$ is an infinitely differentiable function with respect to x as well as t, and $\psi(x,t)$ is also a continuous function of x and t. Therefore

$$\langle Hf, \xi\varphi \rangle = -\frac{1}{2\tau}\int_{-\tau}^{\tau}\int_{0}^{\tau} f(x)\psi(x,t)t\cot\frac{t\pi}{2\tau}\,dt\,dx$$

The above iterated integral is absolutely integrable. Switching the order of integration, we get

$$\langle Hf, \xi\varphi \rangle = \frac{-1}{2\tau}\int_{0}^{\tau}\int_{-\tau}^{\tau} f(x)\psi(x,t)t\cot\frac{t\pi}{2\tau}\,dx\,dt$$

$$= \frac{-1}{2\tau}\int_{0}^{\tau}(P)\int_{-\tau}^{\tau} f(x)[\varphi(x-t)-\varphi(x+t)]\cot\left(\frac{t\pi}{2\tau}\right)dx\,dt$$

$$= \frac{-1}{2\tau}\int_{-\tau}^{\tau} f(x)\,dx\,(P)\int_{0}^{\tau} [\varphi(x-t)-\varphi(x+t)]\cot\frac{t\pi}{2\tau}\,dt$$

$$= \frac{-1}{2\tau}\int_{-\tau}^{\tau} f(x)(P)\int_{-\tau}^{\tau} \varphi(t)\cot\frac{(x-t)\pi}{2\tau}\,dt\,dx$$

$$= \frac{1}{2\tau}\int_{-\tau}^{\tau}\left((P)\int_{-\tau}^{\tau} f(x)\cot\frac{(t-x)\pi}{2\tau}\,dx\right)\varphi(t)\,dt$$

$$= \left\langle \frac{1}{2\tau}(P)\int_{-\tau}^{\tau} f(x)\cot\frac{(t-x)\pi}{2}\,dx, \xi(t)\varphi(t) \right\rangle$$

Thus

$$Hf = \frac{1}{2\tau}(P) \int_{-\tau}^{\tau} f(x) \cot \frac{(t-x)\pi}{2\tau} \, dx$$

$$= \frac{1}{2\tau}(P) \int_{-\tau}^{\tau} f(t-x) \cot \left(\frac{x\pi}{2\tau}\right) dx$$

Example 5. Find the Hilbert transform of the periodic function $\sin t$ that is of period 2π. Here $2\tau = 2\pi$ so $\tau = \pi$.

$$H(\sin t) = \frac{1}{2\pi}(P) \int_{-\pi}^{\pi} \sin t \cot \frac{x-t}{2} \, dt$$

$$= \frac{1}{2\pi}(P) \int_{-\pi}^{\pi} \sin(x-t) \cot \frac{t}{2} \, dt$$

$$= \frac{1}{2\pi}(P) \int_{-\pi}^{\pi} (\sin x \cos t - \cos x \sin t) \cot \frac{t}{2} \, dt$$

$$= -\frac{1}{2\pi} \cos x \int_{0}^{\pi} \sin t \cot \frac{t}{2} \, dt$$

$$= \frac{1}{2\pi} \cos x \int_{0}^{\pi} 2 \cos^2 \frac{t}{2} \, dt$$

$$= -\frac{1}{\pi} \cos x \int_{0}^{\pi} (\cos t + 1) \, dt$$

$$= -\cos x$$

$$H \sin x = -\cos x \tag{7.42}$$

We can similarly show that

$$H \cos x = \sin x$$

To find the Hilbert transform of $\cos x$, we can use the inversion formula

$$H \sin x = -\cos x$$

Therefore

$$H^2 \sin x = -H \cos x$$

Using the inversion formula, we have

$$-\sin x + \frac{1}{2\pi} \int_{-\pi}^{\pi} \sin x \, dx = -H(\cos x)$$

Therefore

$$H(\cos x) = \sin x, \tag{7.43}$$

Note that the inversion formulas (7.42) and (7.43) can also be derived by evaluating the integral $\lim_{N\to\infty} \frac{1}{\pi} \int_{-N}^{N} \frac{e^{it}}{x-t} \, dt$ using contour integration technique and then equating the real and imaginary parts.

Example 6. Solve in $Q'_{2\pi}$ the singular integral equation

$$y + Hy = \sin t \tag{7.44}$$

Operating the operator H on both sides of (7.44), we get

$$Hy - y = H \sin t$$
$$Hy - y = -\cos t \tag{7.45}$$

Eliminating Hy from (7.44) and (7.45), we get

$$2y = \sin t + \cos t$$

or

$$y = \frac{1}{2}(\sin t + \cos t).$$

Example 7. Solve

$$y + Hy' = \sin t \tag{7.46}$$

Operating both sides of (7.46) by H, we get

$$Hy - y' = -\cos t \tag{7.47}$$

Differentiating both sides of (7.47), we get

$$Hy' - y'' = \sin t \tag{7.48}$$

Eliminating Hy from (7.46) and (7.48), we get

$$y + y'' = 0$$
$$y = A \sin t + B \cos t$$

Now we select A and B in such a way that (7.44) is satisfied:

$$A \sin t + B \cos t + H(A \cos t - B \sin t) = \sin t$$
$$A \sin t + B \cos t + (A \sin t + B \cos t) = \sin t$$
$$A \sin t + B \cos t = \frac{1}{2} \sin t$$

So $B = 0$, $A = \frac{1}{2}$, and the required solution is $y = \frac{1}{2} \sin t$.

Example 8. Solve the following integral equation in the space Q_2':

$$Hf' = \sin\left(\frac{t\pi}{2}\right)$$

The above equation can be written as

$$\frac{d}{dt}Hf = \sin\frac{t\pi}{2}$$

or

$$Hf = -\frac{2}{\pi}\cos\frac{t\pi}{2} + C$$

So

$$f = -\frac{1}{2\times 1}\left[(P)\int_{-1}^{1}\left(\frac{-2}{\pi}\right)\cos\left(\frac{(x-t)\pi}{2}\right)\cot\frac{t\pi}{2}\,dt - (P)\int_{-1}^{1}C\cot\frac{t\pi}{2}\,dt\right]$$

$$= -\frac{-1}{\pi}(P)\int_{-1}^{1}\left[\cos\frac{x\pi}{2}\cos\frac{t\pi}{2} + \sin\frac{x\pi}{2}\sin\frac{t\pi}{2}\right]\cot\frac{t\pi}{2}\,dt$$

$$= \frac{1}{\pi}\int_{-1}^{1}\sin\frac{x\pi}{2}\cos\frac{t\pi}{2}\,dt = \frac{4}{\pi^2}\sin\left(\frac{\pi x}{2}\right)$$

7.8. APPROXIMATE HILBERT TRANSFORM OF PERIODIC DISTRIBUTIONS

Let $P_{2\tau}$ be the space of infinitely differentiable periodic functions with period 2τ, and let $P_{2\tau}'$ be the space of periodic distributions with period 2τ when the topology over the space $P_{2\tau}$ is generated by the sequence of seminorms $\gamma_k(\theta) = \sup_t |\theta^{(k)}(t)|$, $k = 0, 1, 2, 3, \ldots$. It is well known that every $f \in P_{2\tau}'$ is identifiable as a continuous and linear functional over $P_{2\tau}$.

We define the approximate Hilbert transform

$$F(x, y) \quad \text{of} \quad f \in P_{2\tau}'$$

by

$$F(x, y) = f(t).q(x - t, y), \qquad y \neq 0$$
$$= \langle f(t), \xi(t)q(x - t, y)\rangle, \qquad y \neq 0$$

Where

$$q(t, y) = \frac{1}{\pi}\lim_{N\to\infty}\sum_{n=-N}^{N}\frac{(t - 2n\tau)}{(t - 2n\tau)^2 + y^2}$$

Clearly for a fixed $y \neq 0$, $q(t, y) \in P_{2\tau}$. It is shown that $\lim_{y \to 0^+} F(x, y) = Hf$, the Hilbert transform of the periodic distribution f, in the weak topology of $P'_{2\tau}$:

$$\lim_{y \to o^+} F(x, y).\theta(x) = \lim_{y \to o^+} \int_{-\tau}^{\tau} F(x, y)\theta(x)\, dx$$

$$= Hf.\theta \qquad \forall \theta \in P_{2\tau}$$

Our result is used in finding the Harmonic function $u(x, y)$, $y > 0$ which is periodic in x with period 2τ such that $u(x, y) \to 0$ as $y \to \infty$ uniformly $\forall x \in R$ and

$$\lim_{y \to o^+} u(x, y) = f$$

where f is a periodic distribution with period 2τ.

7.8.1. Introduction to Approximate Hilbert Transform

If $f \in L^p$, $p > 1$, then its approximate Hilbert transform

$$\frac{1}{\pi} \int_{-\infty}^{\infty} \frac{f(t)(x - t)\, dt}{(x - t)^2 + y^2} \to (Hf)(x)$$

in L^p sense as well as a.e. sense [99]. The main objective of this chapter is to find an analogue of this result for the space of periodic distribution $P'_{2\tau}$ and use our result in solving an associated boundary-value problem. To this end we define a function $q : R \times R \to R$ by

$$q(t, y) = \lim_{N \to \infty} \sum_{n=-N}^{N} \frac{1}{\pi} \frac{t - 2n\tau}{(t - 2n\tau)^2 + y^2}$$

$$= \frac{1}{\pi} \frac{t}{(t^2 + y^2)} + \lim_{N \to \infty} \frac{1}{\pi} \sum_{\substack{n=-N \\ n \neq 0}}^{N} \frac{t - 2n\tau}{(t - 2n\tau)^2 + y^2}$$

Clearly $q(t, y)$ is infinitely differentiable with respect to each of the variables t and y, and each such partial derivative is a continuous function of t and y over the interval $[-\tau, \tau]$, even on the real line. We define the approximate Hilbert transform $F_\eta(x)$ of the periodic distribution f by

$$F_\eta(x) = \langle f(t), \xi(t)q(x - t, \eta)\rangle \tag{7.49}$$

and prove that

$$\lim_{\eta \to o^+} F_\eta(x) = f(x) \qquad \text{in } P'_{2\tau}$$

That is,

$$\lim_{\eta \to o^+} \int_a^{a+2\tau} F_\eta(x)\varphi(x)\, dx = \langle Hf, \xi\varphi\rangle \qquad \forall \varphi \in P_{2\tau} \tag{7.50}$$

Here Hf indicates the Hilbert transform of the distribution f, which is periodic with period 2τ, and a is a fixed real number. Taking $a = -\tau$, we get

$$\lim_{\eta \to 0^+} \int_{-\tau}^{\tau} F_\eta(x)\varphi(x)\,dx = \langle Hf, \xi(t)\varphi\rangle \qquad \forall \varphi \in P_{2\tau}$$

[99, p. 137]. The fact that $F_\eta(x)$ is infinitely differentiable with respect to x for a fixed $\eta > 0$ will also be proved.

Our result will be very useful in solving the boundary-value problem

$$u_{xx} + u_{yy} = 0, \qquad y > 0,\ x \in R$$

$u(x, y)$ is periodic in x with period 2τ and

$$\lim_{y \to 0^+} u(x, y) = f \text{ in the weak topology of } Q'_{2\tau} \tag{7.51}$$

where f is a periodic distribution with period 2τ.

7.8.2. Notation and Preliminaries

The letter f throughout this chapter stands for a periodic distribution with period 2τ, $\tau > 0$ unless stated otherwise. The testing function space $P_{2\tau}$ is the same as defined in [12] and [110]. It consists of infinitely differentiable periodic functions with period 2τ, $\tau > 0$, defined on the real line R. The topology over the space $P_{2\tau}$ is defined by the separating collection of seminorms $\{\gamma_k\}_{k=0}^\infty$ where

$$\gamma_k(\theta) = \sup_t |\theta^{(k)}(t)| \tag{7.52}$$

Therefore a sequence $\{\theta_\nu\}_{\nu=1}^\infty$ tends to zero in the space $P_{2\tau}$ if and only if for each $k = 0, 1, 2, 3, \ldots$, $\gamma_k(\theta_\nu) \to 0$ as $\nu \to \infty$. The space $P_{2\tau}$ is closed under convergence. As a matter of fact $P_{2\tau}$ is locally convex sequentially complete and a Hausdorff topological vector space [101]. Throughout this book D stands for the space of infinitely differentiable complex-valued functions with compact support. We now prove

Lemma 2. Let $\theta(t) \in P_{2\tau}$. Then $\forall x \in [-\tau, \tau]$,

$$|\theta(x)| \le A \left[\int_{-\tau}^{\tau} |\theta(\xi)|\,d\xi + \int_{-\tau}^{\tau} |\theta'(\xi)|\,d\xi \right]$$

where A is a constant independent of θ.

Proof. Our result follows by using a slight modification of the technique used by A. Friedman in [3, p. 83], and it goes as follows:

Set $S = \frac{dT}{dx}$, where T is a distribution of compact support. Now

$$T = \partial \star T = \frac{\partial}{\partial x}[h(x) \star T] = h(x) \star \frac{\partial T}{\partial x}$$

or

$$T = h(x) \star S$$

Take $\alpha \in D$ such that $\alpha = 1$ if $|x| < \frac{1}{2}$ and $\alpha = 0$ if $|x| > 1$. We assume that T is a regular distribution generated by $\alpha(x - y)\varphi(y)$ for a fixed real x lying in the interval $[-\tau, \tau]$.

The function $\theta(x) \in P_{2\tau}$; also $\theta(x) \in E$. The regular distribution generated by $\alpha(x - y)\theta(y)$ also belongs to E', the space of distributions of compact support. So

$$\alpha(x - y)\,\theta(y) = \int_R h(\xi)\frac{\partial}{\partial\xi}[\alpha(x - y + \xi)\,\theta(y - \xi)]\,d\xi$$

Now take $y = x$. We get

$$\alpha(0)\,\theta(x) = \int_R h(\xi)\frac{\partial}{\partial\xi}[\alpha(\xi)\,\theta(x - \xi)]\,d\xi$$

Therefore

$$|\theta(x)| \le B \int_{-k\tau}^{k\tau}[|\theta(\xi)| + |\theta'(\xi)|]\,d\xi$$

Here k is a sufficiently large positive integer such that the support of α is contained in $[-k\tau, k\tau]$. Since θ is periodic with period 2τ, we have

$$|\theta(x)| \le kB\left[\int_{-\tau}^{\tau}|\theta(\xi)|\,d\xi + \int_{-\tau}^{\tau}|\theta'(\xi)|\,d\xi\right] \qquad \square$$

Lemma 3. Let $\theta \in P_{2\tau}$. Then for $p > 1$, there exists a positive constant C independent of θ such that

$$|\theta| \le C\left[\|\theta\|_p + \|\theta^{(1)}\|_p\right]$$

where

$$\|\theta\|_p = \left[\int_{\tau}^{\tau}|\theta(t)|^p\,dt\right]^{1/p}$$

Proof. This is an immediate consequence of Lemma 1 and the Holder's inequality. Note that if $\theta \in P_{2\tau}$, then $\theta^{(k)} \in P_{2\tau}$ for each $k = 1, 2, 3, \ldots$. $\square$

Lemma 4. Let us define the sequence of seminorms $\{\beta_k\}_{k=0}^{\infty}$ over the testing function space $P_{2\tau}$ by

$$\beta_k(\theta) = \left(\int_{-\tau}^{\tau} |\theta(t)^p| \, dt \right)^{1/p}$$

for an arbitrary (fixed) $p > 1$, and let $\{\gamma_k\}_{k=0}^{\infty}$ be the sequence of seminorms defined by (7.52) over the space $P_{2\tau}$.

Then the topology generated by the sequence of seminorms $\{\gamma_k\}_{k=0}^{\infty}$ over the space $P_{2\tau}$ is equivalent to the topology generated by the sequence of seminorms $\{\beta_k\}_{k=0}^{\infty}$ over the space $P_{2\tau}$.

Proof. Let $\theta \in P_{2\tau}$. Then

$$\beta_k(\theta^{(k)}) = \left[\int_{-\tau}^{\tau} |\theta^{(k)}(x)|^p \, dx \right]^{1/p} \leq (2\tau)^{1/p} \gamma_k(\theta) \tag{7.53}$$

Using Lemma 2, we get

$$|\theta^{(k)}(x)| \leq A \left[\int_{-\tau}^{\tau} |\theta^{(k)}(\xi)| \, d\xi + \int_{-\tau}^{\tau} |\theta^{(k+1)}(\xi)| \, d\xi \right]$$

Therefore

$$\sup_x |\theta^{(k)}(x)| \leq (2\tau)^{1/q} \left[\left(\int_{-\tau}^{\tau} |\theta^{(k)}(\xi)|^p \, d\xi \right)^{1/p} + \left(\int_{-\tau}^{\tau} |\theta^{(k+1)}(\xi)|^p \, d\xi \right)^{1/p} \right]$$

where

$$\frac{1}{p} + \frac{1}{q} = 1$$

or

$$\gamma_k(\theta) \leq (2\tau)^{1/q} \left[\beta_k(\theta) + \beta_{k+1}(\theta) \right] \tag{7.54}$$

The result now follows in view of (7.53) and (7.54).

The topology generated by the sequence of seminorms $\{\beta_k\}_{k=0}^{\infty}$ over the space $P_{2\tau}$ is the same as the topology generated by the sequence of seminorms $\{\beta_k'\}_{k=0}^{\infty}$, where

$$\beta_k'(\theta) = \max[\beta_0(\theta), \beta_1(\theta), \dots, \beta_k(\theta)]$$

[110, p. 8]. Therefore there exists a constant $C > 0$ and $\gamma \geq 0$ satisfying

$$|f.\theta| \leq C\beta_\gamma'(\theta) \qquad \forall \theta \epsilon P_{2\tau}$$

If it is not so, we can find a sequence $\{\theta_\nu\}_{\nu=1}^{\infty}$ in $P_{2\tau}$ satisfying

$$|f.\theta_\nu| > \nu\beta_\nu'(\theta_\nu)$$

or

$$\left| f\left(\frac{\theta_\nu}{\nu\beta'_\nu(\theta_\nu)} \right) \right| > 1 \qquad \forall \nu = 1, 2, 3, \ldots$$

Now put $\psi_\nu = \frac{\theta_\nu}{\nu\beta'_\nu(\theta_\nu)}$. Therefore

$$|f.\psi_\nu| > 1 \qquad \forall \nu = 1, 2, 3, \ldots$$

For each $k < \nu$,

$$\beta_k(\psi_\nu) \le \beta'_\nu(\psi_\nu) = \frac{1}{\nu} \to 0 \qquad \text{as } \nu \to \infty$$

For each $k = 0, 1, 2, 3, \ldots$,

$$\beta_k(\psi_\nu) \le \beta'_\nu(\psi_\nu) \to 0 \qquad \text{as } \nu \to \infty$$

Therefore by Lemma 4 we see that $\psi_\nu \to 0$ as $\nu \to \infty$ in $P_{2\tau}$. But

$$|f.\psi_\nu| \to 0 \qquad \text{as } \nu \to \infty$$

a contradiction. This contradiction establishes our assertion. □

7.9. A STRUCTURE FORMULA FOR PERIODIC DISTRIBUTIONS

Theorem 7. Let $f \in P'_{2\tau}$, and $L^p_{2\tau}$ be the space of periodic functions with period 2τ which are L^p integrable over an interval of length 2τ. Then for each fixed $p > 1$, there exist measurable functions $f_k \in L^p_{2\tau}$, $k = 0, 1, 2, \ldots, \nu$ such that

$$f = \sum_{k=0}^{\gamma} (-1)^k D^k f_k$$

That is,

$$f.\theta = \left\langle \sum_{k=0}^{\gamma} (-1)^k D^k f_k, \xi\theta \right\rangle \qquad \forall \theta \in P_{2\tau}, \ \xi \in U_{2\tau}$$

$$= \sum_{k=0}^{\gamma} \int_{-\tau}^{\tau} f_k(t)\varphi^{(k)}(t)\,dt$$

Proof. Let $q > 1$ such that $\frac{1}{p} + \frac{1}{q} = 1$. If $\theta \in P_{2\tau}$, then $\theta \in D_{L^q_{2\tau}}$. That is, θ is a periodic function with period 2τ such that $\int_{-\tau}^{\tau} |\theta(x)|^q\,dx < \infty$.

Then, as stated before, there exist a constant $C > 0$ and a nonnegative integer γ such that

$$|f.\theta| \le C\beta'_\gamma(\theta)$$

or

$$|f.\theta| \le C \left[\beta_0(\theta) + \beta_1(\theta) + \cdots + \beta_\gamma(\theta) \right]$$

Therefore there exist measurable functions $f_\alpha \in L_{2\tau}^p$ such that

$$f.\theta = \sum_{\alpha=0}^{\gamma} f_\alpha.D^\alpha \theta = \sum_{\alpha=0}^{\gamma} (-D)^\alpha f_\alpha.\theta \qquad \forall \theta \in P_{2\tau}$$

The periodicity of each of $f_\alpha^{(i)}$ follows by virtue of the fact that each element of the dual of $P_{2\tau}$ can be identified as a periodic distribution with period 2τ, namely each $f_\alpha^{(i)} \in L_{2\tau}^p$. This proves our claim. $\square$

Corollary 2. If $f \in P_{2\tau}'$, then there exist measurable functions f_α such that $\int_{-\tau}^{\tau} |f_\alpha(x)| \, dx < \infty$ and such that

$$f = \sum_{\alpha=0}^{\gamma} (-D)^\alpha f_\alpha$$

The result follows in view of Theorem 3 and the fact that if $f_\alpha \in L^p[-\tau, \tau]$, then $f_\alpha \in L^1[-\tau, \tau]$, $p > 1$.

Definition. If f is a continuous linear functional over the space $P_{2\tau}$, then its Hilbert transform as a continuous linear functional over $P_{2\tau}$ is defined by

$$Hf \cdot \theta = -f \cdot H\theta \qquad \forall \theta \in P_{2\tau}$$

Thus, if $f \in P_{2\tau}'$, then we define its Hilbert transform Hf by

$$\langle Hf, \xi(t)\theta(t) \rangle = \langle f, -\xi(t)H\theta \rangle \qquad \forall \theta \in P_{2\tau}$$

Clearly we see that $Hf \in P_{2\tau}'$. For if we assume that $\theta(t)$ is an element of $P_{2\tau}$ generated by a $\varphi \in D$ through (7.22) then

$$\langle Hf, \varphi(t) \rangle = Hf \cdot \theta = \langle Hf, \varphi(t + 2\tau) \rangle$$

Theorem 8. Let $F(x, y)$ be the approximate Hilbert transform of a periodic distribution $f \in P_{2\tau}'$, $\tau > 0$, defined by

$$F(x, y) = F_y(x) = \left\langle f(t), \xi(t) \frac{1}{\pi} \left[\frac{x - t}{(x - t)^2 + y^2} \right] \right\rangle$$

$$= \langle f(t), \xi(t)q(x - t, y) \rangle$$

where

$$q(x, y) = \lim_{N \to \infty} \frac{1}{\pi} \sum_{n=-N}^{N} \frac{(x - 2n\tau)}{(x - 2n\tau)^2 + y^2}$$

Then $\frac{\partial^{m+n}}{\partial x^m \partial y^n} F(x, y)$ are all continuous functions of (x, y) for each $m, n = 0, 1, 2, 3, \ldots,$ and

$$\frac{\partial^{m+n}}{\partial x^m \partial y^n} F(x,y) = \left\langle f(t)\,\xi(t)\frac{\partial^{m+n}}{\partial x^m \partial y^n} q(x-t,y)\right\rangle \tag{7.55}$$

$$\lim_{y\to 0+} F(x,y).\theta(x) = \lim_{y\to 0+} \int_{-\tau}^{\tau} F(x,y)\theta(x)\,dx \tag{7.56}$$

$$= \langle Hf, \xi\theta\rangle \qquad \forall \theta \in P_{2\tau}$$

where Hf is the Hilbert transform of the periodic distribution f.

Proof. In view of the proved structure formula for $f \in P'_{2\tau}$, we see that

$$F(x,y) = \left\langle \sum_{\alpha=0}^{\gamma} (-1)^\alpha D^\alpha f_\alpha(t),\ \xi(t)\frac{1}{\pi}\lim_{N\to\infty}\sum_{n=-N}^{N}\frac{x-t-2n\tau}{(x-t-2n\tau)^2+y^2}\right\rangle$$

or

$$F(x,y) = \sum_{\alpha=0}^{\gamma}\int_{-\tau}^{\tau} f_\alpha(t)D_t^\alpha q(x-t,y)\,dt \tag{7.57}$$

Since $\frac{\partial^{m+n}}{\partial x^m \partial y^n} q(x,y)$ is a continuous function of (x,y),

$$\frac{\partial^{m+n}}{\partial x^m \partial y^n} F(x,y) = \sum_{\alpha=0}^{\gamma}\int_{-\tau}^{\tau} f_\alpha(t)D_t^\alpha D_x^m D_y^n q(x-t,y)\,dt \tag{7.58}$$

Therefore (7.55) follows in view of results (7.57) and (7.58).

To prove (7.56), let $\theta(x) \in P_{2\tau}$. Then

$$F(x,y).\theta(x) = \int_{-\tau}^{\tau} F(x,y)\theta(x)\,dx$$

$$F(x,y).\theta(x) = \sum_{\alpha=0}^{\nu}\int_{-\tau}^{\tau}\int_{-\tau}^{\tau} f_\alpha(t)D_t^\alpha D_x^m D_y^m q(x-t,y)\,dt\,\theta(x)\,dx \tag{7.59}$$

Since $q(x-t,y)$ and all its partial derivatives are smooth functions, we can switch the order of integration in (7.59). Therefore, for a fixed $y > 0$,

$$F(x,y).\theta(x) = \left\langle f(t),\,\xi(t)\int_{-\tau}^{\tau} q(x-t,y)\theta(x)\,dx\right\rangle$$

Our result will be proved by showing that

$$\lim_{y\to 0+}\int_{-\tau}^{\tau} q(x-t,y)\theta(x)\,dx = -H\theta$$

$$= -\frac{1}{2\tau}(P)\int_{-\tau}^{\tau}\cot\left(\frac{x\pi}{2\tau}\right)\theta(x-t)\,dx$$

in the weak topology of $P'_{2\tau}$.

Now

$$D_t^k \int_{-\tau}^{\tau} q(x - t, y)\theta(x)\, dx = \int_{-\tau}^{\tau} q(x - t, y)\theta^{(k)}(x)\, dx$$

and

$$D_t^k(H\theta)(t) = (H\theta^{(k)})(t)$$

Our result will be proved by showing that

$$\lim_{y \to 0+} \int_{-\tau}^{\tau} q(x, y)\theta(x - t)\, dx = -H\theta$$

uniformly for all t in the interval $[-\tau, \tau]$.

$$q(x, y) - \frac{1}{2\tau} \cot \frac{x\pi}{2\tau}$$

$$= \lim_{N \to \infty} \sum_{n=-N}^{N} \left[\frac{x - 2n\tau}{(x - 2n\tau)^2 + y^2} - \frac{1}{(x - 2n\tau)} \right]$$

$$= \lim_{N \to \infty} \sum_{n=-N}^{N} \frac{-y^2}{(x - 2n\tau)[(x - 2n\tau)^2 + y^2]}$$

$$= \frac{-y^2}{x(x^2 + y^2)} - 2y^2 \sum_{n=1}^{\infty} \frac{[y^2 x + x^3 + 12xn^2\tau^2]}{(x^2 - 4x^2\tau^2)[(x - 2n\tau)^2 - y^2][(x + 2n\tau)^2 + y^2]}$$

$$\tag{7.60}$$

The second term in the right-hand side of (7.60) $\to 0$ as $y \to 0$ uniformly for $-\tau \le x \le \tau$.

Therefore, in view of Lemmas 3 and 4 and Theorem 8, our result will be proved if we simply show that

$$\lim_{y \to 0+} \int_{-\tau}^{\tau} \frac{y^2}{x(x^2 + y^2)} \theta(x - t)\, dx = 0 \qquad \forall \theta(x) \in P_{2\tau}$$

uniformly for all $t \in [-\tau, \tau]$. Now

$$(P) \int_{-\tau}^{\tau} \frac{y^2}{x(x^2 + y^2)} \theta(x)\, dx = (P) \int_0^{\tau} \frac{\theta(x) - \theta(-x)}{x} \frac{y^2}{x^2 + y^2}\, dx$$

Define a function $\varphi(x, t)$ as follows:

$$\varphi(x, t) = \frac{\theta(x - t) - \theta(-x - t)}{x}, \qquad x \ne 0$$

$$= 2\theta'(-t), \qquad x = 0$$

Clearly $\varphi(x, t)$ is infinitely differentiable over the interval $[-\tau, \tau]$. Therefore

$$P \int_{-\tau}^{\tau} \frac{y^2}{x(x^2 + y^2)} \theta(x - t) \, dx = \int_0^{\tau} \varphi(x, t) \frac{y^2}{x^2 + y^2} \, dx \qquad (7.61)$$

Denoting each of the expressions in (7.61) by I, we get

$$|I| \leq 2 \sup_{-\tau \leq x \leq \tau} |\theta'(x)| \int_0^{\tau} \frac{y^2}{x^2 + y^2} \, dx$$

$$\leq 2 \sup_{-\tau \leq x \leq \tau} |\theta'(x)| y \frac{\pi}{2} \to 0$$

as $y \to 0_+$ uniformly $\forall t \in [-\tau, \tau]$. This completes the proof of our claim. $\square$

7.9.1. Applications

We will now apply our result in solving a boundary-value problem: Find a harmonic function $u(x, y)$ in the region $y > 0$ such that $u(x, y)$ is periodic in x with period 2τ and that

$$\lim_{y \to 0+} u(x, y) = f \qquad \text{in } Q'_{2\tau}$$

and

$$u(x, y) = o(1) \qquad \text{as } y \to \infty \text{ uniformly } \forall x \in R$$

Here f is a periodic distribution with period 2τ and $Q_{2\tau} = \{\theta : \theta \in P_{2\tau} \text{ such that } \int_{-\tau}^{\tau} \theta(t) \, dt = 0\}$. The topology of $Q_{2\tau}$ is that induced by $P_{2\tau}$. Take

$$u(x, y) = \left\langle -(Hf)(t), \xi(t) \left[\frac{1}{\pi} \frac{x - t}{(x - t)^2 + y^2} \right] \right\rangle \qquad (7.62)$$

$$\overset{\Delta}{=} -(Hf)(t) \cdot \left[\frac{1}{\pi} \frac{x - t}{(x - t)^2 + y^2} \right] \qquad (7.63)$$

where

$$[\theta(t)] = \sum_{n=-\infty}^{\infty} \theta(t - 2n\tau)$$

By Theorem 3, it can be shown that

$$\lim_{y \to 0^+} u(x, y) \cdot \theta(x) = \langle -H^2 f, \xi \theta \rangle$$

$$= \langle f, -\xi H^2 \theta \rangle$$

$$= \left\langle f, \left(\theta - \frac{1}{2\pi} \int_{-\tau}^{\tau} \theta(t) \, dt \right) \xi \right\rangle$$

$$= \langle f, \xi \theta \rangle \qquad \forall \theta \in Q_{2\tau}$$

It can also be shown that

$$\langle u(x, y), \xi(x)\theta(x)\rangle = \int_{-\tau}^{\tau} u(x, y)\theta(x)\,dx \tag{7.64}$$

$$= \left\langle f(t), \xi(t) \int_{-\tau}^{\tau} \left[\frac{1}{\pi} \frac{y}{(t-x)^2 + y^2} \right] \theta(x)\,dx \right\rangle \tag{7.65}$$

Note that the range of integration in the right-hand-side integral in (7.64) can as well be taken from $n\tau$ to $(n + 2)\tau$.

Letting $y \to 0^+$ in (7.62), we get

$$\lim_{y \to 0+} \langle u(x, y), \xi(x)\theta(x)\rangle = \langle f, \xi(x)\theta\rangle \qquad \forall \theta \in Q_{2\tau}$$

The fact that $u(x, y) = o(1)$ as $y \to \infty$ can be proved by using Hölder's inequality. We now solve the above problem slightly differently.

Solution. Given

$$u_{xx} + u_{yy} = 0, \qquad y > 0$$

$$u(x, y) = o(1), \qquad \text{as } y \to \infty \text{ uniformly } \forall x \in R$$

$$\lim_{y \to 0^+} u(x, y) = f \qquad \text{in } Q'_{2\tau} \text{ where } f \in P'_{2\tau}$$

That is, f is a periodic distribution with period 2τ. Take

$$u(x, y) = \frac{1}{\pi} \left\langle f(t), \xi(t) \lim_{N \to \infty} \sum_{n=-N}^{N} \frac{y}{(t - x - 2n\tau)^2 + y^2} \right\rangle$$

As in the previous exercise, we take

$$u(x, y) = \left\langle -(Hf)(t), \xi(t) \frac{1}{\pi} \lim_{N \to \infty} \sum_{-N}^{N} \frac{x - t - 2n\tau}{(x - t - 2n\tau)^2 + y^2} \right\rangle$$

$$= \left\langle f(t), \xi(t) \frac{1}{\pi} \lim_{N \to \infty} \sum_{n=-N}^{N} \frac{y}{(x - t - 2n\tau)^2 + y^2} \right\rangle$$

As before, u is a harmonic function of x, y in the region $y > 0$ and $u(x, y) \to f$ in $P_{2\tau}$ as $y \to 0^+$. The fact that $u(x, y) = o(1)$ as $y \to \infty$ uniformly $\forall\, x \in R$ can also be proved by using the structure formula for f and Holder's inequality. The harmonicity of $U(x, y)$ can be proved by direct differentiation. The fact that $\lim_{y \to 0^+} U(x, y) = f$ in $Q'_{2\tau}$ can be proved by using the structure formula for f.

EXERCISES

1. Let $f \in P'_{2\tau}$ and H be the operator of Hilbert transformation over $P'_{2\tau}$. Prove that

$$\left\langle (Hf), \xi(t) \frac{1}{\pi} \lim_{N \to \infty} \sum_{n=-N}^{N} \frac{x - t - 2n\tau}{(x - t - 2n\tau)^2 + y^2} \right\rangle$$

$$= \left\langle f(t) \lim_{N \to \infty} \xi(t) \frac{1}{\pi} \sum_{n=-N}^{N} \frac{y}{(x - t - 2\pi/n\tau)^2 + y^2} \right\rangle, \qquad \xi \in P_{2\tau}$$

2. Prove that

$$\lim_{y \to 0^+} \left\langle f(t), \frac{1}{\pi} \xi(t) \lim_{N \to \infty} \sum_{n=-N}^{N} \frac{y}{(n - t - 2n\pi)^2 + y^2} \right\rangle = f$$

   in the weak topology of $P'_{2\tau}$.

3. Find a harmonic function $u(x, y)$ that is periodic with period 1 in the region $y > 0$ such that $\lim_{y \to 0^+} u(x, y) = \sum_{n=-\infty}^{\infty} \delta(t - n)$ in P'_1 and $u(x, y) = o(1)$, $y \to \infty$ uniformly $\forall x \in R$. Verify your result.

4. Prove that every element of the dual space of $Q_{2\tau}$ can be identified as a Schwartz distribution with period 2τ.

BIBLIOGRAPHY

[1] Al-Gwaiz, M. A., *Theory of Distributions*, Marcel Dekker Inc., New York, 1992.

[2] Anderson, K., Weighted norm inequalities for Hilbert space and conjugate functions of even and odd functions, *Proc. Amer. Math. Soc.* **56** (1976), 99–107.

[3] Akhiezer, N. I., and Glazman, I. M., *Theory of Linear Operators in Hilbert Spaces*, vols. I and II, Frederick Ungar, New York, 1965.

[4] Arsae, J., *Fourier Transforms and the Theory of Distributions*, Prentice-Hall, Englewood Cliffs, NJ, 1966.

[5] Beltrami, E. J., and Wohlers, M. R., *Distributions and the Boundary Values of Analytic Functions*, Academic Press, New York, 1966.

[6] Beltrami, E. J., and Wohlers, M. R., Distributional boundary value theorems and Hilbert transforms, *Arch. Rational Anal.* **18** (1965), 304–309.

[7] Bergh, J., and Löfström, J., *Interpolation Spaces: An Introduction*, Springer-Verlag, New York, 1976.

[8] Bremermann, H. J., *Distributions, Complex Variables and Fourier Transforms*, Addison-Wesley, Reading, MA, 1965.

[9] Bremermann, H. J., On analytic continuation, multiplication and Fourier transformations of Schwartz distributions, *J. Math. Phys.* **2** (1961), 240–258.

[10] Brychkov, Yu. A., and Prudnikov, A.-P., *Integral Transform of Generalized Functions*, Gordon and Breach, New York, 1989.

[11] Butzer, P. L., and Trebels, W., Hilbert transforms, fractional integration and differentiation, *Bull. Amer. Math. Soc.* **74** (1968), 106–110.

[12] Butzer, P. L., and Nessel, R. J., *Fourier Analysis and Approximation*, vol. 1, Birkhäuser, Basel, 1971.

[13] Calderon, A. P., Singular integrals, *Bull. Amer. Math. Soc.* **72** (1950), 427–465.

[14] Calderon, A. P., and Zygmund, A., On Singular Integrals, *Amer. J. Math.* **78** (1956), 427–465.

[15] Carmichael, R. D., and Richter, S. P., Distributional boundary values in $\mathcal{D}'_{L^p}$, *Rend. Sem. Mat. Univ. Padova* **70** (1983), 55–76.

[16] Carmichael, R. D., and Mitrovic, D., *Distributions and Analytic Functions*, II, Pitman Research Notes in Mathematical Series 206, Pitman, New York.

[17] Carrier, G. F., Krook, M., and Pearson, C. E., *Functions of a Complex Variables*, McGraw-Hill, New York, 1966.

[18] Chaudhry, M. A., and Pandey, J. N., Hilbert problem—A distributional approach, *Can. Math. Bull.* **34** (1991), 321–328.

[19] Cheng, S. Y. A., and Fefferman, R., Some recent developments in Fourier analysis on product domains, *Bull. Amer. Soc.* **12** (1985), 1–44.

[20] Cohn, D. L., *Measure Theory*, Birkhäuser, Basel, 1980.

[21] Carton-Lebrun, C., and Pandey, J. N., The Hilbert transform of distributions, *Technical Report*, Carleton University, Ottawa, 1985.

[22] Carton-Lebrun, C., Product properties of Hilbert transforms, *J. Approx. Theory* **21** (1977), 356–360.

[23] Carton-Lebrun, C., Remarques sur les relations de dispersion, *Bull. Soc. Math. Belgique*, Ser. B, **31** (1972), 209–213.

[24] Carton-Lebrun, C., and Fosset, M., Hilbert transforms for functions of bounded mean oscillations, *Report 80-7*, Université de L'État à Mons, Belgique, 1980.

[25] Carton-Lebrun, C., and Fosset, M., Moyennes et quotients de Taylor dans BMO, *Bull. Soc. Roy. Science de Liege* **2** (1984), 85–87.

[26] Carton-Lebrun, C., An extension to BMO functions of some product properties of Hilbert transforms, *J. Approx. Theory*, 1995, to appear.

[27] Chaudhry, M. A., and Pandey, J. N., The Hilbert transform of Schwartz Distributions, II, *Math. Proc. Camb. Phil. Soc.* **102** (1987), 553–559.

[28] Chaudhry, M. A., and Pandey, J. N., Generalized $(n + 1)$-dimensional Dirichlet boundary value problems, *Applicable Anal.* **17** (1984), 313–330.

[29] Chaudhry, M. A., and Pandey, J. N., Generalized n-dimensional Hilbert transform and applications, *Applicable Anal.* **20** (1985), 221–235.

[30] Coifmann, R. R., and Fefferman, C., Weighted norm inequalities for maximal functions and singular integrals, *Studia Math.* **51** (1974), 241–250.

[31] Coifmann, R. R., Rochberg, R., and Weiss, G., Factorisation theorems for Hardy spaces in several variables, *Ann. of Math.* **2** (1976), 611–635.

[32] Cossar, J., On conjugate functions, *Proc. London Math. Soc.* Ser. 2, **45** (1932), 369–381.

[33] Cotlar, M., Some generalizations of the Hardy Littlewood maximal theorem, *Rev. Math. Cuyana* **1** (1955), 85–104.

[34] Cotlar, M., A unified theory of Hilbert transforms and ergodic theorems, *Rev. Math. Cuyana* **1** (1955), 105–167.

[35] Donnelly, R. K., Contributions to analysis: A new space of generalized functions for integro-differential equations involving Hilbert transforms and a note on the non-Nu clearity of a space of list functions on Hilbert space, Ph.D dissertation, Department of Mathematics, Carleton University, Ottawa, Canada, 1988.

[36] Ehrenrpeis, L., Analytic functions and the Fourier transform of distributions, I, *Ann. Math.* **63** (1956), 129–159.

[37] Fabes, E. B., and Riviere, N. M., Singular integrals with mixed homogeneity, *Studia Math.* **27** (1966) 19–38.

[38] Fefferman, C., Estimates for double Hilbert transform, *Studia Math.* **44** (1972), 1–15.

[39] Fefferman, C., Recent progress on classical Fourier analysis, *Internat. Congress Math.*, Vancouver (1974), 95–118.

[40] Fefferman C., Characterizations of bounded mean oscillation, *Bull. Amer. Math. Soc.* **77** (1971), 587–588.

[41] Fefferman, C., and Stein, E. M., H^p spaces of several variables, *Acta Math.*, **129** (1972), 137–193.

[42] Friedman, A., *Generalized Functions and Partial Differential Equations*, Prentice-Hall, Englewood Cliffs, NJ, 1963.

[43] Gel'fand, I. M., and Shilov, G. E., *Generalized Functions*, vol. 1, Academic Press, New York, 1964.

[44] Gel'fand, I. M., and Shilov, G. E., *Generalized Functions*, vol. 2, Academic Press, New York, 1967.

[45] Gel'fand, I. M., and Shilov, G. E., *Generalized Functions*, vol. 3, Academic Press, New York, 1968.

[46] Gohberg, I. Ts., and Krupnik, N. Ya., Norm of the Hilbert transformation in the L^p space, *Functional Anal. Appl.* **2** (1968), 180–181.

[47] Gunning, R. C., and Rossi, H., *Analytic Functions of Several Complex Variables*, Prentice-Hall, Englewood Cliffs, NJ, 1965.

[48] Hewitt, E., and Stromberg, K., *Real and Abstract Analysis*, Springer-Verlag, Berlin, 1969.

[49] Heywood, P., On a modification of the Hilbert transform, *J. London Math. Soc.* **42** (1967), 641–645.

[50] Horvath, J., Hilbert transforms of distributions in $\mathbb{R}^n$, *Proc. International Congress of Mathematicians*, Amsterdam (1954), 122–123.

[51] Horvath, J., *Topological Vector Spaces and Distributions*, vol. I, Addison-Wesley, Reading, MA, 1966.

[52] Hunt, R., Muckenhoupt, B., and Wheeden, W., Weighted norm inequalities for the conjugate function and Hilbert transform, *Trans. Amer. Math. Soc.* **176** (1973), 227–251.

[53] Jones, D. S., *Generalized Functions*, McGraw-Hill, New York, 1966.

[54] Katznelson, Y., *An Introduction to Harmonic Analysis*, Wiley, New York, 1963.

[55] Kelly, J. L., *General Topology*, Van Nostrand, New York, 1955.

[56] Kober, H., A modification of Hilbert transforms, the Weyl integral and the functional equations, *J. London Math. Soc.* **42** (1967) 42–50.

[57] Kokilashvili, V. M., *Singular Operators in Weighted Spaces*, Colloquia Mathematica Societatis Janos Bolyai, 35 Functions, Series, Operators, Budapest, Hungary, 1980.

[58] Lauwerier, H. A., The Hilbert problem for generalized functions, *Archive Rat. Mech. Anal.* **13** (1963), 157–166.

[59] Lighthill, M. J., *Fourier Analysis and Generalised Functions*, Cambridge University Press, Cambridge, 1970.

[60] McLean, W. and Elliott, D., On the p-norm of the truncated Hilbert transform, *Bull. Austral. Math. Soc.* **38** (1988), 413–420.

[61] Mitrovic, D., A Hilbert distributional boundary value problem, *Mathematica Balkanica* **1** (1971) 177.

[62] Mitrovic, D., A distributional representation of analytic functions, *Mathematica Balkanica* **4** (1974), 437–440.

[63] Mitrovic, D., Some distributional boundary value problems, *Mathematica Balkanica* **2** (1972), 161–164.

[64] Muckenhoupt, B., Weighted norm inequalities for the Hardy maximal function, *Trans. Amer. Math. Soc.* **165** (1972), 207–226.

[65] Muskhelishvili, N. I., *Singular Integral Equations*, Moscow, 1946.

[66] Neri, U., *Lecture Notes in Mathematics 200*, Springer-Verlag, Berlin, 1971.

[67] Barros-Neto, J., *An Introduction to the Theory of Distributions*, Marcel Dekker, New York, 1973.

[68] Newcomb, W. R., Hilbert transforms—distribution theory, *Stanford Electronics Laboratories, Technical Report* 2250-1, February 1962.

[69] O'Neil, and Weiss, G., The Hilbert transform and rearrangement of functions, *Studia Math.* **23** (1963), 189–198.

[70] Okikiolu, G.O., *Aspects of Theory of Bounded Integral Operators in L_p-Spaces*, Academic Press, New York, 1971.

[71] Oppenheim, A. V., and Schafer, R. W., *Digital Signal Processing*, Prentice-Hall, Englewood Cliffs, NJ, 1975.

[72] Orton, M., Hilbert transforms, Plemelj relations and Fourier transforms of distributions, *SIAM J. Math. Anal.* **4** (1937), 656–667.

[73] Orton, M., Hilbert boundary value problems—A distributional approach, *Proc. Royal Soc., Edinburgh*, **76A** (1977), 193–208.

[74] Pandey, J. N., and Hughes, E., An approximate Hilbert transform and its application, *Tôhoku Math. J.* **28** (1976), 497–509.

[75] Pandey, J. N., Hilbert transform of generalized functions, *Carleton Math. Series*, no. 175, 1981.

[76] Pandey, J. N., An extension of the Gel'fand-Shilov technique for Hilbert transforms, *Applicable Anal.* **13** (1982), 279–290.

[77] Pandey, J. N., and Chaudhry, M. A., The Hilbert transform of generalized functions and applications, *Canadian J. Math.* **35** (1983), 479–495.

[78] Pandey, J. N., and Singh, O. P., On the p-norm of the truncated n-dimensional Hilbert transform, *Bull. Austral. Math. Soc.* **43** (1991), 241–250, 478–495.

[79] Pandey, J. N., The Hilbert transform of Schwartz distributions, *Proc. Amer. Math. Soc.* **89** (1983), 86–90.

[80] Pandey, J. N., and Singh, O. P., Characterization of functions with Fourier transforms supported on orthants, *J. Math. Analysis Appl.* **185** (1994), 438–463.

[81] Pandey, J. N., Approximate Hilbert transform of periodic functions and applications, *J. Integral Transform and Special Functions*, 1995 (to appear).

[82] Pichorides, S. K., On the best values of the constants arising in the theorems of M. Riesz, Zygmund and Kolmogorov, *Studia Math.* **44** (1972), 165–179.

[83] Robinson, A., and Laurmann, J. A., *Wing Theory*, Cambridge University Press, Cambridge, 1956.

[84] Rudin, W., *Real and Complex Analysis*, McGraw-Hill, New York, 1966.

[85] Schaefer, H. H., *Topological Vector Spaces*, Springer-Verlag, Berlin, 1970.

[86] Schwartz, L., Causalité et analyticité, *An. Acad. Brasil. Ciêne* **34** (1962), 13–21.

[87] Schwartz, L., *Theorie des distributions*, Hermann, Paris, 1978.

[88] Shilov, G. E., *Generalized Functions and Partial Differential Equations*, Gordon and Breach, New York, 1968.

[89] Singh, O. P., and Pandey, J. N., The n-dimensional Hilbert transform of distributions, its inversion and applications, *Can. J. Math.* **42** (1990), 239–258.

[90] Singh, O. P, *Math. Review*, 93g44007.

[91] Sneddon, I. H., *The Use of Integral Transforms*, McGraw-Hill, New York, 1972.

[92] Sneddon, I. H., and Lowengrub, M., *Crack Problems in the Classical Theory of Elasticity*, Wiley, New York, 1969.

[93] Sokel-Soholowski, K., On trigonometric series conjugate to Fourier series in two variables, *Fund. Math.* **33** (1945), 166–182.

[94] Stein, E. M., *Singular Integrals and Differentiability Properties of Functions*, Princeton University Press, Princeton, 1970.

[95] Stein, E. M., and Weiss, G., *Introduction to Fourier Analysis on Euclidean Spaces*, Princeton University Press, Princeton, 1975.

[96] Terras, A., *Harmonic Analysis on Symmetric Spaces and Applications I*, Springer-Verlag, Berlin, 1985.

[97] Tillmann, H. G., Distributionen als Randverteilungen analytischer Funktionen II, *Math. Z.* **76** (1961), 5–21.

[98] Tillmann, H. G., Darstellung der Schwartzschen Distributionen durch analytische Functionen, *Math. Z.* **77** (1961), 106–124.

[99] Titchmarsh, E. C., *Introduction to the Theory of Fourier Integrals*, Claredon Press, Oxford, 1967.

[100] Titchmarsh, E. C., *The Theory of Functions*, Oxford University Press, Oxford, 1939.

[101] Treves, F., *Topological Vector Spaces Distributions and Kernels*, Academic Press, New York, 1967.

[102] Treves, F., Distributions and general theory of differential operators, *Technical Report I*, University of California, Berkeley, January 1960.

[103] Tricomi, F., *Integral Equations*, Wiley, New York, 1957.

[104] Vladimirov, V. S., *Methods of the Theory of Functions of Many Complex Variables*, MIT Press, Cambridge, MA, 1966.

[105] Weiss, G., and Stein, M. E., On the theory of harmonic-functions of several variables, I. The theory of H^p spaces, *Acta. Math.* **103** (1960), 25–62.

[106] Williams, J. H., *Lebesgue Integration*, Holt, Rinehart and Winston, New York, 1962.

[107] Wu, T.-Y., and Ohmura, T., *Quantum Theory of Scattering*, Prentice-Hall, Englewood Cliffs, NJ, 1962.

[108] Yosida, K., *Functional Analysis*, 6th ed., Springer-Verlag, New York, 1980.

[109] Zemanian, A. H., *Distribution Theory and Transform Analysis*, McGraw–Hill, New York, 1965.

[110] Zemanian, A. H., *Generalized Integral Transformations*, Wiley, New York, 1968.

[111] Zygmund, A., On the boundary values of functions of serveral complex variables, *Fund. Math.* **36** (1949), 201–225.

[112] Zygmund, A., *Trigonometric Series*, 2d ed., Cambridge University Press, Cambridge, 1968.

INDEX

NOTATION INDEX

PURE AND APPLIED MATHEMATICS

A Wiley-Interscience Series of Texts, Monographs, and Tracts

Founded by RICHARD COURANT

Editor Emeritus: PETER HILTON

Editors: MYRON B. ALLEN III, DAVID A. COX, HARRY HOCHSTADT,
   PETER LAX, JOHN TOLAND

*Now available in a lower priced paperback edition in the Wiley Classics Library.

Printed in the United Kingdom
by Lightning Source UK Ltd.
128945UK00001B/90/A